MODERN UTILIZATION OF
ELECTRIC POWER
Including Electric Drives and Electric Traction

MODERN UTILIZATION OF
ELECTRIC POWER
Including Electric Drives and Electric Traction

Pradip Kumar Sadhu
Professor & Head
Department of Electrical Engineering
Indian Institute of Technology (Indian School of Mines)
Dhanbad, Jharkhand

Soumya Das
Assistant Professor
Department of Electrical Engineering
University Institute of Technology, Burdwan University
West Bengal

CBS

CBS Publishers & Distributors Pvt Ltd
New Delhi • Bengaluru • Chennai • Kochi • Kolkata • Mumbai
Hyderabad • Jharkhand • Nagpur • Patna • Pune • Uttarakhand

MODERN UTILIZATION OF
ELECTRIC POWER

ISBN: 978-93-87085-18-3

Copyright © Authors and Publisher

First Edition: 2018

All rights reserved. No part of this book may be reproduced or transmitted in any form or by any means, electronic or mechanical, including photocopying, recording, or any information storage and retrieval system without permission, in writing, from the authors and the publisher.

Published by Satish Kumar Jain and produced by Varun Jain for

CBS Publishers and Distributors Pvt Ltd 4819/XI Prahlad Street, 24 Ansari Road, Daryaganj, New Delhi 110 002, India. Ph: 23289259, 23266861, 23266867   Fax: 011-23243014

Website: www.cbspd.com   e-mail: delhi@cbspd.com; cbspubs@airtelmail.in.

Corporate Office: 204 FIE, Industrial Area, Patparganj, Delhi 110 092, India

Ph: 4934 4934   Fax: 4934 4935   e-mail: publishing@cbspd.com; publicity@cbspd.com

Branches

- **Bengaluru:** Seema House 2975, 17th Cross, K.R. Road,   Banasankari 2nd Stage, Bengaluru 560 070, Karnataka, India

  Ph: +91-80-26771678/79   Fax: +91-80-26771680   e-mail: bangalore@cbspd.com
- **Chennai:** 7, Subbaraya Street, Shenoy Nagar, Chennai 600 030, Tamil Nadu, India.

  Ph: +91-44-26680620, 26681266   Fax: +91-44-42032115   e-mail: chennai@cbspd.com
- **Kochi:** Ashana House, No. 39/1904, AM Thomas Road, Valanjambalam, Ernakulam 682 016, Kochi, Kerala, India.

  Ph: +91-484-4059061-62-64-65   Fax: +91-484-4059065   e-mail: kochi@cbspd.com
- **Kolkata:** 6/B, Ground Floor, Rameswar Shaw Road, Kolkata-700 014 (West Bengal), India.

  Ph: +91-33-2289-1126, 2289-1127, 2289-1128   e-mail: kolkata@cbspd.com
- **Mumbai:** 83-C, Dr E Moses Road, Worli, Mumbai-400018, Maharashtra, India.

  Ph: +91-22-24902340/41   Fax: +91-22-24902342   e-mail: mumbai@cbspd.com

Representatives

- **Hyderabad** 0-9885175004
- **Jharkhand** 0-9811541605
- **Nagpur** 0-9021734563
- **Patna** 0-9334159340
- **Pune** 0-9623451994
- **Uttarakhand** 0-9716462459

Printed at: Mudrak, Delhi, India

PREFACE

"Modern Utilization of Electric Power: *Including Electric Drives and Electric Traction*" has been designed as a textbook for students pursuing studies in BE/BTech; AMIE, diploma and for those preparing for other competitive examinations in India and overseas. It will also be helpful to students preparing for various competitive examinations. It is equally helpful to practicing engineers to understand the theoretical aspects of their profession. This book is easy to read and stimulating in its direct approach.

This book is designed to cover an extensive range of topics under electric power utilization. It consists of eight chapters namely: 1. Control devices for industrial motor; 2. Electric drive; 3. Illumination; 4. Electric heating; 5. Electric welding, 6. Electric traction; 7. Electrolytic processes; 8. Refrigeration and air conditioning.

The presentation of subject matter is very systematic and language of text is lucid, direct and easy to understand. Authors lay no claim to the original research in preparing the book. Liberal use of the materials available in the works of renowned authors has been made. In all modesty, they may claim only that they have tried to approach the huge amount of material available from primary and secondary sources into coherent body of description and analysis.

The authors welcome any constructive criticism of the book and will be grateful for any appraisal by the readers.

Pradip Kumar Sadhu
Soumya Das

ACKNOWLEDGEMENTS

We are fortunate to have received many useful comments and suggestions from students, which have helped in improving the technical content and clarity of the book. We are greatful to all of them. In particular, Saswata Mukherjee, Sabyasachi Samanta, Anirban Kundu, and Rishabh Das.

We are indebted to many readers in the academia and industry worldwide for their invaluable feedback and for taking the trouble to draw our attention to the improvements required and the errors in the first edition.

We also thank the reviewers who took time from their busy schedules to send us suggestions.

Most importantly, it was the help and the advice of CBS Publishers & Distributors Pvt. Ltd., staff that made this whole project a reality. We are greatful to the authorities of "Indian Institute of Technology (Indian School of Mines) Dhanbad" and "University Institute of Technology, Burdwan University" for providing all the facilities for writing this book.

Finally, the authors are greatful to their families for their love, tolerance, patience and support throughout this very time consuming project. Readers of the book are welcome to send their comments and feedback.

Pradip Kumar Sadhu
Soumya Das

CONTENTS

Preface   v

1. CONTROL DEVICES FOR
INDUSTRIAL MOTOR   1

1.1 Introduction 1
1.2 Semiconductor Devices 1
   1.2.1 Diode 1
   1.2.2 Zener Diode 3
   1.2.3 Power Transistor 4
   1.2.4 Thyristor 5
   1.2.5 Silicon-Controlled Rectifier
      (SCR) 6
   1.2.6 Triac 8
   1.2.7 Diac 9
   1.2.8 Gate-turn-off Switch 10
   1.2.9 Unijunction Transistor (UJT) 11
1.3 Thyratron 12
1.4 Ignitron 15
1.5 Amplidyne 16
1.6 Saturable Reactor (Magnetic
   Amplifier) 18
1.7 Pilot Devices 20
   1.7.1 Push Button 20
   1.7.2 Limit Switches 21
   1.7.3 Float Switches 21
   1.7.4 Pressure Switches 21
   1.7.5 Thermostats 22
   1.7.6 Plugging Switches 22
   1.7.7 Master Controller 23
   1.7.8 Transfer Switches 23
   1.7.9 Contactors 23
   1.7.10 Speed Governors 26

2. ELECTRIC DRIVE   28

2.1 Introduction 28
2.2 Factors Governing Selection of
   Motors 28
2.3 Nature of Electric Supply 29

2.4 Nature of the Drive 31
   2.4.1 Group Drive 31
   2.4.2 Individual Drive 32
2.5 Starting Characteristics 32
   2.5.1 Starting Torque of DC
      Motors 32
   2.5.2 Three-phase Induction
      Motors 33
   2.5.3 Methods of Starting 3-phase
      Induction Motors 35
   2.5.4 Limitation of Size 38
   2.5.5 Single-phase Induction
      Motors 38
   2.5.6 Synchronous Motors 39
   2.5.7 AC Commutator Motors 40
2.6 Running Characteristics 40
   2.6.1 DC Motor 40
   2.6.2 The Three-phase Induction
      Motor 40
   2.6.3 The Schrage Motor 40
   2.6.4 The Three-phase Series
      Motor 41
   2.6.5 The Synchronous and
      Synchronous Induction
      Motor 41
   2.6.6 The Single-phase Series
      Motor 41
   2.6.7 The Single-phase Induction
      Motor 41
   2.6.8 The Repulsion Motor 41
2.7 Speed Control 41
   2.7.1 Speed Control of DC Motors 41
   2.7.2a Field Control in Shunt Motors 42
   2.7.2b Field Control in Series Motor 42
   2.7.3 Control of Speed by Change
      of Series Resistance in the
      Armature Circuit 44
   2.7.4 Control of Motor Speed by
      Shunting the Armature by a
      Resistance 44

2.7.5 Booster Control 46
2.7.6 Ward-Leonard Method of Speed Control 46
2.7.7 Metaldyne Control 48
2.8 Electronic Control of DC Motors 48
2.8.1 Speed Relations of DC Motors 48
2.8.2 Thyristor Control of DC Motors 49
2.8.3 DC Motor Supplied by Three-phase Thyristor Rectifiers 51
2.8.4 Field Excitation of DC Motor and Generator from Rectifiers 52
2.8.5 Chopper Control of DC Motors 53
2.8.6 Closed Loop System for Automatic Speed Control of DC Motor 54
2.8.7 Automatic Control 55
2.9 Speed Control of Induction Motors 55
2.9.1 Frequency Method of Speed Control 56
2.9.2 Pole Changing Method of Speed Control 57
2.9.3 By Applying Variable Voltage to Stator 57
2.9.4 By Varying the Rotor Copper Losses 58
2.9.5 Speed Control by Slip Coupling 58
2.10 Electric Braking 59
2.10.1 Advantages and Disadvantages of Electric Braking Over Mechanical Braking 59
2.10.2 Types of Electric Braking 59
2.11 Type of Insulation Employed 63
2.12 Heating of Motor or Temperature Rise 64
2.12.1 Equation for Heating of Motor 65
2.12.2 Heating Time Constant 66
2.12.3 Equation for Cooling of Motor or Temperature Fall 66
2.13 Cooling Time Constant 68
2.14 Duty Cycles 68
2.15 Rating of Machines 71
2.15.1 Effect of Altitude on Rated Output 72
2.15.2 Overload Capacity of Induction Motor 73

2.16 Choice of Rating of Motors 73
2.17 Methods of Ventilation and Cooling of Machines 75
2.18 Load Equalization 77
2.19 Use of Flywheels 77
2.20 Mechanical Features of Electric Motor 81
2.21 Types of Enclosures 81
2.22 Bearings 82
2.23 Type of Mountings 82
2.24 Transmission of Drive 84
2.25 Noise 84
2.26 Cost Consideration 85
2.27 Motors for Particular Service 85

3. ILLUMINATION   109

3.1 Introduction 109
3.2 Nature of Light 109
3.3 Definitions 111
3.3.1 Plane Angle and Solid Angle 111
3.3.2 Light Energy 112
3.3.3 Luminous Flux 112
3.3.4 Luminous Intensity 113
3.3.5 Candela 113
3.3.6 Illumination 113
3.3.7 Luminance (Brightness) 114
3.3.8 Mean Horizontal Candle Power (MHCP) 115
3.3.9 Mean Spherical Candle Power (MSCP) 115
3.3.10 Mean Hemispherical Candle Power (MHSCP) 115
3.3.11 Reduction Factor 115
3.3.12 Lamp Efficiency 115
3.3.13 Specific Consumption 115
3.3.14 Space Height Ratio 115
3.3.15 Utilization Factor (UF) 115
3.3.16 Maintenance Factor (MF) 116
3.3.17 Depreciation Factor 116
3.3.18 Waste Light Factor 116
3.3.19 Absorption Factor 116
3.3.20 Luminous Efficiency 116
3.3.21 Beam Factor 117
3.3.22 Reflection Factor 117
3.3.23 Coefficient of Utilisation 117
3.4 Polar Curve 118
3.5 Rousseau Diagram 118
3.6 Laws of Illumination 119
3.7 Sources of Light 122
3.7.1 Arc Lamps 122
3.7.2 Incandescent Lamps 124

3.7.3 Gaseous Discharge Lamps 128

3.7.4 Fluorescent Tubes 135

3.7.5 Compact Fluorescent Lamp 142

3.7.6 LED 142

3.8 Requirements of Good Lighting 143

3.9 Diffusing and Reflecting Surfaces: Globes and Reflectors 144

3.10 Types of Light Fittings or Luminaries 146

3.11 Methods of Lighting Calculation 148

3.12 Factory Lighting 148

3.13 Street Lighting 149

3.14 Flood Lighting 150

3.15 Photometry 152

3.16 Integrating Sphere 156

4. ELECTRIC HEATING 163

4.1 Introduction 163

4.2 Advantages of Electrical Heating 163

4.3 Heating Methods 164

4.4 Resistance Heating 164

4.5 Direct Resistance Heating 165

4.6 Indirect Resistance Heating 165

4.6.1 Types of Furnaces 166

4.6.2 Requirement of a Good Heating Material 168

4.6.3 Materials of Heating Elements 169

4.6.4 Causes of Failure of Heating Elements 169

4.6.5 Temperature Control of Resistance Furnaces 170

4.6.6 Deign of Heating Element 172

4.7 Radiating Heating 173

4.8 Electric Arc Furnace 174

4.8.1 Direct Arc Furnace 174

4.9 Indirect Arc Furnace 183

4.10 Electron Bombardment Heating 184

4.11 High Frequency Heating 186

4.12 Induction Heating 186

4.12.1 Direct Core-type Furnace 188

4.12.2 Vertical Core-type Furnace 189

4.12.3 Indirect Core-type Furnace 190

4.12.4 Coreless Furnace 190

4.12.5 Sources of High Frequency for Induction Heating 191

4.13 Dielectric Heating 193

4.13.1 Advantages of Dielectric Heating 195

4.13.2 Uses of Dielectric Heating 195

4.14 Choice of Frequency 196

5. ELECTRIC WELDING 201

5.1 Introduction 201

5.2 Classification 201

5.3 Electric Arc Welding 202

5.4 How Weld Metal is Deposited? 203

5.5 Four Positions of Arc Welding 204

5.6 Bare Metal Arc Welding 205

5.7 Coated Electrodes 205

5.8 Types of Joints and Applicable Welds 206

5.9 Requirements of Good Weld 207

5.10 Atomic Hydrogen Welding 208

5.11 Submerged Arc Welding 210

5.12 Inert Gas Metal Arc Welding 212

5.12.1 Tungsten Inert-Gas (TIG) Process 212

5.12.2 Metal Inert-Gas (MIG) Process 213

5.12.3 MAG Welding 214

5.13 Carbon Arc Welding 215

5.14 Electric Supply for Arc Welding 216

5.15 Machines for Arc Welding 217

5.15.1 DC Welding Machines with Motor Generator Set 218

5.15.2 AC Rectified Welding Unit 219

5.15.3 AC Welding Machines 219

5.16 Resistance Welding 221

5.17 Spot Welding 221

5.18 Seam Welding 223

5.19 Projection Welding 224

5.20 Butt Welding 224

5.20.1 Flash Butt Welding 224

5.20.2 Upset Butt Welding 226

5.20.3 Percussion Welding 226

5.21 Electron Beam Welding 227

5.22 Electro Slag Welding 229

5.23 Electro Gas Welding 230

5.24 Plasma Arc Welding 230

5.25 Laser Welding 232

5.26 Ultrasonic Welding 233

5.27 Power Supply for Resistance Welding 234

5.28 Machines for Resistance Welding 234

5.29 Electronic Welding Control 236

5.29.1 Ignitron Contactor 236

5.29.2 Heat Control Unit 237

5.29.3 AC Timer Circuit 238

5.30 Energy Storage Welding 239

5.30.1 Capacitor Discharge Circuit 239

5.30.2 Magnetic Storage Welding Circuit 240

6. ELECTRIC TRACTION   242

6.1 Introduction 242

6.2 Traction Systems 242

6.3 Steam Engine Drive 242

6.4 Diesel-electric Drive 243

6.5 Internal Combustion Engine Drive 243

6.6 Electric Drive 244

6.7 Battery Drive 245

6.8 Hybrid Drive 245

6.9 Flywheel Drive 245

6.10 Tramways 246

6.11 Trolley Bus 246

6.12 Systems of Railway Electrification 247

6.12.1 Direct Current System 248

6.12.2 Single-phase Low Frequency AC System 248

6.12.3 Three-phase Low Frequency System 249

6.12.4 Composite System 249

6.13 Advantages of 25 kV, 50 Hz AC System 250

6.14 Comparison between DC and AC Systems of Railway Electrification from the Point of View of Main Line and Suburban Line Railway Service 252

6.15 Traction Mechanics 253

6.15.1 Units used in Traction Mechanics 253

6.15.2 Types of Services 253

6.15.3 Speed Time Curve 253

6.15.4 Simplified Speed–Time Curve 254

6.15.5 Average Speed and Schedule Speed 258

6.16 Train Movement 258

6.17 Tractive Effort 258

6.18 Tractive Effort-speed Characteristic 262

6.19 Power of the Traction Motor 263

6.20 Specific Energy Consumption 265

6.20.1 Factors Affective Specific Energy Consumption 266

6.21 Mechanics of Train Movement 267

6.22 Coefficient of Adhesion 267

6.22.1 Factors Affecting Slip 269

6.23 Electric Traction Systems—Power Supply 273

6.23.1 Transmission Lines to Substations 273

6.23.2 Substations 274

6.23.3 Feeding and Distribution System on AC Traction 275

6.23.4 Feeding and Distribution System for DC Tramways 276

6.23.5 Electrolysis by Currents through Earth 276

6.23.6 Negative Boosters 276

6.24 Block Diagram of AC Electric Locomotive 277

6.25 Overhead Equipment (OHE) 278

6.26 Current Collection System 280

6.26.1 Conductor Rail System 280

6.26.2 Current Collection Gear for OHE 281

6.27 Traction Motor Connections 282

6.28 Smoothing Reactors 283

6.29 Desirable Characteristics of Traction Motors 284

6.30 Traction Motors 285

6.30.1 Suitability of Series Motor for Traction Duty 285

6.30.2 Series Motor Using Undulating DC 286

6.30.3 Suitability of Shunt Motor for Traction Duty 287

6.30.4 Single-phase Series Motors 288

6.30.5 Three-phase Induction Motor 288

6.30.6 Linear Motor 288

6.31 Traction Motor Control 292

6.31.1 Control of DC Traction Motors 292

6.31.2 Series Parallel Control 293

6.32 Advantage of Series Parallel Starting 296

6.33 Metaldyne Control 296

6.34 Multiple Unit Control 299

6.35 Wheel Arrangement 301

6.36 Bogie Arrangements 302

6.36.1 Monomotor Bogie 303

6.37 Transmission of Drive 304

6.38 Braking 307

6.38.1 Requirements of a Braking System 307

6.39 Types of Braking 308

6.39.1 Mechanical Braking 308

6.39.2 Air Brake System 308
6.39.3 Vacuum Brake System 308
6.40 Hydraulic Brake 310
6.41 Eddy Current Brakes 310
6.42 Magnetic Brakes 310
6.43 Types of Electric Braking 311
6.44 Mechanical Regenerative Braking 311
6.45 Electrical Regenerative Braking 312
  6.45.1 How Electrical Regenerative Braking is Applied 312
  6.45.2 Conditions Necessary to Achieve Electric Regenerative Braking (ERB) 313
  6.45.3 Suitability of DC Shunt, Series and Induction Motors for ERB 314

7. ELECTROLYTIC PROCESSES 327

7.1 Introduction 327
7.2 Electrolysis-Basic Principle 327
7.3 Laws of Electrolysis 328
7.4 Terms Connected with Electrolytic Processes 330
7.5 Applications of Electrolysis 331
7.6 Electrodeposition 332
  7.6.1 Electroplating 333
  7.6.2 Electrodeposition of Rubber 335
  7.6.3 Electrometallisation 335
  7.6.4 Electrofacing 335
  7.6.5 Electroforming 335
  7.6.6 Electrotyping 336
7.7 Manufacture of Chemicals 336
7.8 Anodizing 336
7.9 Electropolishing 337
7.10 Electrocleaning or Pickling 337
7.11 Electroparting or Stripping 338
7.12 Electroextraction 338
7.13 Electrorefining 339
7.14 Power Supply for Electrolytic Processes 339

8. REFRIGERATION AND AIR CONDITIONING 342

8.1 Introduction 342
8.2 Basic Terminology 342
8.3 Applications of Refrigeration 343
8.4 Refrigeration Systems 343
  8.4.1 Vapour Compression Refrigeration System 343
  8.4.2 Absorption Refrigeration System 344
  8.4.3 Thermoelectric Refrigeration System 345
8.5 Coefficient of Performance 346
8.6 Unit of Refrigeration 346
8.7 Refrigerants 346
8.8 Domestic Refrigerator 347
  8.8.1 Troubleshooting of Refrigerator 349
8.9 Water Cooler 350
  8.9.1 Instantaneous Type Water Coolers 351
  8.9.2 Storage Type Water Cooler 353
8.10 Desert Cooler 354
8.11 Air Conditioning 355
8.12 Types of Air Conditioning 355
  8.12.1 Unitary Type 356
  8.12.2 Central Type 356
  8.12.3 Unitary Central Type 356
8.13 Window Air Conditioner 356

Bibliography 361
Index 363

CONTROL DEVICES FOR INDUSTRIAL MOTOR

1.1 INTRODUCTION

Industrial control in its expanded meaning encompasses all the methods used to control and maintain the performance of an electrical system. When applied to machinery it includes the starting, acceleration, reversal and stopping of a motor and its load. For this purpose semiconductor devices are becoming more and more popular these days. It, therefore, becomes very important to understand the principle of their working as various control devices.

1.2 SEMICONDUCTOR DEVICES

A semiconductor is a substance, which has resistivity (10^{-4} to $0.5\ \Omega$m) in between that of conductors and insulators. These materials have very limited number of free electrons and holes, which accounts for only partial conduction. These materials have negative temperature coefficient of resistance. These materials may be of silicon or germanium base and depending upon whether these are doped with donor or acceptor, we get P type N type of crystals. This is explained in the following paragraphs.

1.2.1 Diode

The term diode denotes two electrode device. A semiconductor diode is simply a P-N junction connecting leads or terminals on the two sides of the junction. A diode is a unidirectional device permitting the easy flow of current in one direction but restraining the flow in opposite direction. A major application of diodes is rectification, i.e. conversion of AC into DC. Semiconductor diode is gaining more popularity these days due to its smaller size, cheapness, robustness and higher operating efficiency.

Circuit symbol of a semiconductor diode is given in Fig. 1.1a. The graphical symbol for a diode is shown in Fig. 1.1b. The arrow in the symbol indicates the direction of conventional current flow when the diode is on, i.e. from the positive terminal through the device to the negative terminal. The P-side of the diode is always the positive terminal for forward bias and is designated the anode. The N side is called the cathode and is the negative terminal when the device is forward biased.

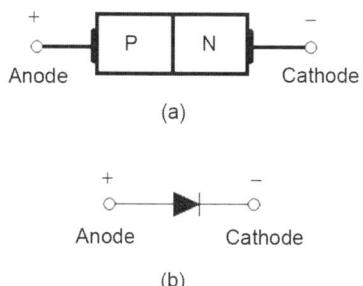

Figs 1.1a and b: (a) Circuit symbol, (b) Graphical symbol.

The semiconductor diode may be either silicon one or germanium one. The silicon junction diode is similar in appearance to the germanium diode but differs in internal properties. Silicon diodes have, in general, higher peak inverse voltage (PIV) and current ratings and wider temperature ranges than germanium diode.

The volt-ampere characteristics of a diode are illustrated in Fig. 1.2. When the junction is forward biased, very little current, called the forward current, flows until the forward voltage exceeds the junction barrier potential (0.3 V for Ge and 0.7 V for Si). The characteristics follow an exponential law. With the increase in forward voltage forward current increases almost linearly and the P-N junction starts behaving as a resistor and when the applied voltage exceeds a certain value, extremely large current would flow and the P-N junction may get damaged due to overheating. A practical diode has a very low forward resistance (of the order 75 Ω in case of Ge and 150 Ω in case of silicon) and has a voltage drop of 0.3 to 0.7 V at all current levels in forward direction.

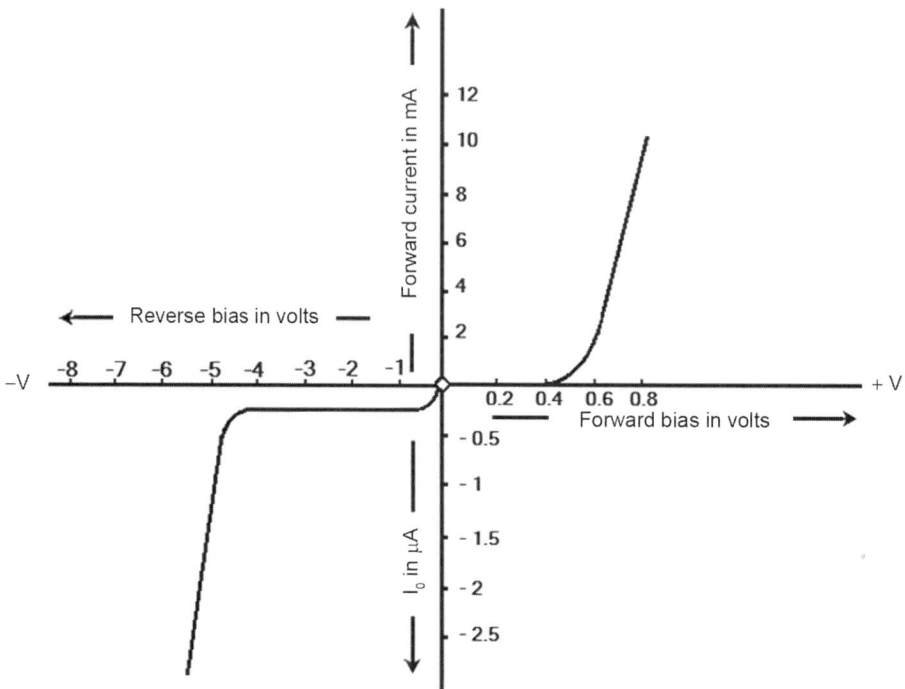

Fig. 1.2: V-I characteristics of a diode.

When an ordinary P-N junction diode is reverse biased, normally only very small reverse saturation current flows. It is almost independent of the voltage applied. However, if the current increases abruptly (Fig. 1.2), it could destroy the junction. If the reverse current is limited by means of a suitable series resistor, the power dissipation at the junction will not be excessive, and the device may be operated continuously in its breakdown region to its normal (reverse saturation) level. It is found that for a suitably designed diode, the breakdown voltage is very stable over a wide range of reverse currents. This quality gives the breakdown diode many useful applications as a voltage reference source.

An ideal diode offers zero impedance to current flow in one direction and infinite impedance in the opposite. The symbol and V-I characteristic of an ideal diode are shown in Figs 1.3a and b. An ideal diode acts like an automatic switch. The switch is closed when the diode is forward biased and is opened when reverse biased.

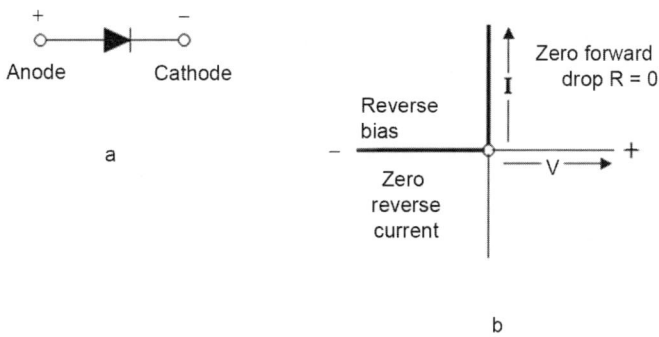

Figs 1.3a and b: (a) Symbol of ideal diode, (b) V-I characteristic of an ideal diode.

The main parameters of a diode are blocking or peak inverse voltage or PIV (about 1,000 V for Si and 400 V for Ge) forward current and maximum operating junction temperature (nearly 125°C for Si and 60°C for Ge). Protection is compulsarily provided against voltage surges, over-current and excessive temperature rise.

A "free-wheeling" diode is commonly employed in motor control system to provide an alternate path for continuity of current in the inductive circuit following the switching-off some power device between the motor and the supply source.

1.2.2 Zener Diode

Zener diode, also known as the breakdown diode, is a P-N junction diode specially designed for operation in the breakdown region in reverse bias condition. The breakdown diode may be silicon or germanium one but silicon is preferred over germanium because of higher operating temperature and current capability. The knee point is also sharper in case of a silicon diode. These diodes are analogous to gas diodes in which large current appears on reaching the breakdown potential.

The symbol of a zener diode is given in Fig. 1.4. This is similar to that of a normal diode except that the line representing the cathode is bent at both ends, i.e. the bar is turned into Z-shape; which stands for zener.

The zener diode, though not a true power controlling device, is quite often use as a voltage control and sensing device in many motor controllers.

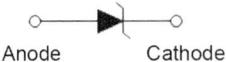

Anode   Cathode

Fig. 1.4: Symbol of a zener diode.

1.2.3 Power Transistor

The transistor is a two junction, three terminal, current controlled solid state device. This device is a 3-layer P-N-P (or N-P-N) structure as illustrated in Figs 1.5a and b. The transistor is capable of amplification and in most respect it is similiar to a vacuum triode. The transistor is only 5 decade old, yet it is replacing vacuum tubes in almost every applications. The reasons are obviously its advantages over vacuum tubes such as compact size, lightweight, rugged construction more resistive to shocks and vibrations, instantaneous operation (no heating required), low operating voltage, high operating efficiency and long-life with essentially no ageing effect if operated within permissible limits of temperature and frequency.

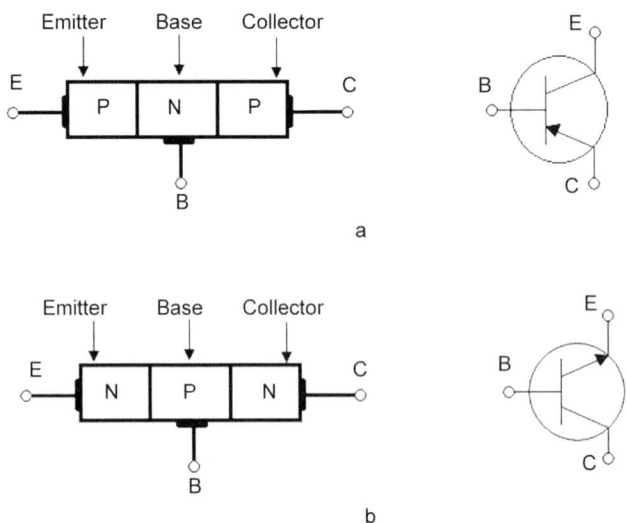

Figs 1.5a and b: (a) PNP transistor, (b) NPN transistor.

Normally, the emitter-base junction is forward biased and collector-base junction is reverse-biased. The collector current equal is slightly less than the emitter current. The collector current can be controlled by base current. Except for small motors, transistors are used in the switching (on/off) mode. The common-emitter configuration is normally employed as it has high power gain.

Power transistors are now available in ratings suitable for motor control. The common-emitter saturation voltage $V_{CE\ (Sat)}$ for typical power transistors varies from 0.3 to 0.7 V. This is invariably less than the on-state anode-cathode voltage drop of an SCR, and therefore, the average power loss in a transistor is apparently less than that of an SCR of an equivalent power rating.

The switching of a power transistor is controlled through the base current. An uninterrupted base current is needed to keep the transistor in the saturated state, i.e., the on-state. In high-current device, a base current of several amperes is required to keep the device in the on state owing to its low current gain (about 10) and, therefore, the power loss in the base circuit may be appreciable.

Power transistors have no surge current capacity and can tolerate only a low rate of change of current. The switching operations of power transistors are normally faster than SCRs, and the commutation problems of SCR are not foreseen in transistors. Power transistor are becoming more and more popular in low to medium power applications, where they complete successfully with thyristors and GTOs. Power transistor applications are in the range of a few kW to several 100 kWs size in voltage fed choppers and inverters with switching frequency up to 10–15 kHz.

1.2.4 Thyristor

Thyristor is the general name given to a family of semiconductor devices having four layers with a control mechanism, although this term is most commonly applied to the silicon-controlled rectifier (SCR). This term is derived from thyratron and transistor because the device combines the rectification action of thyratron and control action of transistor.

Since its inception the thyristor has come to stay as a basic building block in many industrial and power system applications. Its ability to be controlled, compactness, fast response, high reliability, better efficiency, large power handling capacity, high voltage and current ratings, good trigger sensitivity, static operation, large power gain, sturdy construction, long life, very little maintenance and low cost of fabrication—due to advancement in the field of fabrication—have given the thyristor a colourful reception in every field. Today thyristors are finding applications in the control of DC/AC motors; for the improvement of power factor and as switching devices. Thyristors have helped in further cost reduction and in the development of drive system by changing the emphasis from DC motors to AC motors. With cycle converter and inverters, the speed of an AC motor can also be controlled with ease and reliability. Apart from these main applications it finds use as a switching device, particularly in the improvement of power factor of transmission lines and mains. Thyristor can be used as a power switching device with a power handling capacity ranging from a few watts to as high as 4 MW (2,500 A at 1600 V). Some thyristors have a rating as high as 400 A, 10,000 V for use in HV AC transmission lines.

Thyristors, with its large number of advantages and tremendous control capabilities, have numerous applications and have completely replaced the electromagnetic control systems. Thyristor basically serves two functions viz electronic switching an electronic control. Some the applications of thyristors are listed below:

1. Speed control of DC and AC motors.
2. As rectifier for conversion of AC into DC
3. As inverter for conversion of DC into AC
4. As DC chopper or DC to DC converter for converting DC at one level to DC at another level.
5. As cyclo-converter for converting AC of one frequency into AC of another frequency.
6. Control of temperature, level, position and illumination.
7. Power switches (DC and AC circuit breakers).

8. As static switches.
9. Control of induction heating.
10. Relay control.
11. Phase control.

SCRs and triacs having high voltage and current ratings are widely employed for power control applications whereas other members of thyristor family are employed for small power applications and for switching in control and digital circuits.

1.2.5 Silicon-Controlled Rectifier (SCR)

The SCR is a four layer, three-junction and a three-terminal device and is shown in Fig. 1.6a. The end P-region is the anode, the end N region is the cathode and the inner P-region is the gate. The anode to cathode is connected in series with the load circuit. Essentially the device is a switch. Ideally it remains of (voltage blocking state), or appears to have an infinite impedance until both the anode and gate terminals have suitable positive voltages with respect to the cathode terminal. The SCR then switches on and current flows and continues to conduct without further gate signals. Ideally the SCR has zero impedance in conduction state. For switching off or revarting to the blocking state, there must be no gate signal and the anode current must be reduced to zero. Current can flow only in one direction. Now SCRs of voltage rating 10 kV and an rms current rating of 35,000 A with speed of 1 μ s corresponding to a power-handling capacity of 30 MW are available. It can be switched on by low voltage supply of about 1 A and 10 W. Obviously it has tremendous power amplification the order of (3×10^6). Because of its compactness, high reliability and low loss, the SCR has almost replaced the earlier power switching devices-thyratron and a magnetic amplifier.

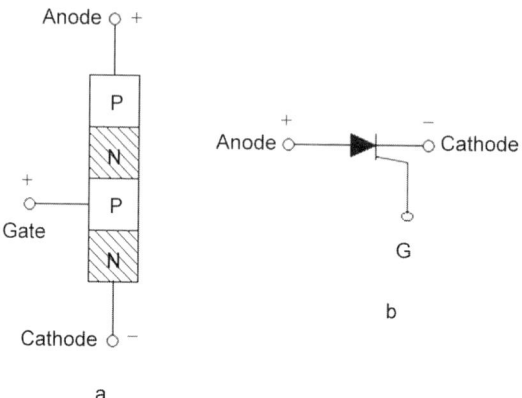

Figs 1.6a and b: (a) Circuit symbol, (b) Graphical symbol of a SCR.

☞ *Volt-Ampere Characteristics*

As mentioned above, the SCR is a four-layer device with three terminals, namely, the anode, the cathode and the gate. When the anode is made positiive with respect to the cathode, junctions J_1 and J_3 are forward biased and junction J_2 is reverse-biased and only the leakage current will flow through the device. The SCR is then said to be in the forward blocking state or in the forward mode or off state. But when the cathode is made positive with respect to the anode, junctions J_1 and J_3 are reverse-baised, a small reverse leakage current will flow through the SCR and the SCR is said to be in the reverse blocking state or in reverse mode.

When the anode is positive with respect to the cathode, i.e. when the SCR is in forward mode, it does not conduct unless the forward voltage exceeds certain value, called the forward breakover voltage, V_{FBO}. In non-conducting state, the current through the SCR is the leakage current which is very small and can be negligible. If a positive gate current is applied, the SCR can become conducting at a voltage, subsequently lower than the forward breakover voltage. The larger the gate current, the lower the break-over voltage. With sufficiently large gate current, the SCR behaves identical to PN rectifier. Once the SCR is switched on, the forward voltage drop across it, is suddenly reduced to a very small value, say about 1 volt. In the conducting or on-state, the current through the SCR is reduced by the external impedance. A typical V-I characteristics of SCR is shown in Fig. 1.7.

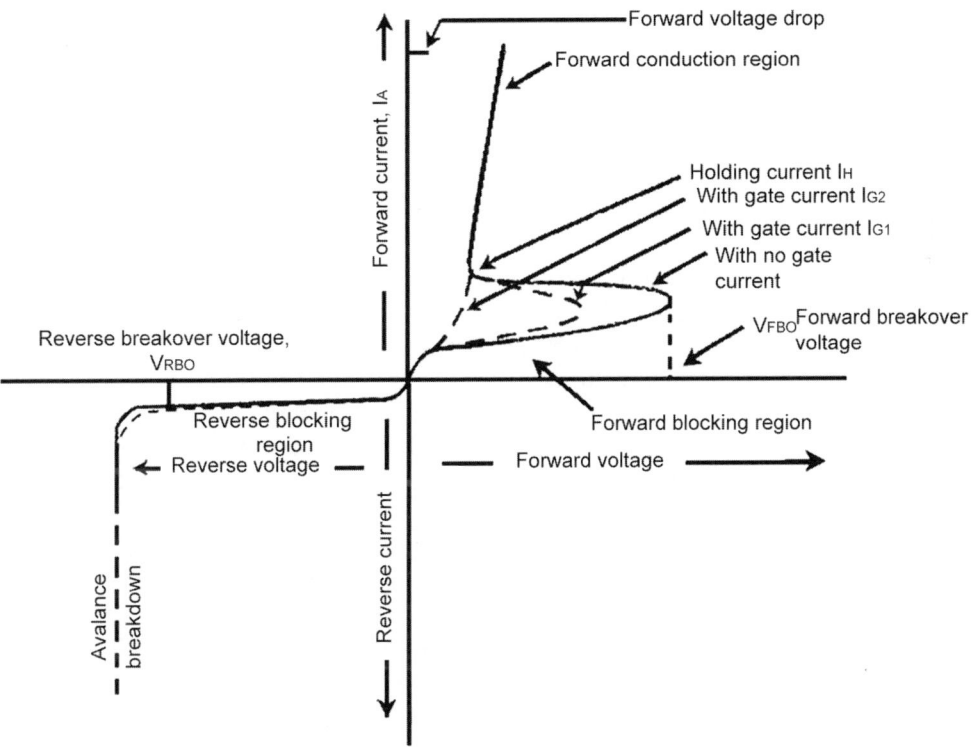

Fig. 1.7: V-I characteristics of SCR.

When the anode is negative with respect to cathode, i.e. when the SCR is in reverse mode or in the blocking state, no current flows through the SCR except a very small leakage current of the order of few micro amperes. But if the reverse voltage is increased beyond the certain value, called the reverse breakover voltage, V_{RBO} avalanche breakdown takes place. Forward breakover voltage V_{FBO} is usually higher than reverse breakover voltage V_{RBO}.

From the foregoing discussion it can be seen that SCR has stable and reversible operating states. The change over from off state to on state called turn-on, can be done by increasing the forward voltage beyond V_{FBO}. A more convenient and useful method of turning on the device employs the gate drive. If the forward voltage is less than forward breakover voltage V_{FBO}, it can be turned on by applying a positive voltage between the gate and the cathode. This method is called the gate control method. Another very important feature of the gate is that once the SCR is triggered to on-state the gate loses its control.

The switching action of gate takes place only when

    i. SCR is forward biased, i.e. anode is positive with respect to cathode.

    ii. Suitable positive voltage is applied between the gate and the cathode.

Once the SCR has been switched on, it has no control on the amount of current flowing through it. The current through the SCR is entirely controlled by the external impedance connected in the circuit and the applied voltage. There is, however, a very small, about 1 V potential drop across the SCR. The forward current through the SCR can be reduced by reducing the applied voltage or by increasing the circuit impedance. There is however, a minimum forward current that must be maintained to keep the SCR in conducting state. This is called the holding current rating of SCR. If the current through the SCR reduced below the level of holding current, the device returns to off-state or blocking state.

The SCR can be switched off by reducing the forward current below the level of holding current which may be done either by reducing the applied voltage or by increasing the circuit impedance. The gate can only trigger or switch-on the SCR, it cannot switched off.

Alternatively the SCR can be switched off by applying negative voltage to the anode (reverse mode), the SCR naturally will be switched off.

Here one point is worth mentioning, the SCR takes certain time to switch off. The time, called the turn-off time, must be allowed before forward voltage may he applied again otherwise the device will switch-on with forward voltage without any gate pulse. The turn-off time is about 15 microseconds, which is immaterial when dealing with power frequency, but this becomes important in the inverter circuits, which are to operate at high frequency.

There are mainly two types of SCRs for use in motor controllers—inverter grade and converter grade. Former ones are used in inverters, cycloconverters, and brushless DC motor stems while the latters are used in choppers, phase-controlled rectifiers, AC regulators, etc. These two grades vary in their time responses. Inverter grade SCRs are costlier than the converter grade.

1.2.6 Triac

The triac is another three-terminal AC switch that is triggered into construction when a low-energy signal is applied to its gate terminal. Unlike the SCR, the triac conducts in either direction when turned on. The triac also differs from the SCR in that either a positive or negative gate signal triggers it into conduction. Thus, the triac is a three terminal, four layer bidirectional semiconductor device that controls AC power whereas an SCR controls DC power or forward biased half cycles of AC in a load. Because of its bidirectional conduction property, the triac is widely used in the field of power electronics for control purposes. Triacs of 16 kW rating are readily available in the market. A schematic symbol of Triac is shown in Fig. 1.8a.

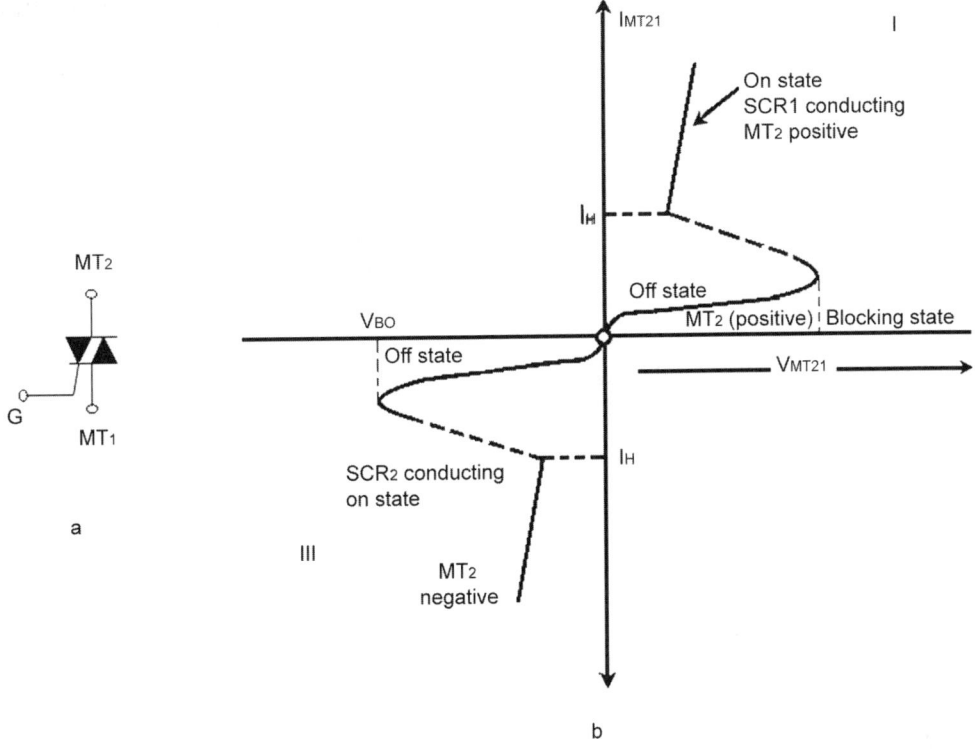

Figs 1.8a and b: (a) Schematic symbol, (b) V-I characteristic of triac.

"Triac" is an abbreviation for three terminal AC switch. "Tri'-indicates the device has three terminals and AC indicates that the device controls alternating current or can conduct in either direction. Though the triac can be turned on without any gate current provided the supply voltage becomes equal to the breakover voltage of the triac but the normal way to turn on the triac is by applying a proper gate current. As in case of SCR, here too, the larger the gate current, the smaller the supply voltage at which the triac is turned on.

Triac can conduct current irrespective of the voltage polarities of terminals MT1, and MT2 with respect to each other and that of gate and terminal MT2.

Typical V-I characteristic of a triac is shown in Fig. 1.8b. The triac has on and off state characteristics similar to SCR but now the characteristic is applicable on both positive and negative voltages. This is expected because triac consists of two SCRs connected in parallel but opposite in directions.

Next to SCR, the triac is the most widely used member of the thyristor family. In fact, in many of control applications, it has replaced SCR by virtue of its bidirectional conductivity. Motor speed regulation, temperature control, illumination control, liquid level control, phase control circuits, power switches, etc. are some of its main applications.

1.2.7 Diac

A Diac is an important member of the thyristor family and is usually employed for triggering triacs. A diac is a two-electrode bidirectional avalanche diode which can be switched from off-state to the on-state for either polarity of the applied voltage. This is just like a triac without gate terminal. Its equivalent circuit is a pair of inverted four layer diodes. Two schematic symbols are shown in Fig. 1.9a. Again the terminal designations are arbitrary since the diac, like triac, is also a bilateral device. The switching from off-state to on-state is achieved by simply exceeding the avalanche breakdown voltage in either direction. Volt ampere characteristics is shown in Fig. 1.9b.

The diacs, because of their symmetrical bidirectional switching characteristics, are widely used as triggering devices in triac phase control circuits used in heat control, universal motor speed control, etc.

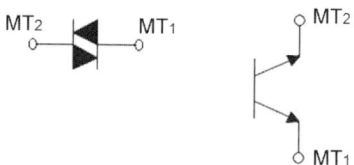

Fig. 1.9a: Schematic symbols of diac.

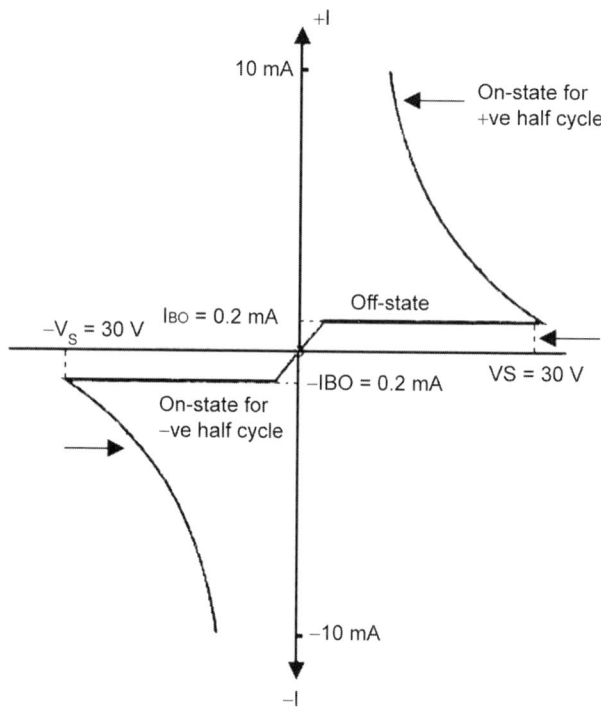

Fig. 1.9b: V-I characteristic of diac.

1.2.8 Gate-turn-off Switch

Gate-turn-off switch is, like a SCR is a four layer, three junction semiconductor device with, three external terminals, namely, the anode, the cathode and the gate. The schematic symbol is shown in Fig. 1.10.

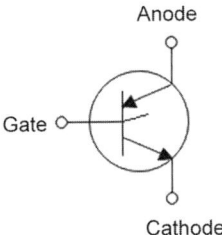

Anode

Gate

Cathode

Fig. 1.10: Schematic symbol of GTO.

The main advantage of the GTO over the SCR or SCS is that it can be turned on or off by applying the proper pulse to the cathode GATE (without the anode gate and associated circuitry required for the SCS). Because of its turn-off capability, there is an increase in the magnitude of the required gate current for triggering. For an SCR and GTO of similar maximum rms current ratings, the gate triggering current may be of the order of 30 μA and 20 mA respectively. The maximum rms current and dissipation ratings of GTOs manufactured today are limited to about 3 A and 20 W respectively.

Another distinct advantage of GTO is its improved switching characteristics. The turn-on time is similar to that of an SCR (typically 1 μs), but the turn-off time duration (1 μs) is much smaller than the typical turn-off time of an SCR (5–30 μs). This mits the use of this device in high speed applications.

1.2.9 Unijunction Transistor (UJT)

As shown in Fig. 1.11a it consists of an N type bar of silicon having side contact of P material called emitter. Two contacts base 1 and base ate attached to N material such that emitter is nearer to base 2. Essentially UJT is a P-N diode. Further that the resistance of silicon bar between base 1 and 2 is of the order of 5 to 10 kΩ. Therefore with emitter current absent, the main body of silicon N material works as a potential divider. UJT, therefore, combines PN junction diode with voltage divider as shown in Fig. 1.11b. Base 1 is usually grounded and base 2 is connected to positive. If resistance between B_2 and x is denoted as RB_2, between B_1 and x as R_{B_1} between B_1 and B_2 as V_{BB} and voltage between B_1 and B_2 as V_{BB} then the voltage between x and ground by voltage divider principle is given as

$$V_{B_1} = \frac{R_{B_1}}{R_{BB}} V_{BB}$$

$$= \eta V_{BB}$$

Where η is called intrinsic stand off radio of the UJT and its value lies between 0.52 to 0.62. Now till external voltage V. applied between y and z remains less than "ηV_{BB}", diode remains reverse biased. As soon as Ve is made larger than "ηV_{BB}", the emitter is forward biased and emitter current flows. Since emitter is of material its current carriers are holes which are rejected into N type bar. These holes are attracted to base 1 being less positive than base 2. Increase in the number of current carriers reduces the resistance R_{B_1}. Since R_{B_2} is in series with R_{B_1}, decrease in R_{B_2} will also reduce V_{XB_1}. We thus see that with the increase in emitter current, emitter voltage becomes less,

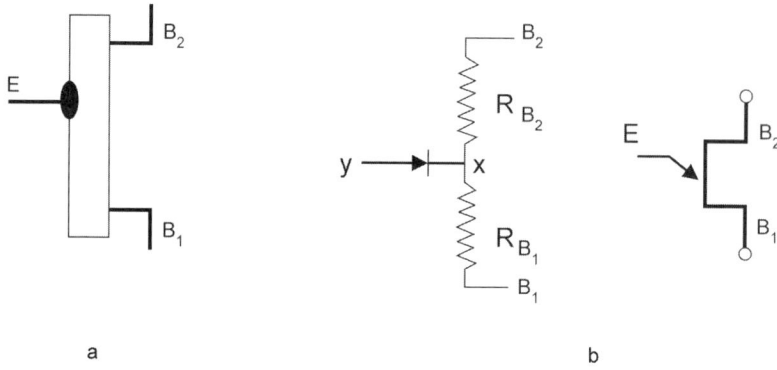

Figs 1.11a and b: (a) UJT, (b) PN junction diode with voltage divider.

indicating that R_{B_1} has negative resistance characteristic at this time. We will summarise the characteristics of UJT as follows:

It has two states of operation off or on. When emitter is reverse biased, i.e. $V < \eta V_{BB}$, UJT is off. When emitter is forward biased $V > \eta V_{BB}$, UJT is on and the value of R_{B_1} becomes very low.

1.3 THYRATRON

Thyratron is a hot cathode gas filled triode. A symbolic representation of thyratron is shown in Fig. 1.12. Normally it is filled with a small amount of inert gas such as argon, hydrogen, neon or mercury vapour. The fast acting devices do not use mercury vapour as the mercury vapour has the disadvantage of characteristics varying with

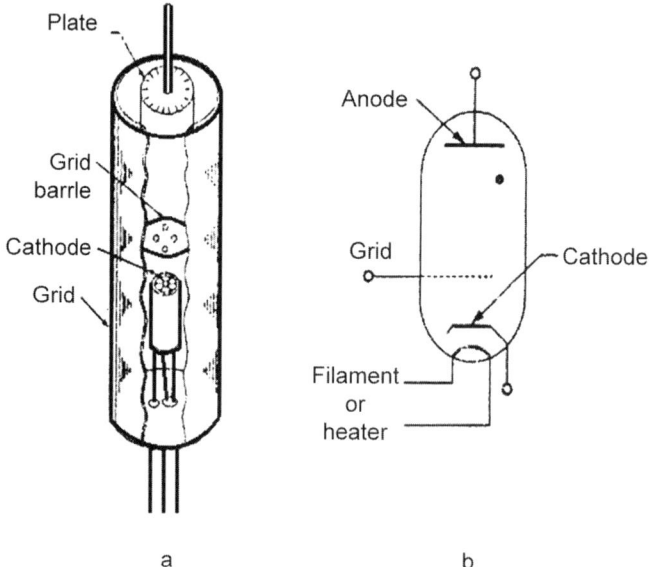

Figs 1.12a and b: (a) Construction, (b) Symbolic representation.

temperature and it takes time in deionization. The characteristics of hydrogen thyratron are stable and the device are fast. Plate is a disc of graphite. The special construction oxide coated cathode is used so that its surface is not damaged due to bombardment of positively charged particles. The control grid has a metal cylinder surrounding the cathode with one or more perforated discs known as grid baffles near the centre. Thus the control grid acts as an electrostatic shield between cathode and anode except for the holes in the grid baffles.

Before the tube fires, its characteristics are similar to those of a vacuum triode, i.e. the plate current (small in magnitude) can be varied by variation of potential applied to the grid. After firing of the tube the grid loses control over plate current completely and plate current is limited by the external resistance in the plate circuit, i.e. the thyratron behaves like a diode. The introduction of grid in between the plate and cathode makes it possible to control the conditions under which the discharge should start.

Since the only function of the grid is to delay or prevent the starting of the current, the grid is naturally designed to achieve its purpose with as little grid voltage as possible. The grid, therefore, is quite an extensive structure. The critical grid voltage varies with the plate voltage to some extent, and also with the design of the grid.

☞ *Plate Voltage-Plate Current (or Plate) Characteristics*

These characteristics are drawn for various grid voltages. For a given grid voltage when the plate voltage is increased from zero, initially there is no plate current but when the plate is further increased, a voltage is reached, called the breakdown voltage, the current at once shoots up and is limited only by the external resistance in the plate circuit. If the grid is made more negative, the tube will breakdown at higher plate voltage, as illustrated in Fig. 1.13.

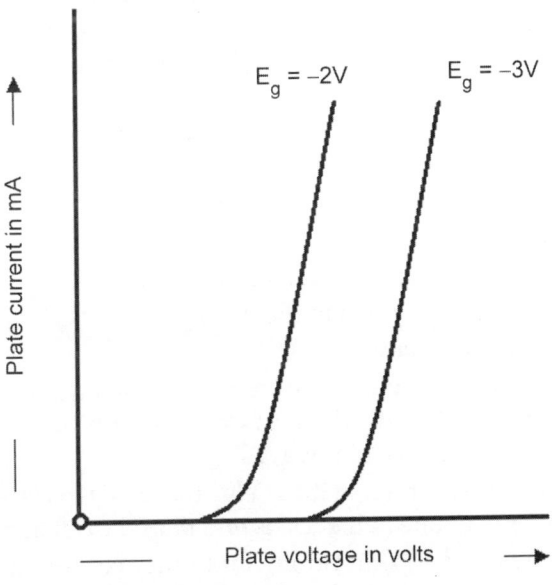

Fig. 1.13: Plate characteristics.

☞ *Grid Control Characteristics*

The curve drawn between plate voltage and critical grid voltage is called the grid control characteristic or starting characteristic or breakdown characteristic. This characteristic can be determined by determining the value of critical grid voltage for different values of plate voltage and plotting them, as shown in Fig. 1.14. The slope of this characteristic is known as control ratio.

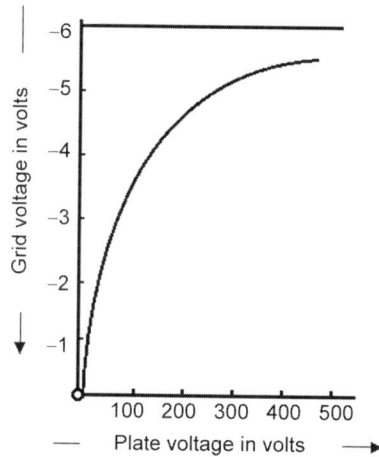

Fig. 1.14: Critical grid voltage for different values of plate voltage.

☞ *Applications*

Thyratron has two main fields of applications:
  i. As an electronic switch to handle heavy currents.
  ii. As a grid controlled rectifier when voltage control is required.

The thyratron, when used as electronic switch, keeps the circuit open from the plate to cathode so long as the grid is kept at high negative potential. As soon as the grid potential is brought to such a value that the tube fires, the circuit is closed.

The thyratron switch has two great advantages over mechanical switches:
  i. It is electrically operated and requires only a minute amount of energy to close it, so that it leads itself readily to automatic operation.
  ii. Operation is very fast.

However, it has got some disadvantages also such as :
  i.   Voltage drop of 8 to 18 volts in the circuit.
  ii. It cannot open DC circuit.
  iii. A single thyratron will pass only the positive half waves. A second thyratron connected in parallel with the first one but with opposite polarity may be used to pass the negative half waves.

A thyratron can open AC circuit, with a delay that a half-cycle, so long as frequency is not so high that the time of one half cycle is less than the deionizing time.

The thyratron finds wide use in timing and other control applications. The fact that its time of firing may be controlled makes the thyratron adaptable in heat control in resistance welding. By delaying or advancing the time during the cycle at which tube fires, the average amount of welding current allowed to flow per cycle may be increased or decreased. In addition to this, the number of cycles that the welding current is allowed to flow may be controlled very accurately.

Thyratron rectifiers are also used in electronic methods of electric motor control for both small and large powers. The main advantages of electronic motor control are that the control can be made to depend on very minute signals such as those obtained from photocell or even a thermo-couple and that where necessary an extremely precise control can be obtained.

Control may be affected in several ways such as:
  i. With direct voltage a as well as to the grid.
  ii. With alternating voltage to the plate circuit and direct voltage to the grid.
  iii. Alternating voltage applied on both the grid and plate.

In the first case the tube will fire and conduct whenever the grid voltage is made less negative than critical grid voltage and will again stop to conduct when the plate voltage is reduce below striking voltage or the plate circuit is opened. In the second case the grid will regain control when the plate voltage becomes negative, hence conduction will be stopped every half cycle and the tube breakdown will occur each half cycle as the plate voltage becomes negative, the point at which the tube fires depends upon the grid voltage. The average current flow can be controlled, therefore, by controlling the length of time during the half cycle that the tube is conducting, i.e. by controlling the breakdown point. In the last case breakdown point is determined by the phase relation of the two voltages as well as by their magnitudes. There will thus be a point in each positive half cycle of the cathode-plate path when the tube will breakdown, if the grid voltage has a small negative value than the critical grid voltage with reference to the plate voltage, which will result in ionization and conduction. By using phase shifting device to control the phase of grid voltage with respect to the cathode plate voltage wave the duration of current flow may be controlled from zero to point represented by practically all the positive half cycles of plate voltage.

The thyratron cannot be employed as a amplifier like a vacuum triode because the grid voltage has no control over the magnitude of plate current after firing of the tube.

1.4 IGNITRON

Ignitron is a gaseous tube having mercury pool cathode, carbon anode and ignitor made of boron carbide. It does not have heating element or grid control as in thyratron. Since this is a cold cathode type tube, electrons are to be made available before tube can start conducting. This function is performed by ignitor which is kept dipped in mercury pool. Boron carbide even when dipped in mercury is not wetted by it, Fig. 1.15. As such there is considerable resistance (10 to 500 ohms) between the mercury pool and ignitor. Symbol of ignition is as in Fig. 1.16. When switch 'S' is closed, full voltage is impressed between ignitor and mercury pool which are very close.

This produces intense electric field which pulls out electrons from mercury pull to start with. If the current flow in ignitor is between 20–40 amps, these electrons of the arc, shock ionize the mercury vapours. Positive ions, so released, are attracted towards cathode mercury pool. Momentum of these mercury ions releases from cathode pool large number of electrons by secondary emission, these electrons are attracted towards anode. When main anode current has started, ignitor current stops as the resistance of

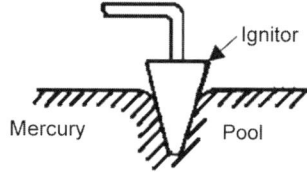

Fig. 1.15: Ignitor dipped in mercury pool.

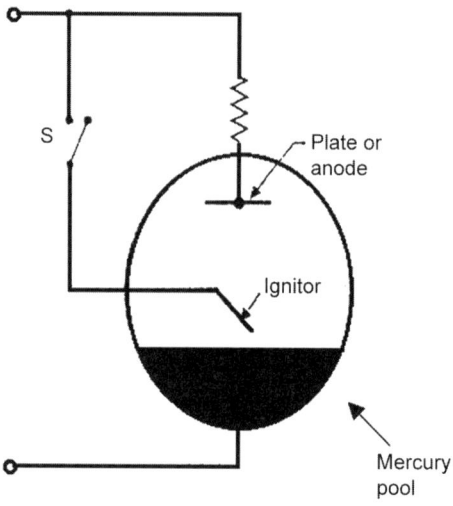

Fig. 1.16: Symbol of ignition.

main path is now very much small as compared to that of ignitor path. This process is to be repeated every time anode becomes positive in AC voltage cycle.

Ignitrons can handle current from 40 to 10,000 amperes. It may also be noted that electronic current flow from ignitor to cathode spoils the ignitor and, therefore, should not be allowed.

1.5   AMPLIDYNE

In a DC shunt or separately excited generator, small power in the field coil controls a great deal more power in the output from generator armature. In this sense generator acts as an amplifier of single stage type. We can have two stage amplifiers by connecting two generators such that armature output of first generator feeds to the field of the second generator as shown in Fig. 1.17. If now the amplification of each stage is 50, one watt fed to the field of first generator will give 50 W armature output. This when fed to the field of second generator would give 2500 W output at the armature of the second generator thereby giving a total gain of 2500 for two stages.

It is possible to combine two machines of two stage amplifier into one explained below. In a separately excited DC generator, shown in Fig. 1.18a, current flow through the field winding now (called control winding) sets up magnetic flux ϕ_F. Rotation of armature in this flux produces an emf across brushes B_1 and B_2.

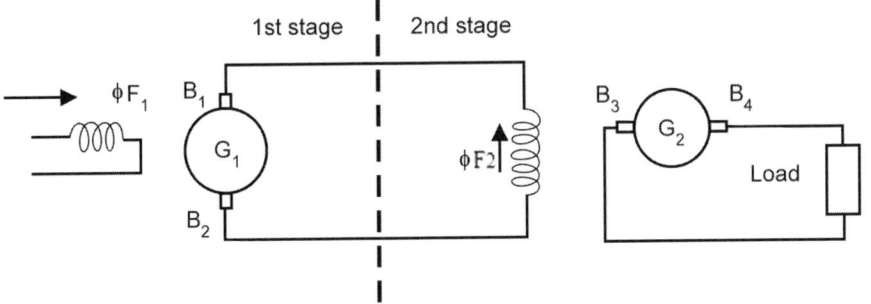

Fig. 1.17: Two-stage amplifiers.

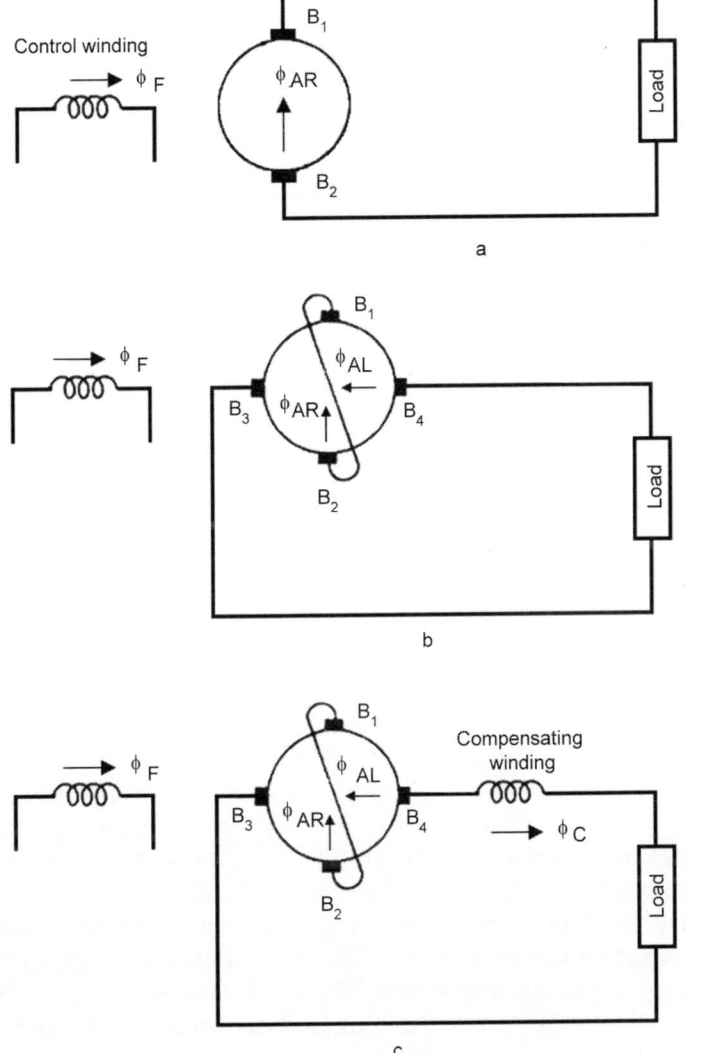

Figs 1.18a to c: (a) Seperately excited DC generator, (b) Brushes B₁ and B₂ as short circuited, (c) Separately excited DC generator with compensating winding.

When load current flows through the armature its sets up armature reaction flux ϕ_{AR}. In Fig. 1.18b we have shown brushes B_1 and B_2 as short circuited instead of being connected to load. Therefore in order that short circuit current across brushes B_1 and B_2 should not be excessive, current through the control winding is to be reduced. Armature reaction flux ϕ_{AR} is produced by the short circuit current flow through armature as usual. This ϕ_{AR} corresponds to ϕ_{F2} as shown in Fig. 1.17. As a result of armature rotating in ϕ_{AR}, emf is produced across brushes B_3 and B_4 across which now is connected load. Load current flow through armature in turn produces armature reaction flux ϕ_{AL}. Its direction is such as to oppose the control field winding flux ϕ_F. This is undesirable as it would interfere with the control action. To overcome this undesirable effect of ϕ_{AL}, a compensating winding is provided wound on the same poles as control winding and is connected in series between armature and load as shown in Fig. 1.18c. It is wound in such a direction that it produces flux C which is equal and opposite to ϕ_{AL} and therefore neutralizes its effect on ϕ_F.

As any DC generator, amplidyne has to be driven by some prime mover which is usually a motor. Motor amplidyne set is used to supply large quantities of controlled electrical power in response to weak signals. It is therefore frequently used to control the speed and torque of large motors.

1.6   SATURABLE REACTOR (MAGNETIC AMPLIFIER)

Saturable reactor is a variable impedance whose reactance can be varied by an auxiliary DC source of low power. Saturable reactor when interposed in series in between AC power source and load as shown in Fig. 1.19. controls the power, voltage or current available at the load terminals.

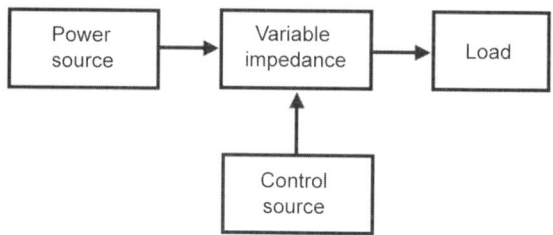

Fig. 1.19: Functioning of saturable reactor.

Saturable reactor in appearances, resembles an ordinary transformer having two winding Fig. 1.20.

Control winding consisting of may turns of thin wire is supplied from a low power DC source through variable resistance and a switch. Output winding of thick wire and having fewer number of turns is connected in series with the load and AC source. If there is no DC supply in the control winding, the output winding has maximum reactance. As a result of this more voltage drop takes in it and therefore voltage applied across load becomes less, If on the other hand DC flows through control winding, reactance of output winding becomes less; in that case less voltage drop takes place in it. The voltage applied across the load as slid, becomes more. We will now understand how the reactance of the output winding is varied inductance (L) of a coil consisting of 'N' number of turns wound over a magnetic core of cross sectional area 'a' and length 'l' having relative permeability of μ_r is given as:

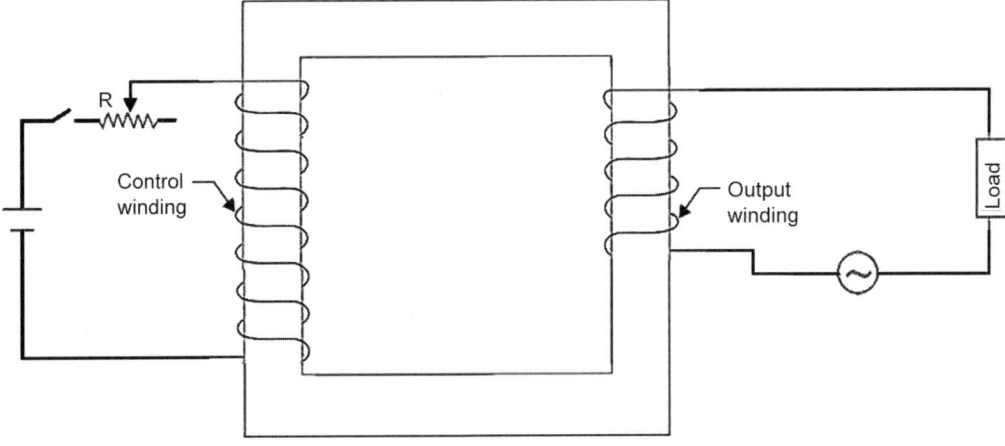

Fig. 1.20: Circuit diagram of saturable reactor.

$$L = \frac{N_2}{\dfrac{l}{\mu_0\mu_r a}} = \frac{N^2\mu_0\mu_r a}{l} \qquad\qquad ...(1.1)$$

Therefore inductance of a coil can he changed by changing the relative permeability of the magnetic path. When current through an air cored coil is increased, magnetic flux produced by it also increases linearly but not very steeply. If the coil is wound on a core of magnetic material, magnetic flux produced by the coil also increases linearly but very steeply. This is because magnetic material offers easy path for the flow of magnetic flux. Steep rise of flux with increase in current signifies high permeability. Steep rise of flux continues up to the point of saturation after which it continues to rise at the same rate as for air cored coil. This is shown in Fig. 1.21.

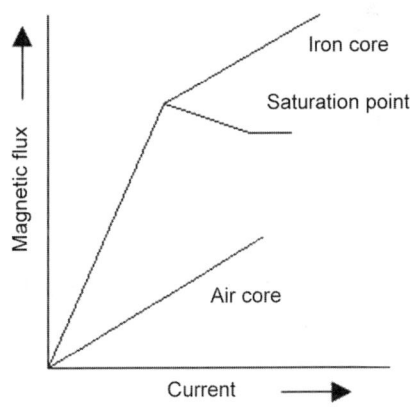

Fig. 1.21: Current *vs* magnetic flux.

Permeability of the magnetic material now becomes less thereby decreasing the inductance of the coil. DC flow through the control winding produces a steady magnetic flux with the result that the output winding has to produce less flux to drive the iron core to saturation. As a result of this the output winding will have reduced reactance and more current will flow through the load. Thus we see that the pre-magnetisation level of the core produced by the control winding determines the current flow through the load. Pre-magnetisation level of the iron core in turn is determined by the amount of DC flowing through the control winding. This current is controlled by varying the resistance R in the DC circuit as shown in Fig. 1.20.

Saturable reactor shown in Fig. 1.20 suffers from a serious drawback. AC flow through the output winding produces alternating magnetic flux which threads with the control winding producing there in AC voltages. This AC induced voltage will interfere with the DC source of the control winding. In order to avoid this, saturable reactor is constructed with a three legged core. Control winding is wound on the central leg and output winding is divided into two halves, each half wound on out side leg in series but with opposing connections so that flux produced by them passes through the outer legs only and not through the central leg thereby eliminating the interference. This can be explained in an other way also. One-half of the output winding produces an emf in the control winding which is cancelled out by the voltage induced by other part, thus producing no interference in the control circuit.

Since control winding consists a many turns, to produce a given amount of flux, it requires relatively small amount of current. Therefore a small power input to the control winding say from a photo cell, thermo-couple or similar device can control relatively large power in the output winding circuit. Saturable reactor is therefore also known as a magnetic amplifier. The advantages that magnetic amplifier possesses over tube or transistor amplifiers are its ruggedness, greater resistance to mechanical shock, greater life expectancy and requires no heating power.

Saturable reactor is used to control the lighting system of a theatre which requires variable voltage. Saturable reactor is used to control the heat of an oven, speed of a DC motor. It may be used as variable inductor in the phase shift circuit to control the firing angle of thyratrons for the speed control of a DC motor.

1.7   PILOT DEVICES

Motor control components may be classified as primary control devices and pilot control devices. Primary control device is one which connects to or disconnects, the motor or load from line. Pilot devices control devices and do not control the load directly. To clarify this we will take up an example of DOL starter. Main contactors and triple pole switch which connect the motor to the line will be classified as primary control device. Start or stop push botton which does not connect or disconnect the motor from line but only energizes or de-energizes the operating coil is an example of pilot device. Pilot devices, in general, will be found to be having sensing element to sense such things as fluid pressure, temperature, liquid level or mechanical pressure applied. Pilot devices convert the above information that they sense to the control of primary control devices. Among various pilot we will consider only most commonly used ones such as push buttons, limit switches, float switches, contactors, pressure switching, thermostats, plugging switches, arc, master controllers, transfer switches and speed governors.

1.7.1 Push Button

This is the most simple pilot device and provides control by pressing a button. The push-button switches are used in remotely operated electromagnetic apparatus and signaling circuits. Push buttons are usually of two types namely maintained contact type in which on pushing of the button the contacts close and remain closed until stop button is pushed and the momentary contact type in which contacts remain closed till push button is held pushed. The contacts of push buttons are usually double break, actuated by plastic buttons. Two sets of contacts are usually provided namely normally open contact 'NOC' and normally closed contacts 'NCC.' So that when the button is pressed one set is opened and the other set is closed. The push button station usually consists of a green button to start and red button to stop. Start push button is provided with normally open contact (NOC) and stop push button is provided with normally closed contact (NCC). The contacts are made silver tipped to withstand numerous of switching operations. Since push buttons are used in control circuits, not used in power circuits, current ratings of button station is, therefore, usually very small (2 A, 500 V for AC working). On pressing start push-button contactor coil is energized, which closes the main contactor in the circuit.

1.7.2 Limit Switches

Limit switches are ordinarily used as pilot devices in control circuits of magnetic starters to govern starting, stopping, or reversing of electric motors. These may be used either as control devices for regular operation or as emergency switches to prevent improper functioning of machinery. These are actuated by the travel of the driven machinery and automatically disconnect the machine after it has travelled through its entire path. These switches can be actuated by cams, levers, push rods and the like, and also through gears. Leaver type actuators may be either of normal roller type or angular type. The switch is either stationary or actuated by some moving member or it moves with the machine and is actuated by a stationary member. These are usually provided with one NOC (normally open contact) and one NCC (normally-closed contact). These switches are usually air break type. These are also used for opening and closing of doors, cinema curtains, lifts, elevators.

1.7.3 Float Switches

This is a type of limit switch and is actuated by a float through a chain, one end of which is connected to the float and the other to the counter-weight. Stops connected the chain move a lever which turns and tilts the mercury tube forming a switch contact. These are designed and used for maintaining a certain level of liquid. The raising or lowering of a float mechanically attached to electrical contacts may start motor driven pump to empty or to fill tanks, whichever is desired. They are also used for opening or closing piping valves the control of fluids.

1.7.4 Pressure Switches

There are basically three types of pressure switches depending upon the means of operation. In the bellow type shown in Fig. 1.22 bellows expand or contract in response to the increase or decrease in pressure. The contacts are held at the end of pivoted lever. Which is acted upon by the bellows. As bellows expand the lever moves against spring I forte. This makes or breaks the contacts depending upon whether they are NO or NC contacts. In the second type, diaphragm is placed in place of belows. Rest of the arrangement is same as for bellows type. In the third type bourdon hollow tube of eleptical shape is employed. Increase in pressure tends to straighten the tube just the same way as in pressure gauge. This action is transformed into rotary motion by linkages which trips mercury switch.

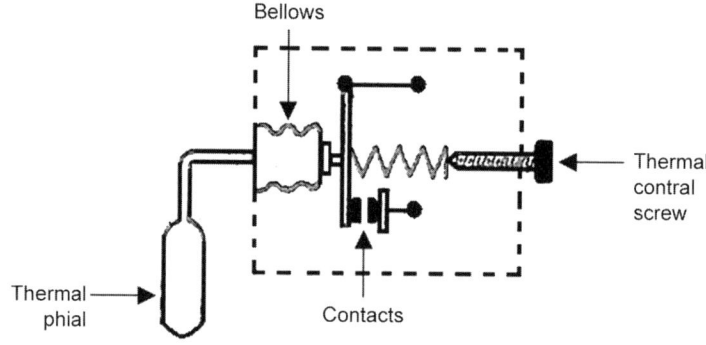

Fig. 1.22: Bellow type pressure switch.

1.7.5 Thermostats

These are switches which make or break a set of contacts in response to temperature changes. Thermostat commonly used in refrigerators is of bellow type shown in Fig. 1.22. It consists of thermal phial containing liquid. This tube is in physical contact with the freezer compartment. Vapour pressure at the liquid inside the phial will depend upon the temperature of the liquid. This pressure acts on the lever through bellows and works pressure switch. Some thermostats employ bimetallic strips to sense temperature and actuate the contacts.

1.7.6 Plugging Switches

We know that DC motors are brought to quick stop by reversing the direction of current flow in either field or armature and that induction motors are reversed by the interchange of any two line leads. As soon as motor comes to stop, it has to be disconnected from the supply. This control function is achieved with the aid of plugging switch. As shown in Fig. 1.23, plugging switch consists of a double control arm held in a central position by springs. In this position two sets of contacts are

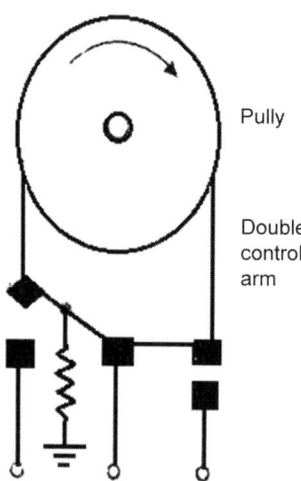

Fig. 1.23: Plugging switch.

open. When forward contactor is closed, motor starts to turn. For clock-wise rotation of motor pulley the belt pulls the contact arm over until right hand contacts close. The arm is now prevented from moving further. Belt now slips over the pulley holding the contacts closed by means of friction. Closure of the right hand contacts energized the reverse contactor. This reverse contactor however cannot close because it is interlocked with forward magnetic contactor. As soon as stop button is pressed, forward contactor opens. Now reverse contactor which remained energised by plugging switch is free to close. In this way motor is connected to the line in reverse direction until it stops. Now the belt friction is not there and control arm comes to central position and opens the circuit to the reverse contactor. We thus see that as soon as stop push button is closed, motor is connected to the supply through reverse contactor. As soon as it comes to rest, it is immediately disconnected from the supply. In this way motor conies to abrupt stop. In case or anticlock rotation of motor, left hand contacts will close and sequence events as described above takes place.

1.7.7 Master Controller

Controller is a device or group of devices that governs in a predetermined manner the delivery of electric power to an electric apparatus with a view to accelerate, decelerate, starts, stop or reverse it as desired. Master controller performs above functions in respect of many electrical apparatus at a time. Control of many electrical apparatus from one place by master controller is made possible by the use of magnetic contractors.

1.7.8 Transfer Switches

Some times it becomes necessary to operates and control an electrical apparatus from either of two masters of pilot devices. For instance multiple unit stock is controlled by two master controllers at ends of a train. For this purpose transfer switches are used which enable electrical apparatus to be connected to either of the masters. Transfer switch merely a double throw switch of required numbers of poles.

1.7.9 Contactors

These are mechanically switching devices having only one position of rest (corresponding to the position of the main contacts either open or closed) operated other than by hand, capable of making, carrying and breaking overload conditions, the speed of make and break being independent of the operator. Following are the various types of contactors:

a. **Electromagnetic contactor:** Here the moving elements leave the position of rest when electromagnet is energized and acts directly on the mechanism of the contactor.

b. **Pneumatic contactor:** Here the moving element leaves the position of rest when compressed air is fed without the use of electrical means, to a device which acts on the mechanism of the contactor.

c. **Electropneumatic contactor:** This is similar to pneumatic contactor except that compressed air is fed by means of electrical controlled valves.

d. **Latched contactor:** A contactor the moving elements of which leave the position of rest when the operating means are energized, but which are prevented by means of a latching arrangement from returning to the position of the rest when the operating means are de-energized. The latching and the release of the latching may be mechanical, magnetic, electrical or pneumatic.

Contactors may be air break or oil immersed type depending upon whether main circuit is made and broken in air or oil. In addition to main contacts, there may be control contacts and auxiliary contacts of low current rating. Control contacts are included in the control circuit of the contactor. For instance normally closed contacts are actuated by overload relays are control contacts. In control schemes certain operations have to be carried out in proper sequence to prevent faulty operation of the equipment and to protect the system against overloads short circuits, etc. This requires auxiliary contacts known as electrical interlocks which make the operation of a piece of apparatus dependent upon certain predetermined conditions being fulfilled.

1.7.9.1 Electromagnetic Contactor or Relay

In this device contacts are made to touch or separate by the action of an electromagnet. In Fig. 1.24, is shown one set of contacts, normally separated by the action of a spring upon the soft iron armature. These contacts are said to be normally open (NO) or make contacts. Other set of contacts normally touch due to the action of the spring. These contacts are called normally closed (NC) air break contacts.

When current from the control circuit flows through the coil, the relay gets energized and the magnetic field produced attracts the armature to create minimum reluctance for the magnetic flux path. As a result of this NO contacts are made to touch and NC contacts are made to separate. In the first case controlled circuit is closed and in the second case it is opened. When the flow of current through the coil ceases, the relay is de-energized and reverse happens under the actions of the spring. In addition to above two types of contacts, it is possible to control several circuits simultaneously by other contact arrangements such as 'break, make' 'make-before-break' and 'break-before-make, break'. Various relay contact configurations are shown in Fig. 1.25. In Fig. 1.25a is shown configuration of NO or make contacts. When relay coil is energized contact 2 moves up to touch contact 1. In Fig. 1.25b is shown configuration of NC or break contacts. When relay coil is energized contact '1' moves up and breaks with the contact 2. In Fig. 1.25c contact 2 breaks with 3 and then touches contact 1. This is break make arrangement. In Fig. 1.25d contact 3 first makes with 1 and then separates it from 2. This is called make before break contacts. In Fig. 1.25e on relay energization,

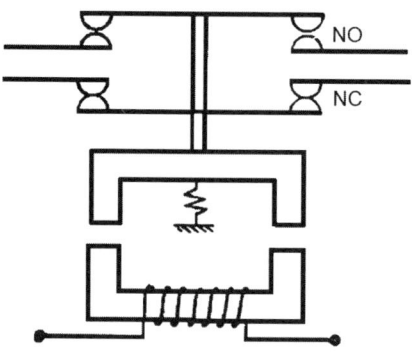

Fig. 1.24: Electromagnetic contactor.

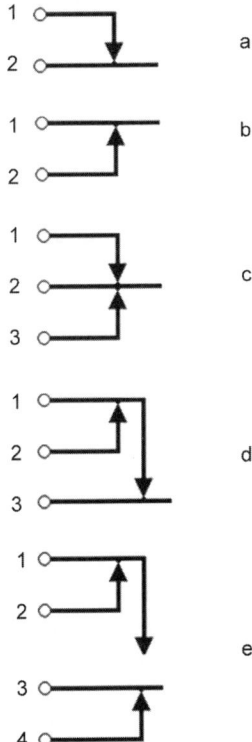

Figs 1.25a to e: Various relay contact configurations (a) Make (NO), (b) Break (NC), (c) Break make, (d) Make before break, (e) Break before make, break.

contact 3 breaks with contact 4 first. It then touches contact 1 separating it from contact 2. It is possible to combine these various arrangements of contacts in many different pile ups or stacks to perform various operations.

Relay coil is of two types. It may carry very small current. In that case coil may consist of many turns of fine wire. The armature and contact points are also light and current of controlled circuit is also relatively small. On the other hand, relay operated from heavier source will have coil with few turns of heavy wire. It may have also heavy armature and contact points and the controlled circuit may also carry large current. Contact points, across which electric spark takes place on closing or opening, are made of material which has high melting point and high resistance to oxidation such as silver, tungsten, palladium, etc. Even then in case of heavy current contact, a condenser in series with resistance is connected across the contact points as shown in Fig. 1.26. Function of condenser is to reduce the sparking when contacts break because now the voltage across the contacts will have to charge the capacitor. When contacts make this capacitor voltage produces current surge which causes spark across them. Resistance in series with capacitor reduces this sparking and thereby minimising the pitting and oxidation of contact points.

Relay coil may be energized by DC or AC With DC no special problem arises. However with AC, alternations of magnetic field cause armature chattering. This affects the controlled circuit also. This problem is overcome in following three ways.

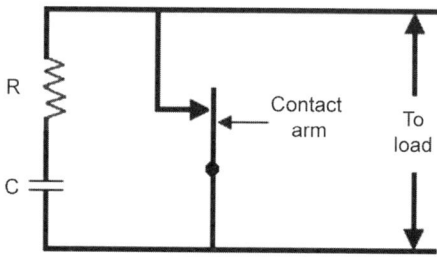

Fig. 1.26: A condenser in series with resistance.

1. To rectify the AC by means of small semiconductor diodes before applying it to the relay coil.
2. To connect condenser of large capacity across the relay coil. This condenser gets charged simultaneously when coil is energized. During the period in between half cycles, when current goes from one alteration to the next the capacitor discharges through the coil, keeping it energized and preventing chatter.
3. To use shading band round haft of the cross section of core of the coil at the top. Two portions of fluxes emitted from the core have time phase difference. The portion of flux emitted from the portion surrounded by shading band lags behind the flux through unshaded portion. Thus at no time magnitude of flux becomes zero. This therefore prevents chattering.

1.7.10 Speed Governors

Speed governors, actuated by centrifugal force, are employed to protect machinery against over speed or under speed or to perform some functions at given speed of motor or machine. For instance speed governor in conjunction with controller can be used for two speed AC motors. When slow speed is desired, high speed winding is disconnected from the supply. Motor now speeds down till at the required speed, governor connects the slow speed winding to the supply.

QUESTIONS

1. Why are solid-state devices becoming more and more popular in industry nowadays?
2. What is diode? Give its V-I characteristic and uses.
3. What do you understand by the word "Thyristor"? Give its applications.
4. What is an SCR? Give its volt-ampere characteristic and typical parameters.
5. With a neat diagram explain the constructional features of a triac. Sketch its V-I characteristics. How does it differ from an SCR? Mention its applications.
6. What is zener diode and where is it mostly used?
7. Explain the function of a transistor used (I) as a switch (II) as a rheostat and (Ill) as an amplifier.
8. What is diac? Give its schematic symbol, V-I characteristic and uses.
9. Where do you recommend the use of free wheeling diode?
10. What do you understand by the word unijunction transistor UIT and for what applications does it find use?
11. What is thyratron? How is electric motor control affected by thyratron?
12. What is the construction of an ignitron? What usual application?

13. How does an amplidyne differ from an ordinary DC machine? What is usual application?
14. In what way a saturable reactor differs from an ordinary reactor? What is its usual application?
15. What is GTO? Give its schematic symbol and its advantages over SCR.
16. What do you understand by NO and NC contacts?
17. How arcing across relay contacts is reduced?
18. How chattering of armature of electromagnetic relays operated from AC is prevented?
19. Where do you recommend the use of a limit switch?
20. Where do you recommend the use of a pressure switch?

2

ELECTRIC DRIVE

2.1 INTRODUCTION

An electric drive is often described as a form of machine equipment fabricated to convert electric energy into mechanical energy and provide electrical control of the drive.

Electrical energy is gaining increasing popularity in industrial and commercial sectors. Electric drive for industrial purposes is now well accepted everywhere. Electric drive has more or less replaced all other drives in industry. The advantages of electric drives are listed below:

1. It is simple in construction and is very cost effective.
2. Its speed control is easy and very smooth.
3. It is neat, clean and free from any smoke or flue gases.
4. It can be installed at any place thus affording more flexibility in the layout.
5. It can be controlled wirelessly.
6. As it is compact, it requires less space.
7. It can be started immediately without any loss of time.
8. It has comparatively longer life.

However, electric drive system has two inherent disadvantages:

1. It comes to stop as soon as there is failure of electric supply and
2. It cannot be used at far off places which are not served by electric supply.

However, the above two disadvantages can be compensated by deploying diesel-driven DC generators and turbine-driven 3-phase alternators which can be utilized either in the absence of or on the failure of normal electric supply.

2.2 FACTORS GOVERNING SELECTION OF MOTORS

Electric motors are predominantly very trustworthy machines, and require very little maintenance but choice of motor should be precise. So an engineer should engage his utmost care and attention to the selection of motor. One usually should take following factors into consideration before final selection of the motor:

1. **Nature of electric supply:** Whether AC or pure DC or rectified AC supply is to be utilized for motor.

2. **Nature of the drive:** Whether motor is to drive individual machines or a group of machines.

3. **Electrical characteristics of motors**
   a. Starting characteristic.
   b. Running characteristic.
   c. Speed control.
   d. Braking characteristic.

4. **Size and rating of motors**
   a. Requirement for continuous, intermittent or variable load cycle.
   b. Overload capacity.

5. **Mechanical considerations**
   a. Type of enclosures.
   b. Type of bearings.
   c. Transmission of drive.
   d. Noise level.
   e. Heating and cooling time constant of the motor.

6. **Cost**
   a. Capital cost.
   b. Running cost.

Proper choice of motor in the first place, saves lots of worries later on. Moreover due to practical difficulties it is not always possible to satisfy all the above considerations. For instance, either due to non-availability of particular type of motor or its prohibitive cost, we may select some other type of motor. In such circumstances, it is the experience and insight which an engineer has to fallback upon in the final selection of motor.

2.3   NATURE OF ELECTRIC SUPPLY

Wherever AC supply is available, choice should always be made for AC motors. AC motors have the advantage of possessing few working parts, requiring less maintenance and replacement of spares, giving an uninterrupted long service life. These motors, however, suffer from heavy starting current and fixed speed. As against this DC have many working parts such as commutator, brush goes, etc. All of which tend to give trouble. Advantages of DC motor are in no way loss prominent than maintenance troubles and can be found in ease of starting and speed control. Therefore; for certain applications, DC motors have privileged position such as in traction, rolling mills and lift services where frequent starting, stopping, reversing and speed control are required. In case of AC motors, we should try to use three phase motors in the interest of better efficiency, high starting and uniform running torques and balanced loading of supply. Single phase motors are limited to small loads only.

Nature DC supply available, to some extent affects the size of motor required. DC supply available from a DC generator will be more or less smooth as shown in Fig. 2.1.

As against this, DC supply available from rectifiers is undulating as shown in Fig. 2.2. This undulating DC supply, according to Fourier principle, is equal to DC component and AC harmonics. Mechanical power produced by motor is due to DC component only whereas armature heating will be produced by rms value of the total

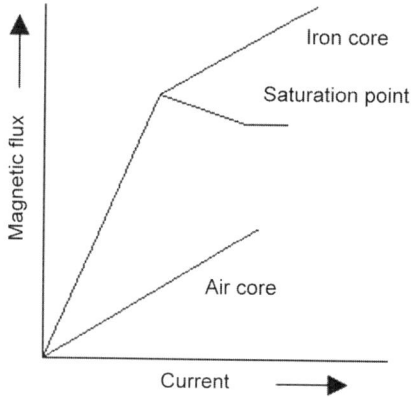

Fig. 2.1: Nature of DC supply.

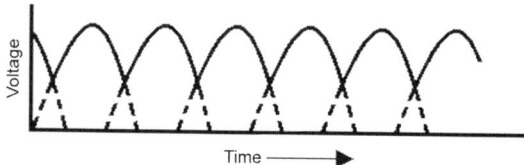

Fig. 2.2: Output of rectifiers.

undulating current. Therefore, when a DC motor is driven from a rectifier slightly larger motor size is required for a given HP requirement.

AC power generation is at conventional frequency of 50 Hz. However this standard frequency is not capable of meeting the requirements of certain industries, railway signalling and defence service. As such there has been recent trend towards developing high frequency power generation. Relative advantages and disadvantages of high frequency and power frequency equipment are given below:

1. Output of a motor is given by equation 2.1

$$P \propto D^2 Ln \qquad \qquad ...(2.1)$$

where 'P' stands for output, 'D' for the diameter, 'L' for length and 'n' for the speed of motor in rps.

In case of alternator $P = VI \cos \phi$

$$4K_1 K_2 K_3 f T \phi_m I \cos\phi \qquad \qquad ...(2.2)$$

It is observed from both of equations 2.1 and 2.2 that as speed or frequency increases, output of the equipment increases. This results in the reduction in size and weight of the equipment. For same power output high frequency generator may be 1/10th the weight of 50 Hz generator. On account of high, power to weight ratio, relative cost of the equipment decreases frequency increases. Decrease in the weight of the equipment plays vital role in transporting and handling the equipment at high altitude for defense services.

2. High frequency equipment needs less maintenance, low operational cost and higher efficiency. It has long span of life squiring minimum repairing cost.

3. High frequency equipment gives more reliable service in arduous conditions of high altitude, low temperature and high humidity.

4. For the production of high frequency power, size of generator being small the prime mover will also be of smaller size. As a result problem of high noise will be reduced.

5. High frequency equipment has the problem of increased losses due to hysteresis and eddy currents. However by the use of proper low loss materials, proper flux density and adequate cooling arrangements, problems of high frequency can be overcome. Third and fifth harmonics can be eliminated by the use of proper LC filter circuits.

2.4   NATURE OF THE DRIVE

Practically, a motor may drive a number of machines at a time, called group drive or each machine may be fitted with a separate motor called individual drive.

2.4.1 Group Drive

In group drive, a single motor drives a number of machines through belts from a common shaft. It is also called line shaft drive. In switch over from non-electric drive to electric drive, easiest way was to replace the engine by means of motor and retaining the rest of power transmission system.

Each type of electric drive has its own advantages and disadvantages. The group drive has following advantages:

1. It leads to saving in initial cost because one 150 kW motor costs much less than ten 15 kW motors needed for driving 10 separate machines.

2. Since all ten motors will seldomly be required to work simultaneously, a single motor of even 100 kW will be sufficient to drive the main shaft. This diversity in load reduces the initial cost still further.

3. Since a single large motor will always run at full-load, it will have higher efficiency and power factor in case it is an induction motor.

4. Group drive can be used with advantage in those industrial processes where there is a sequence of continuity in the operation and where it is desirable to stop these processes simultaneously as in a flour mill.

The disadvantages of group drive not less conspicuous which are listed below:

1. Any fault in the driving motor renders all the driven equipment idle. Hence, this system is unreliable.

2. If all the machines driven by the line shaft do not work together, the main motor runs at reduced load. Consequently, it runs with low efficiency and with poor power factor.

3. Considerable amount of power is lost in the energy transmitting mechanism.

4. Flexibility of layout of different machines is lost since they have to be so located as to suit the position of the line shaft.

5. The use of line shaft, pulleys and belts, etc. makes the drive look quite untidy and less safe to operate.

6. It cannot be used where constant speed is required as in paper and textile industry.
7. Noise level at the worksite is quite high.

Group drive is all the same employed in processes where stoppage of one operation necessitates stoppage of sequence of operations as in case of textile mills.

2.4.2 Individual Drive

Here individual machine is connected with its own motor. This method, of drive has more flexibility of layout, smoothness in speed control, clean appearance and safe working conditions. Output of one motor does not affect the sequence of operations of other machines. The only disadvantage is that it is very costly in the initial stage. However, over a number of years and by the way of improved performance, it would be more economical to use individual drive. Modern trend, therefore, is to use as much individual drive as possible.

2.5   STARTING CHARACTERISTICS

The starting torque exerted by a motor should be large enough to accelerate the motor and its load to the rated speed in a reasonably short in. Some motors may have to start against full-load torque, e.g. motors driving grinding mills or oil expellers. In the case of lifts and hoists, the motors have to start frequently with high acceleration.

At the time of starting a motor two torques come into play: the torque required to overcome the static friction and the torque necessary to accelerate the motors and its load to the desired speed. The torques required for static friction cannot be easily determined. The torque for acceleration depends upon the load torque itself. The load torque may:

i. increase with speed, i.e. may be proportional to (speed)^2 as in the case of a fan or centrifugal pump or
ii. remain constant with speed as in the case of a hoist.

The starting gear should, therefore, be able to limit the starting current taken by a motor to a safe value consistent with the production of the necessary starting torque.

2.5.1 Starting Torque of DC Motors

The torque in DC motor is given by

$$T = \frac{1}{2\pi}\phi I_a Z \frac{P}{A}$$

where ϕ = the flux per pole in webers
   I_a = the armature current
   Z = the number of armature conductors
   P = number of poles
   A = number of parallel paths in the armature winding.

The torque, therefore, depends upon the product of flux and armature current and is independent of speed, i.e. $T \propto I_a$.

In the case of a shunt motor, both the armature and the field are connected in parallel across constant voltage mains. The current taken by the field is, therefore, constant and hence the flux will be maintained constant so long as the field current remains constant. Therefore the torque in a shunt motor varies as the armature current. The torque-armature current curve is a straight line passing through the origin. Full-load current will produce full-load torque and if we double the current, the torque will also be doubled and vice versa.

But in case of the DC series motor, the current in the series winding and the armature is the same. The flux is dependent directly on the same. The flux is dependent directly on the value of the current the motor draws. Torque is, therefore, proportional to the square of the armature current, i.e. $T \propto I_a^2$. The torque-current curve is, therefore, a parabola. But the flux varies as the current only up to the limit of saturation of the magnetic circuit and the torque-current curve is parabolic in shape only up to the limit of saturation. Beyond the saturation point since "ϕ" does not vary appreciably the torque current curve is almost a straight line (Fig. 2.3). A DC series motor is, therefore suitable for drives starting with heavy loads, i.e. electric train, hoist and lifts, etc.

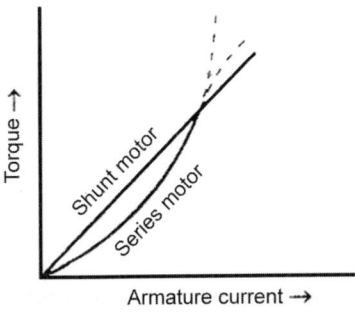

Fig. 2.3: Variation of torque with armature current.

2.5.2 Three-phase Induction Motors

In a three-phase induction motor, if

r_1 = stator resistance per phase

r_1' = rotor resistance per phase referred to the stator

x_1 = stator reactance per phase

r_2' = rotor reactance per phase referred to the stator

s = fractional slip

V = stator applied voltage per phase,

then torque, T, in (N–m) is given by

$$T = k\frac{V^2\left(r_2'/s\right)}{\left(r_1 + r_2'/s\right)^2 + \left(x_1 + x_2'\right)^2}$$

and stator current per phase

$$I = \frac{V}{\sqrt{\left(r_1 + r_2'/s\right)^2 + \left(x_1 + x_2'\right)^2}}$$

If k is made unity, the torque is expressed in synchronous watts per phase. At starting,

$s = 1$

therefore, starting torque,

$$T_s = \frac{V^2 r_2'}{\left(r_1 + r_2'/s\right)^2 + \left(x_1 + x_2'\right)^2}$$ synchronous watts per phase and starting current per phase

$$I_s = \frac{V^2}{\sqrt{\left(r_1 + r_2'\right)^2 + \left(x_1 + x_2'\right)^2}}$$ amperes.

The starting torque is a maximum if the rotor resistance is made equal to its leakage reactance. It is, therefore, usual to start a slipring induction motor with a variable resistance in its rotor circuit to have a good starting torque and to cut the resistance in steps as the motor speeds up. The resistance in the rotor circuit also serves the purpose of limiting the starting current taken from the supply.

In the squirrel-cage induction motor, it is not possible to insert a variable resistance for starting in the rotor circuit on account of the cage construction. The motor is started by the application of a reduced voltage for which purpose star-delta or auto-transformer starters may be employed. The motor has a low starting torque.

Induction motors are normally used where constant torque is required, e.g. in paper machinery, textile machinery, compressors, conveyors, etc. Squirrel-cage motors are more reliable, cheaper and easier to use whereas phase wound motors are expensive and maintenance is complicated. The former are used for low and medium HP while the latter are used for high HP.

Motors with double cage have a high starting torque. The outer cage is made of high-resistance metal bars and the inner cage is of copper bars. The inductance of the inner winding is higher than that of the outer high resistance winding. At the instant of starting, the motor induced currents are at the line frequency and the inner cage has a high reactance ($2\pi fL$) with the result that the rotor currents remain confined to the outer cage despite its high resistance. The starting torque is, therefore, high. During normal running, the reactance of the inner cage decreases ($2\pi sfL$), and the rotor currents are confined to the inner cage which is a low resistance winding. This gives a high efficiency of the motor.

Fig. 2.4 shows the speed-torque curves of a single cage and double cage motor.

An important relation proved in most textbooks of electrical machines in the case three-phase induction motors is:

$$\frac{\text{Starting torque}}{\text{Full-load torque}} = \left(\frac{\text{Starting current}}{\text{Full-load current}}\right)^2 \times \text{Full-load slip}$$

i.e. $\dfrac{T_s}{T_1} = \left(\dfrac{I_s}{I_{fl}}\right)^2 \times S_{fl}$

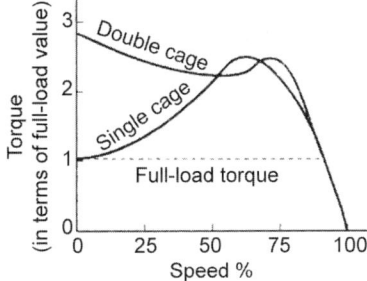

Fig. 2.4: Speed-torque curves of a single cage and double cage induction motor.

2.5.3 Methods of Starting 3-phase Induction Motors

It is desirable to start AC motors at full voltage to attain simplicity and economy in the starting gear. Induction motors can be designed and built to enable them to be started on full voltage. But in case of cage motors the starting current could be large enough to produce considerable voltage drop in the distribution system which may adversely affect other apparatus and also cause light flicker.

Induction motors are, hence, started on reduced voltage. The various methods of starting are discussed below:

1. Resistor Starting

A series resistor is used in each line and can be arranged in a manner so that the resistance is reduced to zero in steps and the motor current may increase to the full value gradually and transients are avoided. The torque efficiency which is:

$$\frac{\text{Torque developed by the motor/full-load torque}}{\text{Current of motor/full-load current}}$$

is lower at reduced voltage start than for full voltage starting. A motor having a starting torque equal to twice the full-load current (these values being at full voltage starting) has a torque efficiency of $2/6 = 0.33$. If the same motor is started at 60 percent voltage by using resistors, the line current will be ($0.6 \times 6 \times$ full-load current) and the starting torque will be $[(0.6)2 \times 2 \times$ full-load torque]

The torque efficiency will then be

$$\frac{(0.6)^2 \times 2}{0.6 \times 6} = 0.2$$

2. Reactor Starting

It is not as widely used as resistors or autotransformer starting, though the method is similar to resistor starting. The acceleration is very smooth in this case though the accelerating time is comparatively longer.

3. Autotransformer Starting

There are taps on the autotransformer so that the motor can be started at a lower voltage. Taps are for 50, 65 and 80 percent of line voltage; the 50 percent tap are provided only in the case of sizes above 50 HP. In the second method of operation (Fig. 2.5), the large transient current is reduced since the motor is always connected through the autotransformer winding to the line.

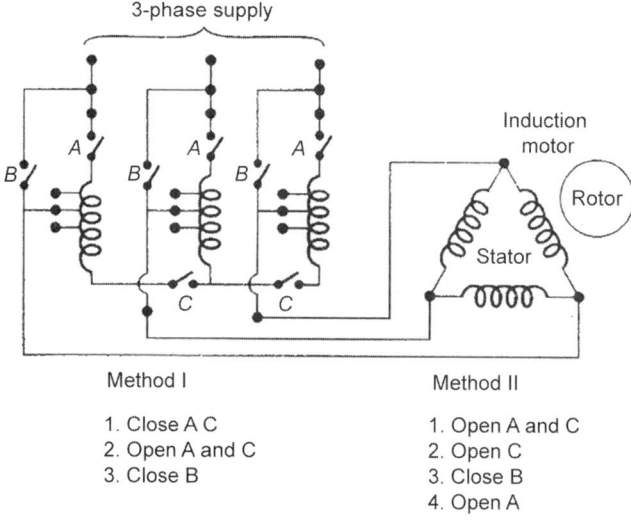

Method I
1. Close A C
2. Open A and C
3. Close B

Method II
1. Open A and C
2. Open C
3. Close B
4. Open A

Fig. 2.5a: Induction motor starting with autotransformer.

We can determine the reduction in the starting current when using autotransformer starter. Fig. 2.5b shows a starting of autotransformer. Consider an autotransformer with a transformation ratio k, i.e. $\dfrac{\text{primary voltage}}{\text{secondary voltage}} = k$. Consider also that the motor has a starting torque equal to twice the full-load torque and starting current equal to six time the full-load current. If the motor is started at full-load voltage, phase voltage $E_{ph} = E/\sqrt{3}$ and the starting current = 6I. When the motor is started through an auto-transformer, the phase voltage is E_{ph}/k and starting current = $6I/k$.

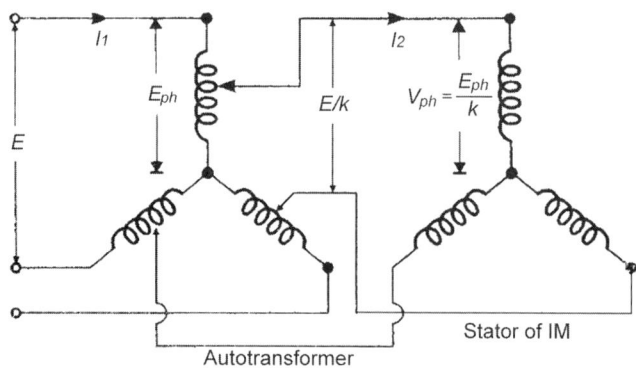

Fig. 2.5b: Autotransformer starting.

Also we have

$$\frac{E_{ph}}{V_{ph}} = k = \frac{N_1}{N_2} = \frac{I_2}{I_1}$$

where N_1 = no. of primary turns

N_2 = no. of secondary turns

I_1 = primary current

I_1 = secondary current

$I_1 = 6I/k = kI_1$

$I_1 = (6/k^2)I$

Therefore line current is reduced inversely as the square of the ratio of transformation. Since the torque is proportional to the square of applied voltage, the starting torque is proportional to V_{ph}^2 or $(E_{ph}/k)^2$. If T is the full voltage starting torque, starting torque at reduced voltage $= T/k^2$.

The torque efficiency $= \dfrac{2}{6/k^2} = 0.33k^2$

4. Star-delta starting

While starting the stator of the cage motor is connected in star and in the running position it is connected in delta, so that $\dfrac{1}{\sqrt{3}}$ of the line voltage is impressed on each phase at the time of starting. A star delta starter is shown schematically in Fig. 2.6. The starting line current of the motor with star delta starter is also reduced to $\dfrac{1}{\sqrt{3}}$ full voltage starting line current. The starting torque which is proportional to $(E/\sqrt{3})^2$ is reduced to one-third (E being the line voltage).

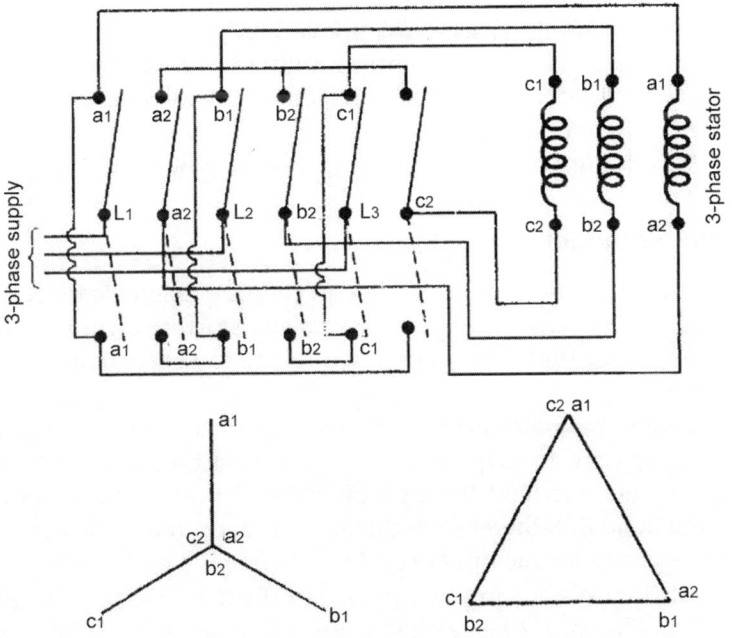

Fig. 2.6: Star delta starting.

The torque efficiency in this method is $\dfrac{2}{6\sqrt{3}} = \dfrac{1}{\sqrt{3}} = 0.576$ for the motor considered under resistor starting.

5. Starting by the use of slip-ring starter

In this method reduced voltage is not applied to the motor . The full voltage is applied to the stator but resistance is inserted in each phase of the wound rotor (Fig. 2.7). Since the stator and rotor can be regarded as the primary and secondary of a transformer, the resistors on the secondary limit the currents in the rotor windings and since in a transformer $\dfrac{I_1}{I_2} = \dfrac{V_2}{V_1}$, the starter current I_1 is also reduced.

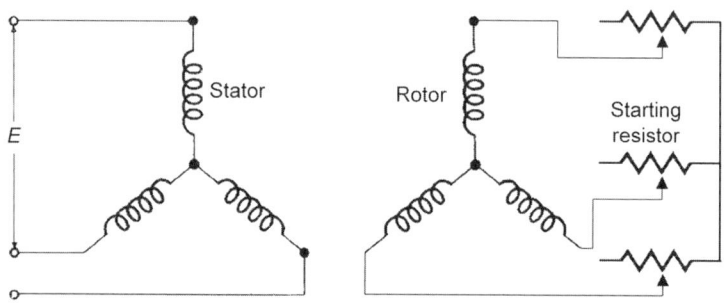

Fig. 2.7: Induction machine starting by resistors in rotor circuit.

2.5.4 Limitation of Size

From the above discussion it is seen that the starting torque in case of cage motor varies with the type of starting method employed. In fact the starting difficulty of the cage-type motors has hindered their application only to loads requiring more than about 40 or 50 HP with not more than about 50 percent full-load torque. The slip-ring motor can be used beyond this limit.

We shall now discuss briefly the starting of other types of motors.

2.5.5 Single-Phase Induction Motors

A single phase induction motor does not have a rotating magnetic field. It has only got a pulsating field and therefore does not possess any starting torque. So it needs to be started with external support. The following three methods are employed to make the motor self-starting.

    a. **Pole shading:** The motor is of squirrel cage type and the stator pole shaded by heavy copper wire or strip (Fig. 2.8). The current induced in the shading coil causes the magnetic field through the shaded portion of the pole face to lag behind the main flux thereby producing a rotating magnetic field. Such motors have low starting torque but are quite cost effective in small sizes.

    b. **Phase-splitting:** We obtain a two-phase supply from a single phase line by using a capacitor. The motor has a cage rotor and a stator containing two separate windings located in the same manner as for a two-phase stator. One of the windings is connected directly to supply and the other through the capacitor.

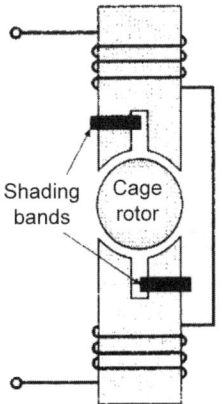

Fig. 2.8: Shaded pole motor.

A rotating magnetic field is obtained in the air gap. There are two types of capacitor motors; the "capacitor start" motor, in which case the capacitor is in circuit only during the starting period and is disconnected at a predetermined speed by a centrifugal switch, the other type is the "capacitor start and run" motor where the capacitor is connected permanently and improves the power factor of the motor (Fig. 2.9).

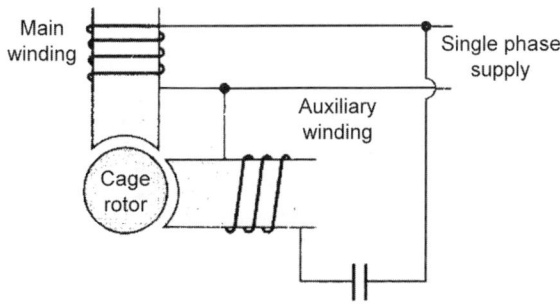

Fig. 2.9: Capacitor motor.

c. **Repulsion motor starting:** The rotor has a repulsion motor winding and therefore starts as a repulsion motor giving a high starting torque. As it runs to speed a centrifugal device short-circuits the commutator bars and lifts the brushes, converting the motor into a plain squirrel cage one.

2.5.6 Synchronous Motors

Synchronous motor is not self-starting as there is no starting torque. It has to be run up to synchronous speed by another motor and synchronised to the supply. To make it self-starting a cage winding is provided on the poles. It starts as a plain squirrel cage motor and when it has attained nearly synchronous speed at no load, the DC excitation is switched on and the rotors pulls into synchronism. Starting torque between 50 and 100 percent full-load torque can be obtained with twice full-load current. The "synchronous induction motor" has a cylindrical rotor with a slip-ring induction motor winding. It starts as a slip-ring induction motor with good starting torque and when it has almost reached synchronous speed, DC is passed through the rotor winding making rotor pull into synchronism.

2.5.7 AC Commutator Motors

These are started by the application of reduced voltage or by shifting the brushes.

2.6 RUNNING CHARACTERISTICS

The running characteristics of a motor include the speed-torque or speed-current characteristics, losses, efficiency and power factor at various loads. Power factor consideration crops up in the case of AC motors only.

2.6.1 DC Motor

In case of a DC shunt motor the speed is pretty much constant with load; although there is only a slight fall in speed when the load comes up. The speed torque characteristic is a slightly drooping straight line.

For a DC series motor, the speed is normally high at low loads and decreases as the motor is loaded (Fig. 2.10). The speed torque characteristics is a sharply drooping-curve.

In a compound motor, the speed-torque characteristics may be made to lie anywhere between the pure shunt and the pure series by suitably adjusting the series and shunt windings.

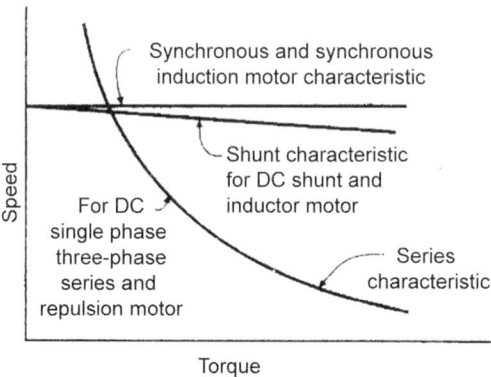

Fig. 2.10: Speed torque characteristics of different motors.

2.6.2 The Three-Phase Induction Motor

It possesses shunt characteristics. The power factor is very poor at low loads but improves as the load increases. The power factor, however always remains less than unity.

2.6.3 The Schrage Motor

It is an AC commutator motor and gives a speed range of 321 by shifting the brushes round the commutator. It is available up to sizes of about 400 HP and possesses shunt characteristics. The running power factor is high.

2.6.4 The Three-Phase Series Motor

It possesses series characteristics and is suitable for haulage and waste work.

2.6.5 The Synchronous and Synchronous Induction Motor

The synchronous motor is a constant speed motor: the speed is fixed by the frequency of the supply. It is not, however, self-starting. It is started by an auxiliary motor and then synchronised to the supply. This disadvantage is eliminated in the synchronous-induction motor where the machine starts as plain induction motor and when its speed is very near the synchronous speed the DC excitation to the rotor is switched on the motor pulls into synchronism.

By varying the field excitation of these types of motors the power factor may be made unity or even be made leading. And over excited synchronous motor works at a leading power factor while an under excited motor works at a lagging power factor. It is therefore, used in improving the power factor in a system using many inductions motors.

2.6.6 The Single-Phase Series Motor

It possesses the series characteristics. In large size it finds an exclusive application or traction work. In the fractional horse power size it is used in domestic appliances like vacuum cleaners and refrigerators, etc.

2.6.7 The Single-Phase Induction Motor

It has shunt characteristics. Since it has no starting torque, additional means have to be provided to make it self-starting. The repulsion start and the capacitor-start motor are the most common modifications of the single phase induction motor.

2.6.8 The Repulsion Motor

It has series characteristic and closely resembles the series motor in construction. The armature is short circuited on itself.

2.7   SPEED CONTROL

Control of speed for an industrial drive depends upon the nature of work being carried out. A certain operation may require a continuously varying speed; another one may only require to fixed speeds. Sometimes creeping speed may be necessary to adjust the work. For most industrial drives, however, a control speed within ± 20 percent may be suitable.

2.7.1 Speed Control of DC Motors

The speed of DC motor is given by the expression

$$N \propto \frac{V - I_a.R}{\phi} \text{ or } N = k\frac{V - I_a R}{\phi}$$

where, N = speed in rpm
ϕ = flux per pole

V = supply voltage

R = resistance in the armature circuit

I_a = current drawn by the motor armature

Two methods of speed variation are possible:

1. by the variation of flux ϕ, i.e. field control; and
2. by changing resistance in armature circuit.

2.7.2a Field Control in Shunt Motors

The flux per pole is varied by inserting and extra resistance in the field circuit. Variation of the flux per pole changes the speed of the motor (Fig. 2.11.).

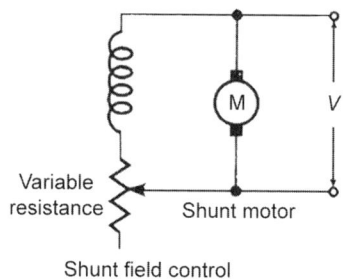

Shunt field control

Fig. 2.11: Speed control of DC motor.

Some limitations of the methods are discussed below:

◈ With the regulating resistance in the field circuit zero, the flux per pole is maximum which gives the lowest speed of the motor. Motors are usually designed to work are a speed slightly less than the rated speed when the regulating resistance is zero. It is obvious that any lower speed than this cannot be achieved by this method.

◈ The speed of motor N is proportional to VI/ϕ whereas the full-load torque T is proportional to TN or VI_a. The horse power delivered by the motor is proportional to TN or VI_a which is constant. Therefore, these method can be utilised only where the horse power of the load remains constant.

There is a limit to which the field can be weakened to obtain high speed. At such a speed the motor will tend to draw large current to develop the same torque. But this will result in the main field ampere—turns becoming much smaller than the armature mmf. The armature reaction will demagnetise and distort the main flux making the operation of the motor unstable. In motors where a wide speed range required, this difficulty is overcome by having a light series winding connected cumulatively to provide stable operation.

Speed variation by this method is limited to a ratio of 5 to 1.

2.7.2b Field control in Series Motor

Three method are used for changing the flux per pole in series motors. These are:

1. Diverter field control
2. Tapped field control
3. Series-parallel field control

1. **Diverter field control:** A shunt is employed in parallel with the series field to divert the current partially in the series field thus causing field weakening (Fig. 2.12a). Speeds higher than normal are attained when the diverter is used.

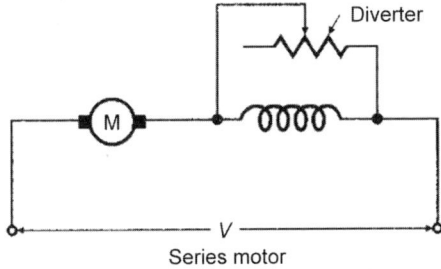

Fig. 2.12a: Diverter field control.

2. **Tapped field control:** Tapping are provided on the field winding and current may be passed through different number of turns thereby changing the field ampere-turns (Fig. 2.12b). This method is commonly used for series motor used in traction work.

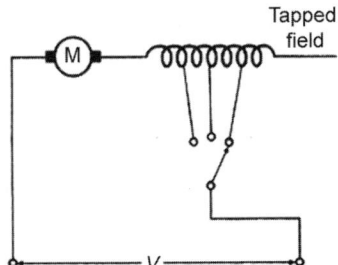

Fig. 2.12b: Tapped field control.

3. **Series-parallel field control:** The field winding is designed in two sections which may either be connected in series or in parallel (Fig. 2.12c). The field ampere turns are reduced to half the value in parallel connection as compared to those in the series connection. The speed therefore becomes about twice the initial. Though the method is simple and inexpensive, only two speed are possible.

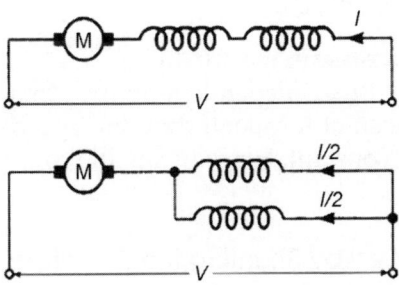

Fig. 2.12c: Series-parallel field control.

2.7.3 Control of Speed by Change of Series Resistance in the Armature Circuit

Since $N \propto (V - I_a R)/\phi$, the speed of a DC motor can also be changed by altering R, resistance in the armature circuit. The torque of a motor is proportional to the product of the flux, ϕ and the armature current, I_a. In case of a shunt motor, since ϕ is constant, N will be proportional to $V - I_a R$. If constant torque is required I_a must be unchanged. But since speed is to be varied R has to be varied. Increase of R (for constant torque and therefore constant armature current) will give decreasing values of speed. The minimum value of R is I_a, the resistance of the armature itself. Fig. 2.13a shows motor speed-torque characteristics.

$$R = r + R_a$$

For a theoretical value of R= 0, the horizontal straight line passing through 100% speed ordinate is limiting value. For any other value of R which may be $R_1 = r_1 + R_a$ or $R_2 = r_2 + R_a$, $R_2 > R_1$ or $R_3 = r_3 + R_a$, $R_3 > R_2$, etc. the curve are shown in Fig. 2.13b.

If this method is employed for a load demanding constant torque at all speed, the armature current must be constant and in turn input of the motor (i.e. armature) is also constant. But the output decreases with the decrease in speed and hence deteriorating the efficiency of the motor at lower speeds. The loss of power takes place in the controlling resistance r. In case of fans and centrifugal pumps where the load torque decreases with decreasing speeds, this method can be quite convenient and economical for short-term periods. Creeping speed may also be obtained by this method.

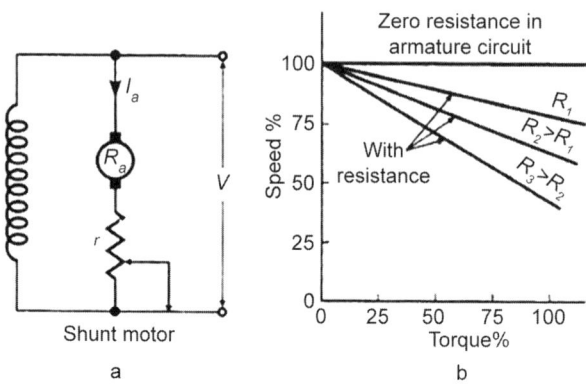

Shunt motor

a

b

Figs 2.13a and b: (a) Speed control of DC shunt motor by resistance in the armature circuit, (b) Speed torque characteristic of DC shunt motor.

In a series motor, an increase in the armature circuit resistance will decrease both speed and torque. Since the flux is dependent on the armature current, the torque is proportional to I_2. For a constant torque, if different speeds are required, I has to be constant which will make ϕ constant. For reducing the speed r is to be increased (Refer Figs 2.14a and b).

2.7.4 Control of Motor speed by Shunting the Armature by a Resistance

The arrangement of varying the speed of a DC motor by changing the series resistance in the armature circuit is a time not applicable as the speed of the motor rises if the

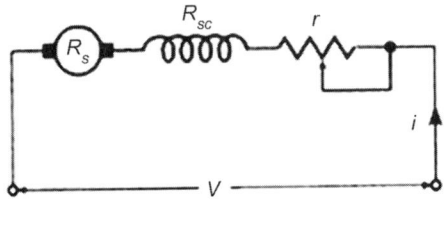
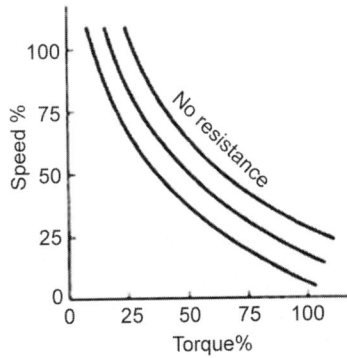

Figs 2.14a and b: (a) Speed control of DC series motor by resistance in the armature circuit, (b) Speed torque characteristic of DC series motor.

load is reduced. We can see from the equation $N \propto (V - I_a R)/\phi$ that as I_a diminishes N increases. To eliminate this drawback, the armature is shunted by a variable resistance. A series resistance also used as shown in Figs 2.15a to c. By adjusting P and Q a number of speed torque curves can be obtained.

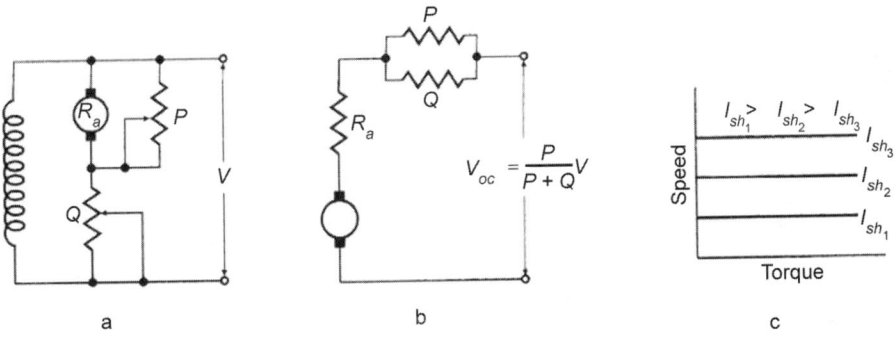

Figs 2.15a to c: (a) Circuit diagram, (b) Equivalent circuit based on the application of Thevenin's Theorem, (c) Shunted armature method of speed control.

If we apply Thevenin's Theorem to the circuit in Fig. 2.15a, we get P and Q in parallel, i.e. short circuit resistance R_{sh} by short-circuiting the source supply and removing the branch (i.e. armature) through which we wish to find the value of current flowing. Therefore the open circuit voltage across the armature is $V_{oc} = [P/(P + Q)]V$. Fig. 2.15b shows the equivalent circuit based on Thevenin's Theorem. The circuit is given by

$$I = \frac{V_{oc}}{R_a + R_{sh}}$$

where

$$V_{oc} = \frac{P}{P+Q}V$$

and

$$R_{sh} = \frac{PQ}{P+Q}$$

The efficiency of this method is poor and heavy current may be drown from the supply at certain speed.

2.7.5 Booster Control

In Fig. 2.16, M_1 is the primary and central motor employed for driving the load. DC supply is fed to the motor terminal for excitation. Here B is used as the separately excited booster. The voltage and polarity of this booster can be controlled by the booster field (BF). By tapping the voltage from the parallel resistances, the current through the booster field can be reversed which in turn reverses the polarity of the booster armature voltage. Thus, required boost or buck can be supplied to M_1. Now M_2 is the motor which drives the booster armature. The speed can be maintained over a wide range depending on the size of the booster. This method is however, suitable for small motors, otherwise the size of the booster becomes practically not feasible.

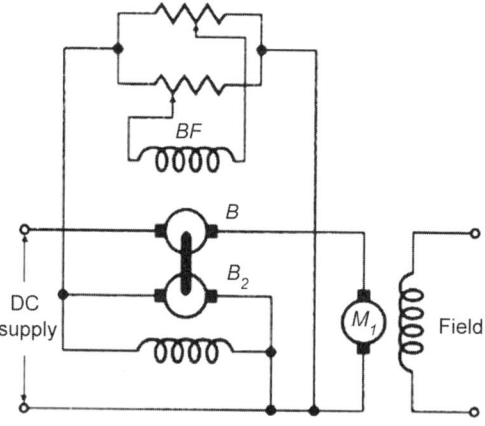

Fig. 2.16: Speed control by booster.

2.7.6 Ward-Leonard Method of Speed Control

The primary adjustable voltage armature control method of speed control achieved by means of an adjustable voltage generator is known as the Ward-Leonard system. This total set-up system consists of driving the motor with a constant excitation and applying a variable voltage to its armature to provide the required speed. From a motor-generator or from simply a convertor set, we can obtain the variable supply voltage. The scheme is illustrated in Fig. 2.17. M_1, is the working motor, powered by the generator G, which is driven by a synchronous or induction motor M_2. The excitation current, needed for the motor M_1 and the generator G is supplied by the exciter E mounted on the same shaft as the generator. The Ward-Leonard set is initiated by starting the driving motor. The field rheostat R of the generator is gradually brought-out of the circuit as the generator picks up the speed, and the work motor begins to rotate. The variable voltage across the terminals of the generator across the motor is obtained by varying the exciting current of the generator G by means of shunt regulator R. The direction of rotation of motor armature can be reversed by reversing the direction of exciting current of the generator G with the help of reversing switch RS. The converter set runs always in the same direction.

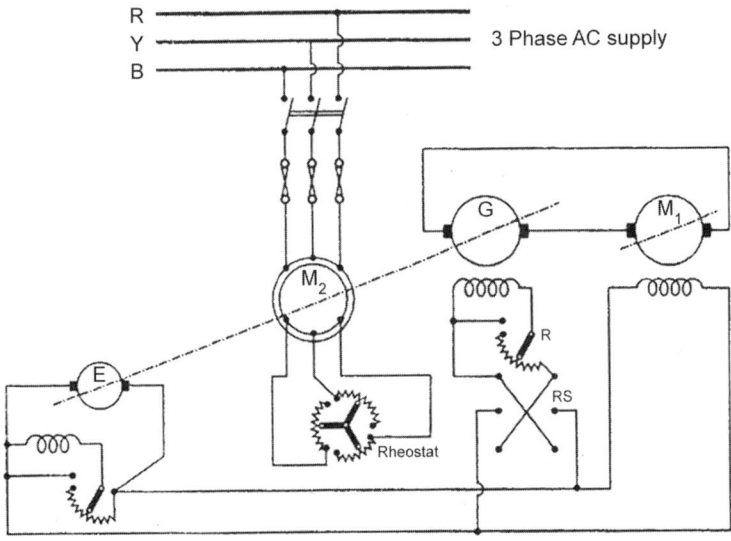

Fig. 2.17: Ward-Leonard method.

Braking of motor M_1, may be carried out by reducing the generator excitation so that its emf is less than the counter-emf of motor M_1. Under these conditions, motor M_1 begins to operate momentarily as a generator, generator G as a motor and AC driving machine M_2 as a generator.

As a result kinetic energy of motor M_1 and its load is returned to the supply mains and braking action on the motor M_1 takes place.

☞ *Advantages*

1. Very fine speed control over the whole range from zero to normal speed in both directions can be obtained.
2. Uniform acceleration can be obtained.
3. Speed regulation is good.

☞ *Disadvantages*

1. Two extra machines are required, so arrangement is costly.
2. Low overall efficiency of the system, especially at light loads.

☞ *Applications*

This system of speed control is best suited where almost unlimited speed control in either direction of rotation is required such as in steel rolling mills, paper machines, elevators, cranes mine hoists, diesel-electric propulsion of ships, etc.

☞ *Ward Leonard Ilgner System*

It is a modified version of Ward Leonard system. It incorporates a heavy flywheel, which is mounted on the shaft coupling the driving motor M_2 and generator G. In operation with a flywheel, the driving motor needs to have a drooping speed—load characteristic, i.e. its speed must increase with the load on the shaft. The primary function of the flywheel is to reduce the fluctuations in the power demand from the supply mains as explained below:

An increase in the load on the shaft causes the work motor to draw more current from the generator so more power is required to drive the later and if there were no flywheel, the driving motor would take all the additional power from the supply line, thus causing sharp fluctuations in it. The heavy flywheel, however, stores a large amount of kinetic energy. When an increase in the generator load causes the driving motor to slowdown, some of the kinetic energy of the flywheel goes to sustain the peak load on the shaft of the work motor. When the load on the work motor decreases, the driving motor picks up speed and the flywheel stores up kinetic energy.

This system of speed control is employed where the load on the motor shaft sharply varies such as in mine hoists, rolling mills, etc.

2.7.7 Metaldyne Control

This is applicable in traction work and is nowadays not extensively used.

2.8   ELECTRONIC CONTROL OF DC MOTORS

Electronic motor control is one of the most extensively used methods in industry these days. It is generally employed where some of the following objects are needed to be achieved: holding the speed or torque at a definite fixed value, limiting the torque automatically, smooth and constant acceleration, facility for reversal under load conditions automatic electric breaking, etc. Since such control involves a huge expense, it is employed where the above-mentioned features are not desirable but are only must for operation. In industry some of the applications are in the area of precision machine tools like lathes and grinders. Also in operations where accuracy and precise control is needed in control devices.

2.8.1 Speed Relations of DC Motors

In a DC motor the speed in rpm is given in the given equation:

$$N = \frac{V - I_a R_a}{k\phi}$$

where V is the voltage applied to the armature, I_a is the current through the armature, R_a is the resistance of the armature and ϕ is the magnetic flux.

On rearranging the equation, we get

$$V - I_a R_a - k\phi N = 0$$

In this equation $k\phi N$ represents the back emf, R_b of the motor. We may now show the equivalent circuit of a separately excited DC motor as given if Fig. 2.18. Though the armature has an inductance L_a, it takes no part in the motor operation under steady state conditions. A voltage equal to $L_a \dfrac{di}{dt}$ will be induced in it whenever there will be change in the armature current.

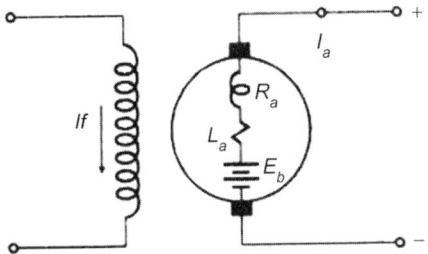

Fig. 2.18: Equivalent circuit of a separately excited DC motor.

2.8.2 Thyristor Control of DC Motors

Speed and acceleration of a DC drive can be controlled by varying the voltage applied to the motor from very low value at start to the full value at rated speed. In the conventional method, this is achieved by utilising a tap changing transformer feeding power to the motor through a set of rectifiers [recently semiconductor rectifiers are being increasingly be used]. The same result is achieved using a thyristor (silicon controlled rectifier). The thyristor output voltage is DC and is a function of firing angle (α). Thus, with the change of 'α', the DC output voltage of the thyristor changes thus controlling the speed of the DC motor.

Main advantage of the thyristor control is the absence of on-load bulky tap changer. However, for thyristor operation, proper firing circuitry must be utilized. Different types of thyristor firing circuitry exit and for this any power electronics book may be referred to.

Fig. 2.19 shows a half-wave thyristor circuit for the armature voltage control of a separately excited DC motor. The voltage waveforms of AC and DC sides have been shown in Fig. 2.20. Full-wave bridge rectifiers are also common for controlling the applied voltage on the motor. For positive half-cycle in the AC side, thyristor 1 and diode 3 conduct while for negative half-cycle for AC side thyristor 2 and diode 4 conducts.

The circuit of a full-wave rectifier controlled DC motor has been shown in Fig. 2.21 while the voltage waveform is shown in Fig. 2.22.

In all AC phase control circuit of thyristors, the firing circuit starts conduction of the SCR while the 'stopping' of conduction (named 'commutation') takes

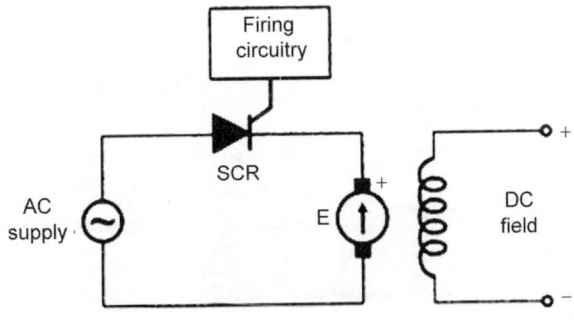

Fig. 2.19: Half-wave thyristor rectifire.

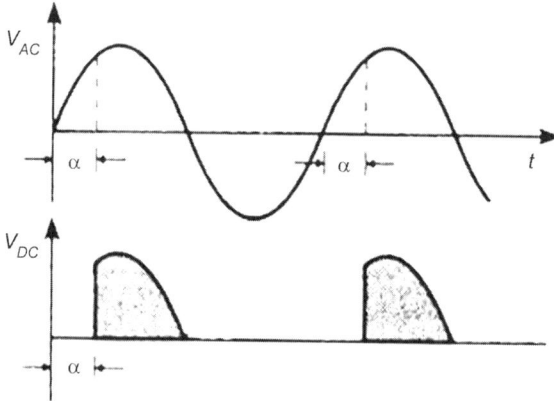

Fig. 2.20: AC and DC sides voltage waveforms.

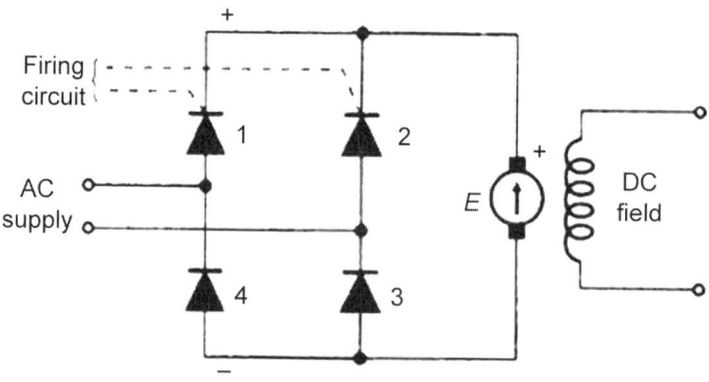

Fig. 2.21: Circuit of a full-wave rectifier controlled DC motor.

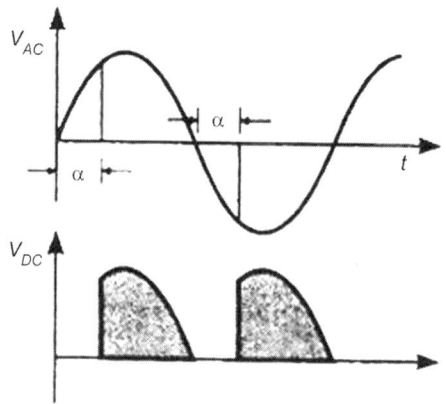

Fig. 2.22: Voltage waveform in AC and DC sides of the full-wave thyristor bridge.

place automatically at the current zero of the AC wave. In half-wave circuit, the conduction remains stopped for negative half-cycle while for full-wave circuit, commutation takes place from one set of thyristor-diode assembly to another set as shown in Fig. 2.22.

2.8.3 DC Motor Supplied by Three-phase Thyristor Rectifiers

Fig. 2.23 shows the circuit diagram of a motor with separate excitation, the armature being supplied from thyristors rectifiers. Thyristors T are used as rectifiers in this case. Each phase of the transformer secondary produce current in each thyristor for one-third of a cycle. The back emf E_b of the motor is of such a polarity that it opposes the flow of this current through the motor armature. This means that the conduction in cannot take place till the instantaneous voltage of the transformer secondary is greater than E_b (back emf). Since E_b is dependent on motor rpm, for a certain torque or current, the instant of firing in each cycle changes as the motor runs up to the rated speed from standstill. We shall now discuss briefly some relevant aspects of the control process.

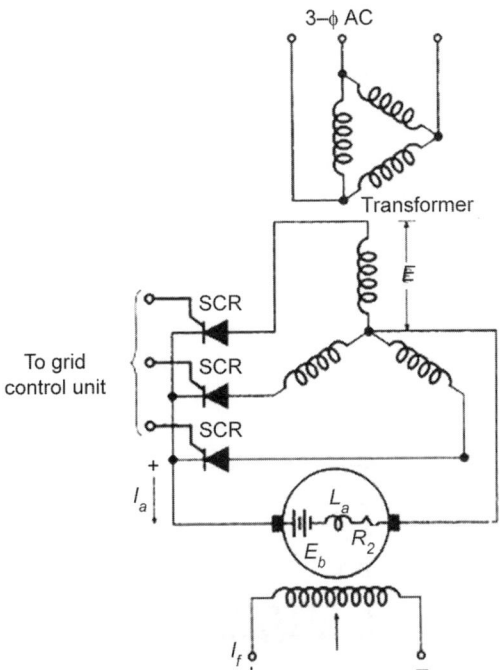

Fig. 2.23: DC motor operated from thyristor rectifiers.

i. **Continuity in conduction.** Neglecting the effect of L_a, it is easy to see that conduction in the SCRs will take place continuously if the applied rectified voltage is never less than E_b. On the output side, the rectifiers deliver fluctuating unidirectional current which has two values: (a) average value and (b) rms value. The average value of the rectifier current determines the power fed into the motor armature whereas the rms value of the current will determine the heating of this armature. Since the rms value is higher than the average value, where a DC motor is supplied by a rectifier the size and rating of the motor has to be a little larger for a particular horsepower requirement.

ii. **Discontinuity in conduction.** If the back emf (E_b) of the motor is greater than the instantaneous emf of a secondary phase of the transformer, current through the SCR will stop and so also through the motor armature. Also the current would stop if the firing of a particular thyristor is delayed till the pervious one has stopped conducting. Discontinuity in conduction is not a desirable feature and is not required to be achieved. But very often it happens that E_b, rises above the instantaneous voltage of a secondary phase of the transformer and discontinuous occurs. Discontinuity in current also results in torque fluctuation and in increasing the heating of motor armature.

iii. **Armature inductance.** Armature induction plays rather a helpful role. It does not allow the current to rise to such a value as it would if there were no induction. Also it minimise discontinuous conduction. Where the value of L_a is not sufficient to achieve these two purposes, extra induction may be added in series with the armature circuit to achieve the desired effects.

2.8.4 Field Excitation of DC Motor and Generator from Rectifiers

Direct current required for field excitation of motors or generators can also be obtained from rectifiers. Generally the shunt field system of a DC machine is highly inductive apart from its having a high resistance also. The ratio L/R is therefore high. We shall now consider how a thyristor rectifier may be used for supplying current to the field system. Referring to Fig. 2.24 the DC field is arranged to be supplied from the thyristor rectifier.

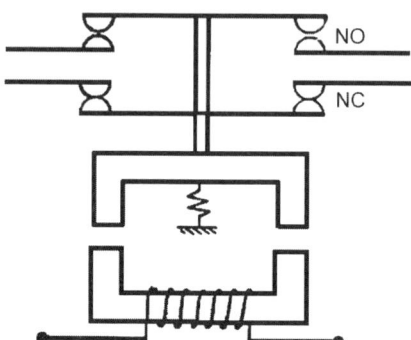

Fig. 2.24: Thyristor half-wave rectifier supplying a hunt field.

If the diode is absent, the current I, ignoring the voltage drop in the tube, is given by the equation

$$E_m \sin \omega t = L \frac{di}{dt} - Ri = 0$$

Initial condition is that i = 0 at t = 0. The current waveform is given in Fig. 2.25.

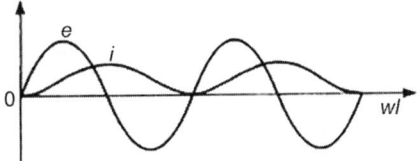

Fig. 2.25: Voltage and current relating to the circuit of 2.24 without the diode.

During the positive half-cycle, on account of the highly inductive nature of field, the current rises slowly and during the negative half-cycle for the same reason the current decays very slowly. Further as the current is increasing voltage $\left(L\dfrac{di}{dt}\right)$ is induced having such a polarity that the cathode (K) becomes positive with respect to the anode (A) in the diode and it does not conduct so long $\left(L\dfrac{di}{dt}\right)$ as the current increases.

When the current starts decreasing the polarity of the voltage reverses, the cathode of the diode becomes negative with respect to the plate and starts conducting, i.e. it acts as a short-circuit across the field system and now the thyristor is not conducting. The following equation is applicable:

$$-L\frac{di}{dt} - Ri = 0$$

From the two differential equations that we have, I can be determined.

The diode helps the stored energy of the field system to circulate during non-conduction time of the SCR. This makes the current through the motor nearly continuous. This diode is named freewheeling diode.

2.8.5 Chopper Control of DC Motors

When the supply is DC, chopper is used for the speed control of the DC motor. A chopper is a thyristor based circuitry which periodically interrupts the DC supply so that the average appearing across the motor is the supply voltage multiplied by the ratio of 'ON' time to time of the complete cycle.

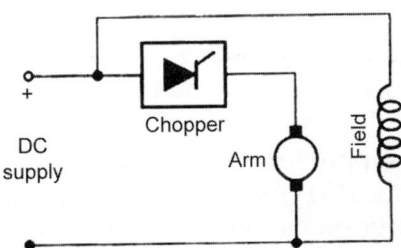

Fig. 2.26a: Chopper control of DC motor.

Fig. 2.26a shows the schematic diagram of the chopper control of armature and Fig. 2.26b exhibits the voltage waveforms at the input and at the motor terminals. It is to be noticed that as there is no current zero in the DC.

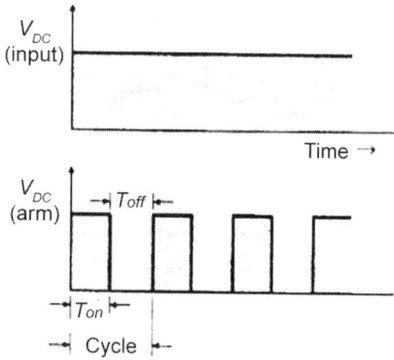

Fig. 2.26b: Voltage waveforms.

Input wave hence natural commutation does not occur and the thyristor is to be forced commutated. Separate hardware is to be incorporated for this purpose.

During start period, the 'ON' period is low making the voltage across the armature low. Slowly the motor starts and the 'ON' period is elongated. At rated speed, the 'ON' and 'OFF' periods are fixed and the motor runs stably. Speed controls can be achieved by varying 'ON' and 'OFF' time periods.

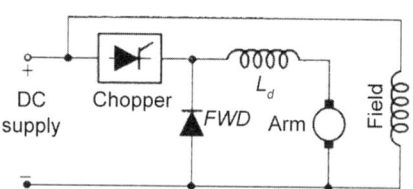

Fig. 2.27: Modified chopper control.

A free-wheeling diode (FWD) may be attached as shown in Fig. 2.27. Such that current remains continuous during 'OFF' period. This eliminates any speed drop or fluctuation in the circuitry. The inductance (L_d) in the armature circuit keeps the energy stored during on cycle releases it during 'OFF' cycle.

2.8.6 Closed Loop System for Automatic Speed Control of DC Motor

A half-wave thyristor rectifier (SCR) is widely used for converting the input AC voltage to DC voltage for the armature. The thyristor is coupled with a gate firing circuit and an error voltage detector.

Filtered DC voltage supply (utilising Diode D and capacitor C) is used as a reference voltage through the POT. As the speed of the motor is directly proportional to the back emf of the armature, the voltage across the armature is checked against the set DC value and error voltage detector Fig. 2.28 exhibits the scheme.

Instead of using this circuitry a tachogenerator can be employed which is already coupled with the armature of the motor mechanically. The output voltage of the tachogenerator is taken (which is proportional to the speed of the motor) and compared with a pre-set reference value. The difference in the result gives us the error voltage. Depending on the error voltage, the firing can be adjusted in the thyristor gate control

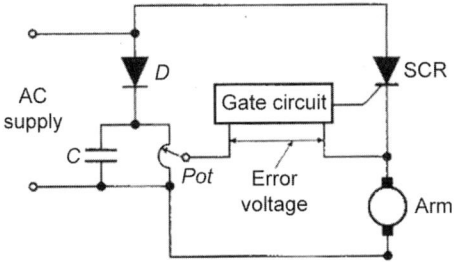

Fig. 2.28: Schematic of closed loop control circuitry of DC motor.

circuit which therefore controls the average output voltage of the thyristor and hence its speed.

A microprocessor can be used in order to compare the dynamic voltage signal (from the output of the tachogenerator) with the pre-set reference value and then it can control the firing angle of the thyristor assembly through a suitable hardware circuit.

2.8.7 Automatic Control

Electronic control of motors can also be made automatic utilizing closed loop speed control scheme and features like the following can be included in such control:

1. Regulation of speed
2. Regulation of voltage across armature
3. Compensation for armature resistance drop
4. Current limitation
5. Reversal of speed
6. Regeneration braking
7. Rheostatic (or dynamic) braking.

Commercial circuit, like this, has been developed by various manufacturers incorporating the above features. Any good book on Industrial Electronics may be referred to for complete description of such a circuit.

2.9 SPEED CONTROL OF INDUCTION MOTORS

Speed of induction motor is given by the following equation:

$$N_r = N_s(1-S) = \frac{120f}{P}(1-S)$$

$$\because N_s = \text{Synchronous speed} = \frac{120f}{P}$$

Thus, the speed of an induction motor, broadly speaking, can be regulated by three methods, i.e. by varying any of the above quantities, viz. Frequency, number of poles or sleep.

Thus speed change can be varying the frequency. Induction motor has drawbacks of developing low starting torque, drawing heavy starting current and having no easy means of continuous easy speed control. On the other hand, assets of induction motor are trouble free operation, less maintenance, high voltage operation consequently needing reduced amount of current and automatic regeneration. In addition to these, because of extreme mechanical simplicity in the construction, the rotor can resist centrifugal forces better than DC armature and, therefore for a given amount of iron and copper, more power can be produced.

2.9.1 Frequency Method of Speed Control

The above equation suggests that speed of the induction motor is directly proportional to the frequency of the supply voltage.

☞ Control from Variable Frequency Supply

Induction motor normally operates at a high efficiency and power factor is also kept high and it operates at speeds near to its synchronous speed. The difference between the actual speed and synchronous speed is called slip, which represents losses in the rotor. Thus in induction motor, operating from constant frequency supply, slip has to be small if efficiency is to be high. In other words, motor should operate at high speed (near to synchronous speed). If, however, synchronous speed itself is brought down near to actual slow operating speed, motor will still be working at high efficiency. This is achieved in variable frequency supply. Another advantage of feeding low frequency supply to induction motor at starting is that it does not take heavy starting current. This is proved as follows:

If suitable variable frequency supply is made available, induction motor can develop high starting torque without excessive rotor currents when it is supplied with low frequency voltage supply say 1/2 to 9 cycles. Intersection of the stable region of torque speed curve with load torque curve (Fig. 2.29) gives the operating speed. As the frequency of supply is reduced, torque curve shown dotted will move more towards left. This increases the starting torque (T_s). Another advantage of variable frequency supply is that as the motor speed falls, the frequency supply is that as the motor speed falls, the frequency of supply is reduced.

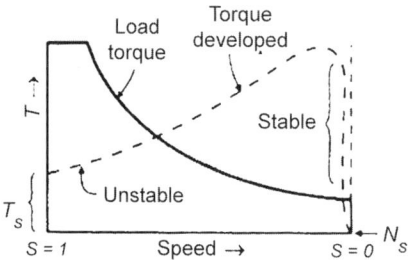

Fig. 2.29: Electromagnetic torque and load torque profiles.

This avoids the operation of motor on unstable portion of the speed torque curve. Induction motors, apart from the advantage of avoiding commutator maintenance can be of smaller size as they can be run at higher speeds and higher temperatures than DC motors and in addition they are more efficient. Induction motor is excellent for the industrial purpose as it has excellent power weight ratio, great mechanical strength, suitability for high speeds, no sliding contacts and high starting torque without overload. But all these advantages are available only through close control of flux and frequency.

☞ *Control by Variable Frequency Inverter Employing Thyristor*

With a conventional rotary converter, it was not possible to obtain low frequencies say ½ to 9 cycles, so that attempts for developing high starting torque resulted in heavy rotor currents. However with the development of silicon controlled rectifier (SCR), used as inverter, the frequency of three phase supply can be adjusted from 0–150 cycles.

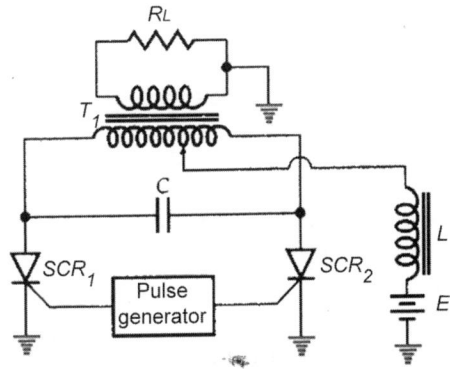

Fig. 2.30: Schematic of variable frequency inverter.

Fig. 2.30 shows a single phase parallel inverter circuit using two SCRs. To start with, assume SCR_1 as conducting and SCR_2 as off. Left plate of condenser C will be at negative potential and right plate will be at positive potential at double the voltage of DC source due to auto transformer action. A trigger from the pulse generator to SCR_2 will switch it on. Now condenser C will send current through SCR_2 and block SCR_1. DC source will send current through inductance L, transformer T and SCR_2. Current pulse flowing through transformer will produce polarity in the secondary of transformer of opposite sign to that produced previously when SCR_1 was conducting. Condenser C will now be charged with right band plate negative and left hand plate positive to double the DC source voltage. Now when SCR_1 is triggered, condenser C will again discharge but this time through SCR_1 to block SCR_2. In this way cycle repeats. For 1:1 turn ratio of transformer, the peak value of AC square wave is half the voltage of DC source and frequency of AC supply depends upon the frequency of the firing of SCRs.

2.9.2 Pole Changing Method of Speed Control

This method is applicable to squirrel cage motors only, as their rotors can adjust themselves to any number of poles. This method of speed control is used for driving drilling machines which requires different speeds for drilling different metals. This is also used for lifts where regenerative breaking can be applied by pole changing. On increasing the number of poles, synchronous speed becomes less than the actual running speed and motor now works as an induction generator. Some motors have two stator windings wound for different poles.

2.9.3 By Applying Variable Voltage to Stator

In the Fig. 2.31 two speed torque curves of a motor with different applied voltages are shown. Torque developed is proportional to square of applied voltage. Hence speed

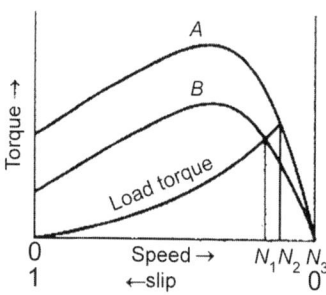

Fig. 2.31: Torque speed characteristics with voltage control.

torque curve A will be for higher voltage and B for lower voltage. Intersection of load torque line with the torque developed, gives us speed N_1 with higher voltage applied and N_2 speed with lower voltage applied. With constant torque loads, speed control by this method gives limited variation of speed. However for loads whose torque varies as the square of speed, this method gives wide ranges of speed.

2.9.4 By Varying the Rotor Copper Losses

For constant load torque applications, slip is directly proportional to the copper losses in induction motor. By increasing the rotor copper losses, slip will be increased and hence speed can be decreased. Rotor losses can be regulated by following methods.

1. **By inserting external resistance in the rotor circuit:** In induction motor, for a given torque to be developed, rotor current remains constant. Therefore if resistance is inserted in the rotor circuit, it will increase the rotor copper losses and, therefore, slip. This method, besides being wasteful for low speeds, requires heavy three-phase controller to dissipate the losses.

2. **By cascade control:** Instead of wasting the energy in rotor resistance and creating a problem of disposal of heat, it can be taken out of the rotor and fed to another motor which is capable of taking power at low frequency. This motor is usually mechanically coupled with the main motor. Thus power taken from the rotor is converted to mechanical energy. This arrangement would give increased output at reduced speeds. This rotor power at low frequency can also be converted to power at supply frequency by means of frequency converter and fed back to the line. Arrangement of connecting auxiliary machine to the rotor of main motor to perform any of the above functions is called cascading.

2.9.5 Speed Control by Slip Coupling

This method allows driver shaft to run faster than the driven shaft by an amount which can be controlled. There will be some power loss in the coupling itself. One form of slip coupling consists of stator and rotor. Rotor is squirrel cage type and is keyed to the driven. In ordinary induction motor rotating magnetic field is produced electrically by the three phase currents. As against this, in slip coupling, magnetic field is mechanically driven by the driver. This will exert torque on the rotor which will be transmitted to the driver. In induction motor the magnitude of the torque is dependent upon the magnitude of the rotating field or applied voltage. In slip coupling, also the magnitude of the torque at which slip occurs can be varied by the excitation of the slip coupling which is conveyed to it through two slip-rings. Torque slip characteristics of sleep coupling are essentially same as that of the induction motor. Slip coupling can, therefore, make it possible to have variable speed drive from constant speed driver.

2.10 ELECTRIC BRAKING

There are two types of applications where braking torque is applied to the apparatus driven by electric motor. These are those applications where, (a) kinetic energy of the moving parts is only involved, (b) wherein addition to kinetic energy of moving parts, potential energy, usually gravitational which can drive the apparatus possibly at an excessive speed, is involved. Braking torque can be applied either by mechanical brakes or electrodynamically. Electrodynamics braking can be applied by a separate eddy current brake or driving motor can be made to work as brake. In this article we will not deal with mechanical braking. Before we deal with electric braking it will be desirable to know the advantages and limitations of this method of braking.

2.10.1 Advantages and Disadvantages of Electric Braking Over Mechanical Braking

1. In mechanical braking; due to excessive wear on brake drum, it needs timely and costly replacement. This is not needed in electrical braking and so electrical braking is more cost effective than mechanical braking.
2. Due to wear and tear of brake liner frequent adjustments are needed thereby making the maintenance costly.
3. Mechanical braking gives metal dust as a by-product, which can damage bearings. Electrical braking has no such problems.
4. If mechanical brakes are not correctly adjusted it may result in shock loading of machine or machine parts in case of lift, trains which may result in discomfort to the occupants.
5. Electrical braking is smooth.
6. In mechanical braking the heat is produced at brake liner or brake drum, which may be a source of failure of the brake. In electric braking the heat is produced at convenient place, which in no way is harmful to a braking system.
7. In regenerative braking electrical energy can be returned back to the supply which is not possible in mechanical braking.
8. Noise produced is very high in mechanical braking. Only disadvantage in electrical braking is that it is ineffective in applying holding torque.

2.10.2 Types of Electric Braking

There are three types of electric braking as applicable to common types of electric motors in addition to eddy-current brakes.

1. Plugging or reverse current braking.
2. Rheostatic or dynamic braking.
3. Regenerative braking.

In many cases, provision of an arrangement for stopping a motor and its driven load is as important as starting it. For example, a planing machine must be quickly stopped at the end of its stroke in order to achieve a high rate of production. In other cases, rapid stops are essential for preventing any danger to operator or damage to the product being manufactured. Similarly, in the case of lifts and hoists, effective braking must be provided for their proper functioning.

☞ Plugging

In plugging, the connections of the armature terminals are reversed so that the motor tends to rotate in the opposite direction thus providing the necessary braking effect. However, the power supply must be cut-off when the motor comes to rest otherwise it will start rotating in the reverse direction. Plugging may be used with DC, induction and synchronous machines.

    a. **Plugging with DC motors**: The current in the armature reverses as the armature is reversed with respect to the field . Under normal running condition, the back emf E is opposite to the direction of the armature current but during braking, the back emf E and the armature current are in the same direction. At the instant of reversal of armature connections a voltage equal to $V + E$ is available across the armature circuit, V being the supply voltage. Since E is very nearly equal to V, the impressed voltage is becoming twice and is approximately $2V$. This will cause a great rush of current in the armature circuit. To prevent this, the starting resistance is reinserted in the armature circuit as shown in Fig. 2.32.

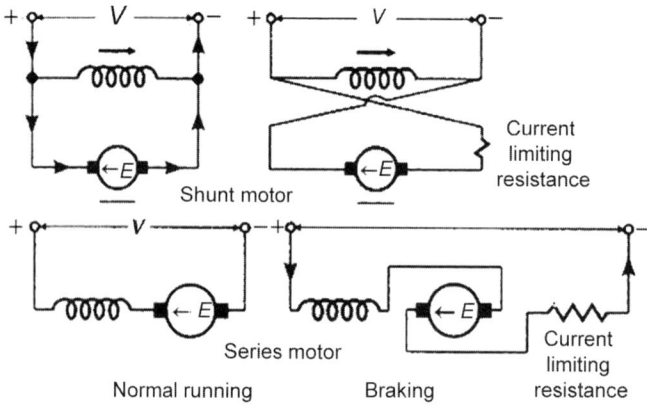

Fig. 2.32: Plugging of DC motors.

    It should, however, be noted that during braking, in addition to the kinetic energy of the motor being dissipated in the resistance, some energy is being drawn from the supply. There is, therefore, a waste of energy.

    The manner in which electric braking torque (EBT) depends on speed is found from the relation given below:

Electric breaking torque (EBT) $\propto I$

i.e., EBT $= k_1 \phi I$

but $I = \dfrac{V + E}{R}$

where, V = supply voltage,

        E = back emf,

        R = total resistance in the armature circuit.

Therefore,

$$EBT = k_1\phi\frac{V+E}{R}$$

$$E = k_2 N\phi$$

∴

$$EBT = k_1\phi\frac{V+k_2 N\phi}{R}$$

$$= \frac{k_1\phi V}{R} + \frac{V+k_1 k_2 N\phi^2}{R}$$

$$= k_3\phi + k_4 N\phi^2$$

In the case of a series motor, ϕ is proportional to the current and the value of the torque can only be determined from the magnetization curve. For a shunt motor, however, ϕ is constant.

Therefore,

$$EBT = k_3\phi + k_4 N\phi^2 = k_5 + k_6 N$$

When the motor is driving a load, the load will exert an additional breaking torque and the total breaking will be that due to the electric breaking and the load.

b. **Plugging with induction motors:** If any two supply phases are interchanged with each other the direction of rotation of the magnetic field reverses and, therefore, the torque on the rotor also reverses providing a breaking action. Supply, however, has to be cut-off when the motor comes to rest, otherwise the rotor would start building up the motion in the reverse direction. The rotor and the stator currents tend to be abnormally high and a resistance may have to be inserted in the rotor pr stator circuit for the purpose of protection.

c. **Plugging with synchronous motors:** If the DC excitation of the synchronous motor is reversed, the DC and AC fields will rotate in opposite direction and there can be no braking effect. But in case of motor fitted with damper windings the eddy currents induced in them provide braking.

☞ *Rheostatic or Dynamic Braking*

The motor is disconnected from the supply and worked as a generator driven by the kinetic energy of the rotor and the load. A resistance is connected across the motor terminals; the kinetic energy of rotation is converted into electric energy and is dissipated in the resistance.

a. **DC motors shunt:** The armature is disconnected from the supply and connected across resistance. The motor now works as a separately excited generator and a braking torque is applied by the current delivered to the resistance. If however, the supply fails, the braking action vanishes as the excitation disappears. This drawback is sometimes removed by fitting a series winding in the armature circuit which is connected during the braking period only. Due to the action of this winding, the motor self-excites as a series generator and the current delivered by the armature provides braking action (Figs 2.33a and b).

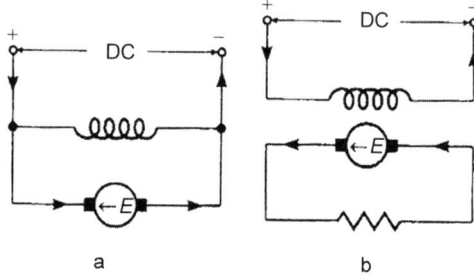

Figs 2.33a and b: Rheostatic braking of shunt motor: (a) Normal running, (b) Resistance braking.

b. **DC motor-series:** The after being disconnected from the supply is made to excite as a series generator. For this it is necessary that the total resistance in the motor circuit should less than the critical resistance, so that the generator may self-excite. Also in order that the flux may build-up, the connection of the armature with respect to the field have to be reversed (Figs 2.34a and b).

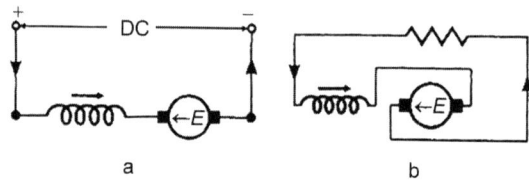

Figs 2.34a and b: Rheostatic braking of shunt motor: (a) Normal running, (b) Braking.

Normally, the starting resistance is employed for braking purposes. Since torque is proportional to the product of flux and current, we have in a series motor electric braking torque $= k_1 \phi I$ and braking current $I = \dfrac{E}{R}$ where E is the induced emf in the armature and R is the total resistance in the motor circuit.

$\therefore$

$$EBT = k_1 \phi \frac{E}{R} = \frac{k_1}{R} \phi E = \frac{k_1}{R} \phi (k_2 N \phi)$$

$$= \frac{k_1 k_2}{R} \phi^2 N = k_3 N \phi^2$$

For a shunt motor ϕ is constant.

$\therefore$

$$EBT = k_4 N$$

c. **Synchronous motors:** The field excitation is maintained and the motor after being disconnected from the supply is connected to resistances in star or delta. It now works as an alternator and the kinetic energy is dissipated in the form of losses in the resistance.

d. **Induction motors:** The stator is disconnected from the supply and direct steady current is passed through its windings. A flux is produced. When the short-circuited rotor conductors cut this steady flux emf is induced in them which provide the necessary braking effect. If the rotor is wound, the braking torque can be controlled by the insertion of suitable resistance in the rotor circuit.

☞ *Regenerative Braking*

In regenerative braking the motor is run as generator by the kinetic energy of the load which is returned to the mains as electrical energy. There is, therefore, an overall saving in energy.

a. **DC motors-shunt:** If the emf generated by the motor is greater than the supply voltage, power will be fed back into the supply. The emf in a shunt motor depends upon its excitation and speed. If the field is disconnected from the supply and the field current is increased by exciting it from another source, the induced emf will exceed the supply voltage and the motor will feed energy into the supply. The speed of the motor, however, falls to value corresponding to the field current at any instant. The condition is shown in Fig. 2.35a.

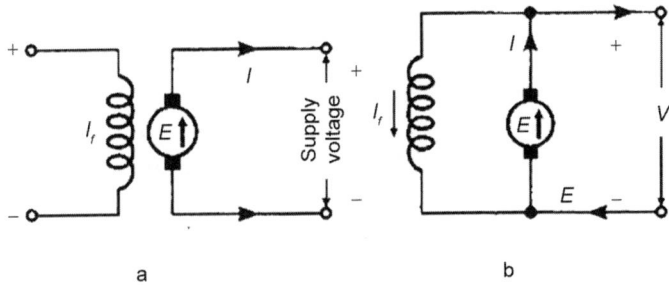

a   b

Figs 2.35a and b: (a) Regenerative braking with DC shunt motor, (b) Regeneration at constant speed

There is another way in which regeneration takes place resulting in braking effect Fig. 2.35b. If the field excitation does not change but the load causes the speed to increase, the induced emf may become greater than the supply voltage and power will be fed back into the supply. The regenerative effect, however, will prevent any tendency of the speed to increase further. This itself constitutes a form of braking effect since in the absence of regeneration, the speed would increase continuously.

b. **DC motor series:** Regenerative braking with series motors is employed mainly in traction work. This is dealt within detail in the chapter on Electric Braking.

c. **Induction motors:** When an induction motor runs at a speed above the synchronous, it works as an induction generator feeding power back into the supply. No extra devices need be employed. It may be noted that regenerative braking of induction motors is hardly useful for stopping the motor but it helps in keeping the load at a speed slightly above the synchronous and returns valuable power to supply.

2.11 TYPE OF INSULATION EMPLOYED

The qualities that an insulating material should possess are: high breakdown voltage, high heat conductivity, high mechanical strength, high flexibility, high resistance against vibration, high resistance against surface creeping currents and low dielectric loss. These qualities deteriorate after sometime but the speed at which these qualities go down depends largely upon the endured temperature and its duration. Insulating materials have therefore been classified according to the maximum temperature they can endure for a reasonably long-time say 15 to 30 years under normal conditions. Electrical machinery is designed for a limit in temperature rise rather than for a maximum temperature. The continuous rating of a machine is that rating for which final temperature rise is equal to or just below the limit in temperature rise for insulating material used in the construction of the machine. When the machine is over-loaded for such a long-time that, it is final temperature rise is higher than the limit, this may cause damage to it. This damage in worst cases will result in an immediate thermal breakdown of the insulating material which will cause short circuit in the motor, thus putting an stop to its functioning. Short circuit may also give rise to a fire. In less severe cases immediate thermal breakdown of the insulating material may not occur but the quality of insulation will deteriorate such that thermal breakdown with future overloads or even normal loads might soon occur, thus shortening the useful life of the machinery and apparatus.

Different insulating materials have different limiting temperatures. Some of the insulating materials are given below as classified in ISS 1271.

◆ **Class Y:** Cotton, silk, paper, similar organic materials neither impregnated nor immersed in liquid dielectric. The limiting temperature is 90°C.

◆ **Class A:** Cotton, silk, paper and similar organic materials impregnated or immersed in liquid dielectric also. Limiting temperature is 105°C.

◆ **Class E:** Certain synthetic organic films and other materials having the same thermal stability. Limiting temperature is 120°C.

◆ **Class B:** Mica, asbestos and similar inorganic materials combined with binding cement. Limiting temperature is 130°C.

◆ **Class C:** Mica, without binding cement porcelain, glass quartz and other similar fire proof material. Limiting temperature is above 130°C and is restricted only by the specific physical, chemical or electrical properties of the material concerned.

The limiting temperatures are never exceeded in practice, otherwise it will result in deterioration and breakdown of the insulation and will shorten the life of the motor.

2.12 HEATING OF MOTOR OR TEMPERATURE RISE

The rise in temperature of a motor results from the heat generated by the losses and an expression for this temperature rise is obtained by equating the rate at which heat is being generated by these losses to the rate at which heat is being absorbed by the motor for raising the temperature of motor and in dissipation from the surfaces exposed to cooling media.

So long as the temperature of machine rises, the generated heat will be stored in body and the rest will be dissipated to cooling medium depending upon the temperature difference. This is called unstable or transient situation.

If the temperature of body rises, it has to store heat. The amount of heat, i.e. stored depends upon the heat capacity of the body. If the temperature of the machine remains constant, i.e. it does not rise, then no further storage of heat takes place and all the heat, i.e. generated must be dissipated. So rate of heat generation in motor equals rate of heat dissipation from the cooling surface. This is called a stable situation.

2.12.1 Equation for Heating of Motor

Let, W → Heat generated in motor due to power loss in watts.

G → Weight of motor (kg)

S → Average specific heat in (Watt - Sec.) to raise the temperature of unit weight through 1°C.

$G \times S$ → Heat required to raise the temperature of motor through 1°C (Watt - Sec.)

θ → Temperature rise above cooling medium in °C.

θ_f → Final temperature rise in °C.

A → Cooling surface area of motor.

λ → Rate of heat dissipation from the cooling surface.

[(Watts/Unit area/°C rise in temperature.) above cooling medium]

$A\lambda$ → Rate of heat dissipation in Watts/°C rise in temperature for a motor.

☞ *Assumptions*

1. Loss 'W' remains constant during temperature rise.
2. Heat dissipation is proportional to the temperature difference between motor and cooling medium.
3. Temperature of cooling medium remains constant.

   {Rate of heat generation in motor} = {Rate of heat absorption by the motor} + {Rate of heat dissipation from cooling surface}

$$\rightarrow \qquad W = GS\frac{d\theta}{dt} + A\lambda\theta \qquad\qquad ... (2.3)$$

$$\text{or} \qquad W - A\lambda\theta = GS\frac{d\theta}{dt}$$

$$\rightarrow \qquad \frac{W}{A\lambda} - \theta = \frac{GS}{A\lambda}\frac{d\theta}{dt}$$

$$\rightarrow \qquad \frac{d\theta}{\left(\dfrac{W}{A\lambda} - \theta\right)} = \frac{dt}{\dfrac{GS}{A\lambda}}$$

By integrating, $\qquad \log_e\left(\dfrac{W}{A\lambda} - \theta\right) = \dfrac{A\lambda}{GS}t + C \qquad\qquad$ (2.4)

At $t = 0$, $\theta = \theta_1$ [Initial temperature rise, i.e. difference between the temperature of cooling medium and temperature of motor, during starting]. If starting from cold position, $\theta_1 = 0$

Substituting the values of t and θ in above equation.

$$C = \log_e\left(\frac{W}{A\lambda} - \theta_1\right)$$

$$\therefore \qquad (2.4) \rightarrow \log_e\left[\frac{\left(\dfrac{W}{A\lambda} - \theta\right)}{\left(\dfrac{W}{A\lambda} - \theta_1\right)}\right] = -\frac{A\lambda}{GS}t$$

By taking antilog,

$$\frac{\left(\dfrac{W}{A\lambda} - \theta\right)}{\left(\dfrac{W}{A\lambda} - \theta_1\right)} = e^{-\frac{A\lambda}{GS}t}$$

$$\therefore \qquad \theta = \frac{W}{A\lambda} - \left(\frac{W}{A\lambda} - \theta_1\right)e^{-\frac{A\lambda}{GS}t} \qquad\qquad (2.4.1)$$

When, the final temperature rise of θ_f is reached, all the heat generated is dissipated from the cooling surface so that, equation (2.3) becomes $W = A\lambda\,\theta_f$ or $\theta_f = \dfrac{W}{A\lambda}$

And $\qquad\qquad \dfrac{GS}{A\lambda} = $ Heating time constant

$$\therefore \qquad\qquad \frac{A\lambda}{GS} = \frac{1}{T}$$

Then equation (2.4.1) becomes;

$$\theta = \theta_f - \left(\theta_f - \theta_1\right)e^{\frac{-t}{T}}$$

If starting from cold, then $\theta_1 = 0$

$$\therefore \qquad\qquad \theta = \theta_f\left(1 - e^{\frac{-t}{T}}\right)$$

2.12.2 Heating Time Constant

Heating time constant of motor is defined as the time required to heat up the motor up to 0.633 times its final temperature rise.

$$\theta = \left(1 - e^{\frac{-t}{T}}\right)$$

At $t = T$, $\theta = 0.633\,\theta_f$

After time $t = T \quad \theta$ reaches to 63.3 % of θ_f

$\qquad\qquad t = 2\,T \;\; \theta$ reaches to 86.5 % of θ_f

$\qquad\qquad t = 3\,T \;\; \theta$ reaches to 95 % of θ_f

$\qquad\qquad t = 4\,T \;\; \theta$ reaches to 98.2 % of θ_f

$\qquad\qquad t = 5\,T \;\; \theta$ reaches to 99.3 % of θ_f

T = Heating time constant.

$\qquad$ = 90 min for motors up to 20 HP.

$\qquad$ = 300 min for larger motors.

2.12.3 Equation for Cooling of Motor or Temperature Fall

If rate of heat generation is less than rate of heat dissipation, cooling will take place.

$\qquad \therefore$ {Rate of heat generation in motor} + {Rate of heat absorption by motor} = {Rate of heat dissipation from cooling surface}

$$\rightarrow \qquad\qquad W = GS\frac{d\theta}{dt} + A\lambda'\theta$$

λ' = Rate of heat dissipation during cooling surface

$$W - A\lambda'\theta = -GS\frac{d\theta}{dt}$$

$$\rightarrow \frac{W}{A\lambda'} - \theta = -\frac{GS}{A\lambda'}\frac{d\theta}{dt}$$

$$\theta - \frac{W}{A\lambda'} = \frac{GS}{A\lambda'}\frac{d\theta}{dt}$$

$$\rightarrow \frac{d\theta}{\left(\theta - \dfrac{W}{A\lambda'}\right)} = \frac{dt}{\dfrac{GS}{A\lambda'}}$$

$$\therefore \quad \int \frac{d\theta}{\left(\theta - \dfrac{W}{A\lambda'}\right)} = \int \frac{dt}{\dfrac{GS}{A\lambda'}}$$

$$\log_e\left(\theta - \frac{W}{A\lambda'}\right) = \frac{A\lambda'}{GS}t + C$$

At $t = 0$ Let $\theta = \theta_0$ Difference of temperature between cooling medium and motor (Temperature rise at which cooling starts.)

$$\therefore \quad C = \log_e\left(\theta_0 - \frac{W}{A\lambda'}\right) \text{Put this value of } C \text{ in the above equation.}$$

$$\therefore \qquad \log_e \frac{\theta - \dfrac{W}{A\lambda'}}{\theta_0 - \dfrac{W}{A\lambda'}} = \frac{A\lambda'}{GS}t$$

If θ'_f is final temperature drop (above that of cooling medium), then at this temperature whatever heat is generated will be dissipated.

$$\therefore \qquad W = A\lambda'\theta'_f \qquad \rightarrow \theta'_f = \frac{W}{A\lambda'}$$

$$\therefore \qquad W = A\lambda'\theta'_f \qquad \rightarrow \theta'_f = \frac{W}{A\lambda'}$$

$$\therefore \qquad \log_e\left\{\frac{\theta - \theta'_f}{\theta_0 - \theta'_f}\right\} = -\frac{t}{T'}$$

Where T' is cooling time constant $= \dfrac{GS}{A\lambda'}$

$$\frac{\theta - \theta'_f}{\theta_0 - \theta'_f} = e^{-\frac{t}{T'}} \rightarrow \left(\theta - \theta'_f\right) = \left(\theta_0 - \theta'_f\right)e^{-\frac{t}{T'}}$$

$$\theta = \theta'_f + \left(\theta_0 - \theta'_f\right)e^{-\frac{t}{T'}}$$

If motor is disconnected from supply, there will be no losses taking place and so final temperature reached will be ambient temperature. Hence $\theta'_f = 0 \; (W = 0)$

$$\therefore \qquad\qquad \theta = \theta_0 . e^{-\frac{t}{T^1}}$$

If $t = T^1$, then $\theta = \theta_0 . e^{-1}$

$$\theta = \frac{\theta_0}{e} = 0.368 \, \theta_0 ;$$

$$\therefore \qquad\qquad \theta = 0.368 \, \theta_0$$

2.13 COOLING TIME CONSTANT

Cooling time constant is defined as the time required to cool machine down to 0.368 times the initial temperature rise above ambient temperature.

By putting different values of T' in

$$\therefore \qquad\qquad \theta = \theta_0 . e^{-\frac{t}{T^1}}$$

After time $t = T'$ θ has fallen to 36.8% of θ_0
$\qquad\qquad t = 2T'$ θ has fallen to 13.5% of θ_0
$\qquad\qquad t = 3T'$ θ has fallen to 5% of θ_0
$\qquad\qquad t = 4T'$ θ has fallen to 1.8% of θ_0
$\qquad\qquad t = 5T'$ θ has fallen to 0.7% of θ_0

2.14 DUTY CYCLES

The nominal duty of a drive motor is the duty corresponding to the service conditions and performance marked on its nameplate.

There are three types of duties, viz. continuous duty, short-time duty and intermittent duty.

The heating and cooling curves for continuous-duty motor are shown in Fig. 2.36a. Continuous duty is that duty when the on-period is so long that the motor attains a steady-state temperature rise.

The heating and cooling curves for short-time duty motor are given in Fig. 2.36b. The short-time duty motor operates at a constant load for some specified periods which

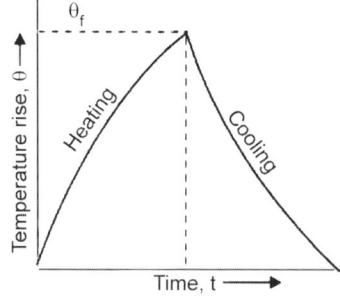

Fig. 2.36a: Heating and cooling curves for continuous-duty motor.

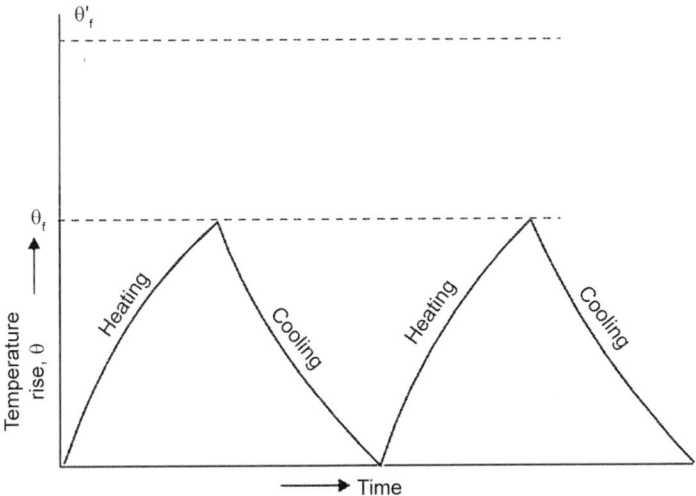

Fig. 2.36b: Heating and cooling curves for short-time duty motor.

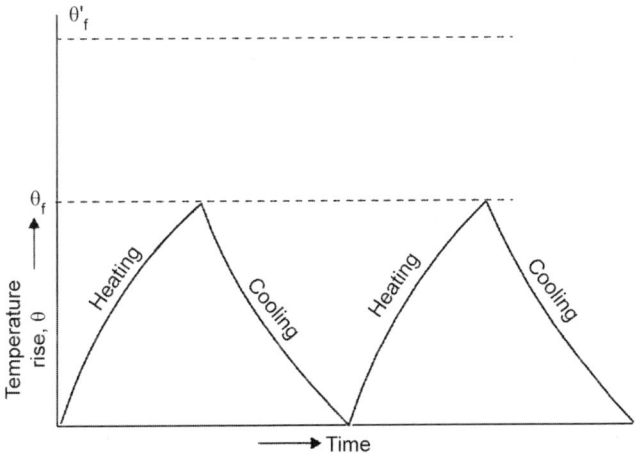

Fig. 2.36c: Heating and cooling curves for intermittent periodic duty motors.

is then followed by a period of rest. The period of run (or load) is so short that machine cannot attain its steady temperature rise while the period of rest is too long that the motor temperature drops to the ambient temperature.

The heating and cooling curves for intermittent periodic duty motors are illustrated in Fig. 2.36c. On intermittent or duty the periods of constant load and rest with machine de-energized alternate. The loading periods are too short to allow the motor to attain its final steady-state value while periods of rest are too small to allow the motor to cool down to the ambient temperature. Intermittent rating of a machine is defined as the load which is applied during a certain fraction of time of a load cycle and the temperature rise limit is not exceeded.

When a machine is intermittently loaded, it will cool down during the time it is off, and temperature will rise when it is on, as illustrated in Fig. 2.37.

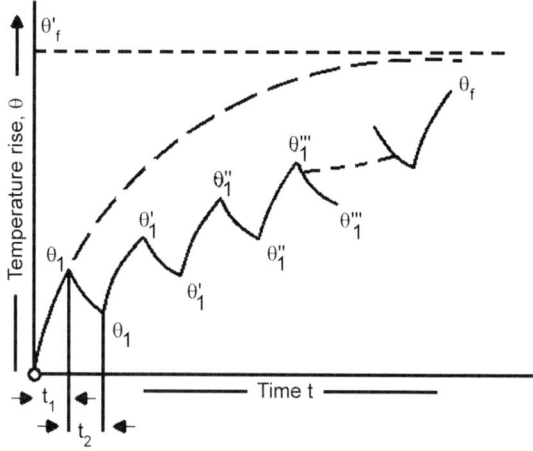

Fig. 2.37: Heating and cooling curves for short intervals.

Continuous duty motors are employed to drive fans, compressors, generators, etc. They may be in operation for many hours and even days in succession. Short-time duty motors are used in navigation-lock gates, railway turn tables, bascule bridges and the like. Intermittent duty motors are employed in cranes hoists, lifts, rolling mills, some metal working machines, etc. The duty cycles for various motors duty are shown in Figs 2.38a to c.

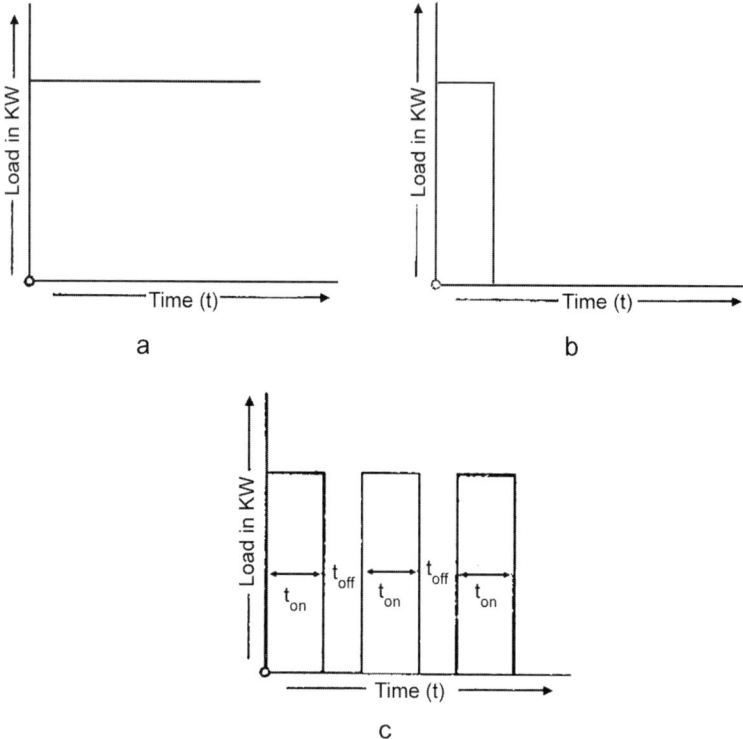

Figs 2.38a to c: (a) Continuous-duty cycle, (b) Short-time duty cycle, (c) Intermittent duty cycle.

2.14.1 *Maximum temperature reached with intermittent loads*

When a machine is intermittently loaded it will cool off during the time it is off and its temperature will rise when it is on as shown in Fig. 2.37. Let θ_1, θ_1' etc. be the temperatures after heating and θ_2, θ_2' etc. be the temperatures after cooling as shown in Fig. 2.37.

$$\theta_1 = \theta_f' \left(1 - e^{-\frac{t_1}{T}} \right) = \theta_f' \left(1 - e^x \right) \qquad \text{where } x = -\frac{t_1}{T}$$

$$\theta_2 = \theta_1 e^{-\frac{t_2}{T'}} = \theta_1 e^y = \theta_2' = \theta_f' \left(\theta - e^x \right) e^y \text{ where } y = -\frac{t_2}{T'}$$

$$\theta_f' = \theta_f' - \left(\theta_f' - \theta_2 \right) e^x$$

$$= \left(1 - e^x \right) \theta_f' + \theta_2 e^x$$

$$= \left(1 - e^x \right) \theta_f' + \theta_f' \left(1 - e^x \right) e^y e^x$$

$$= \theta_f' \left(1 - e^x \right) \left(1 + e^x e^y \right)$$

$$\theta_2' = \theta_1' e^y$$

$$= \theta_1' e^y \left(1 - e^x \right) \left(1 + e^x e^y \right)$$

$$\theta_1'' = \theta_f' \left(1 - e^x \right) + \theta_2' e^x = \left(1 - e^x \right) \theta_f' + \theta_f' e^x e^y \left(1 - e^x \right) \left(1 + e^x e^y \right)$$

$$= \theta_f' \left(1 - e^x \right) \left[1 + e^x e^y + e^{2x} e^{2y} \right]$$

For n times intermittency, we have,

$$\theta_1^{1-n} = \theta_f' \left(1 - e^x \right) \left\{ 1 + e^x e^y + \ldots + e^{(n-1)x} e^{(n-1)y} \right\}$$

$$= \theta_f' \left(1 - e^x \right) \left\{ \frac{1 - e^{nx} e^{ny}}{1 - e^x e^y} \right\}$$

If $n = \infty$ both e^{nx} and e^{xy} will be zero as x and y are negative. If θ_f is the maximum temperature rise with intermittent load, then

$$\theta_f = \theta_f' \left(\frac{1 - e^x}{1 - e^x e^y} \right)$$

$$= \theta_f' \left(\frac{1 - e^x}{1_{-x} \left(x + y \right)} \right)$$

$$= \theta_f' \frac{1 - e^{-\frac{t_1}{T}}}{1 - e^{-\left(\frac{t_1}{T} + \frac{t_2}{T^1} \right)}}$$

2.15 RATING OF MACHINES

A nameplate attached to the outer portion of an electrical machine represents the data pertaining to its rating. A machine rating normally specifies the voltage, current, speed, excitation, power factor, efficiency, power output, etc.

The rating of a machine is such which should contain all necessary information so that, if the machine is operated within the limits of all the factors mentioned in its rating, the machine will function satisfactorily and will operate well within the safe limit. And the life span of the machine will also increase. Therefore, the rating of a machine must represent all the necessary information which will safeguard the application of the machine from conditions of operation which (i) would result in unsafe mechanical or electrical strains upon any part of its structure or (ii) would result in excessive deterioration of the mechanical or electrical characteristics of the materials of which the machine is constructed. For providing this information, the rating of an electric machine should include the output, voltage, speed, and any other information that may be necessary for the proper operation of the machine.

Generators and motors are always rated in terms of kW output at a given speed and voltage. The size and rating of an electric machine for any service is mainly governed by the factor 'temperature rise'. The maximum temperature, to which an electric machine is allowed to reach, is limited by the type of insulation used.

The maximum temperature rise permissible with insulation A and insulation B are 40°C and 50°C respectively. Overloads are generally permissible for short period of time but when machines are required to carry greater loads than those specified, they must he kept under inspection to see that the temperature does not rise too much and that severe sparking at the commutator does not occur.

The type of service to which a machine is subjected is of great importance. The machines operating continuously at rated (or near rated) load are physically larger than those working at intermittent loads. Also electric machines that are not enclosed and are, in addition, well cooled by fans are likely to have higher ratings than covered up machines or those machines located where air does not circulate freely through and over them. High-speed machines and machines employing mica, glass tapes, and the new silicons as insulation can generally be physically smaller, in given ratings, than the low-speed machines employing standard insulations. The output of DC machines is also limited by the factor 'commutation sparking'. This factor often limits the output, even though heating may not have proceeded to permissible values.

The machines according to ISI specifications are classified as follows:

a. **Continuous rating:** This is an output, which a machine delivers continuously without exceeding the permissible temperature rise. It can deliver 25 percent overload for two hours.

b. **Continuous maximum rating:** Similar to continuous rating but not allowing any overload. It is used for motors of capacity larger than 1.84 kW (2.5 hp) per rpm.

c. **Short-time rating:** This is an output which an electric machine can deliver for a specified period (say 1 hr., ½ hr., ¼ hr. etc.) without exceeding the specified temperature rise.

2.15.1 Effect of Altitude on Rated Output

At high altitudes, the density of cooling air decreases. As a result, cooling capacity is reduced. The effect is negligible for elevations not exceeding 1,000 m. The output is reduced as stated in Table 2.2.

Table 2.2	Reduction of rated output at higher altitudes
Altitude above sea level in metres	*Permissible output in percentage of the rated value*
1000	100
2000	95
3000	87
4000	78

2.15.2 Overload Capacity of Induction Motor

Motors are designed to carry overloads without any adverse effect as mentioned below.

Single phase induction motors are usually not designed to carry overload. However 20–25% overload in torque intermittently may be permitted in split phase AC motors and 40–45% overload in torque intermittently for capacitor start motor.

2.16 CHOICE OF RATING OF MOTORS

The choice of ratings of motors for the services requiring fairly constant power such a fans, air-compressors, blowers, pumps, motor-generator sets, etc. is every simple. In such cases the power required is determined and motor having continuous rating of power required is chosen. It is, however, worthwhile to take considerable trouble to find out exactly how much power is actually required by the driven machinery, since if the motor it will be overloaded and will, therefore, overheat and deteriorate, while if it is too large the efficiency and power factor will probably be less than those of a motor of the correct size. If there is no possibility of overloading then a motor with a continuous maximum rating may be selected, which will be comparatively smaller in size and cheaper in cost. Such a motor may operate at a higher temperature (55°C in place of 40°C).

For the constant torque loads, the size of the motor can be determined as follows:

In case of rotary motion, rating of motor required is given as

$$P = T \times \frac{2\pi N}{60} \times \frac{1}{\eta} \times \frac{1}{1000} \text{kW} = \frac{TN}{9550} \text{kW}$$

where T is the load torque in Nm, N is the speed in rpm and η is the product of the efficiency of the driven load and that of the transmitting device. Efficiency varies considerably with the type of drive, bearings, etc. and its value can best be dictated by experience in application.

In case of linear motion, rating of the motor required is given as

$$P = \frac{F \times 9.81 \times v}{\eta} \times \frac{1}{1000} \text{kW} = \frac{F \times v}{102} \text{kW} \tag{2.5}$$

where F is the force caused by the load in kg and v is the velocity of motion of load in m/s.

The above expression [equation (2.5)] can be applied in case of hoisting mechanism, lifts or elevators with slight modification and that is due to counterweight (balancing the weight of the cage or car as well as one-half of the useful load). The modified expression is

$$P = \frac{F \times v}{2 \times 102\eta} \text{ kW}$$

v in case of normal passenger lift cabins lies in the range of 0.5–1.5 m/s.

In case of pumps, the rating of the motor required,

$$P = \frac{Q \times \rho \times h}{102\eta} \text{ kW}$$

where Q is the delivery of pumps in m³/s, ρ is the density of liquid pumped in kg/m³ and h is the gross head (static head + friction head) in m. Efficiency η lies in the range of 80 to 90% in case of reciprocating pumps and 40 to 80% in case of centrifugal pumps.

Similarly, the rating of a fan motor is given as

$$P = \frac{Qh}{102\eta} \text{ kW}$$

where Q is the volume of air or any other gas in m³/s and h is the pressure in mm of water or in kg/m². Efficiency may be taken to be 0.8 for larger power fans and 0.6 in. case of smaller power fans.

The selection of size of motor is also simple for the services where the motor is run for a short time so that the temperature rise is less than the maximum permissible value and then it is allowed to cool down to the ambient temperature. For such services a motor rated at the required output power for the given period is chosen.

In cases where the load fluctuates over a given cycle, as in rolling-mills or colliery winders, the size of the motor is determined accurately by finding the heating and cooling curves of motors under consideration and plotting heating curve for each motor, when working on the given cycle. The smallest motor which delivers the load according to given cycle without exceeding the specified temperature rise may be choosen.

Other methods for determination of rating of a motor for continuous duty and variable load are given below.

1. **Method of average losses:** In this method it is assumed that the temperature rise attained by the motor with fluctuating loading conditions over a certain period of time will be the same as that attained by the motor with a certain load of constant magnitude. This will be true provided that the average losses in the motor for the same time are the same for both the conditions of operation.

   This method consists of determining average losses in the motor when it operates according to the given load diagram and then comparing with the losses corresponding to the continuous duty of the machine when operated at its nominal rating.

2. **Equivalent current method:** This method is based upon the assumption that the actual variable current may be replaced by an equivalent current I_{eq} which produces the same losses in the motor as the actual current.

$$I_{eq} = \sqrt{\frac{I_1^2 t_1 + I_2^2 t_2 + I_3^2 t_3 + ...I_n^2 t_n}{t_1 + t_2 + t_3 + ... + t_n}} \qquad (2.6)$$

The heating and cooling conditions in self-ventilated machines depend upon its speed. At low speeds the cooling conditions are poorer than at normal speeds. Therefore, if the work cycle involves slow-speed operation, it must be taken into consideration when using equation (2.6).

The equivalent current is compared with the rated current of the motor selected. The equivalent current may be less than or equal to the rated current of the machine.

3. **Equivalent torque method:** This method is based upon the assumption that the motor current is proportional to the torque and heating is proportional to the square of current (i.e. heating is proportional to the square of torque).

The equivalent torque is determined in the same manner as the equivalent current, i.e.

$$T_{eq} = \sqrt{\frac{T_1^2 t_1 + T_2^2 t_2 + T_3^2 t_3 + ...T_n^2 t_n}{t_1 + t_2 + t_3 + ... + t_n}} \qquad (2.7)$$

4. **Equivalent power method:** At constant speed or where the variations in speed are small, the equivalent power is given by the relationship,

$$P_{eq} = \sqrt{\frac{P_1^2 t_1 + P_2^2 t_2 + P_3^2 t_3 + ...P_n^2 t_n}{t_1 + t_2 + t_3 + ... + t_n}} \qquad (2.8)$$

A motor having power rating P_{eq} is selected.

The equivalent current method is the most accurate out of the four methods discussed above. This method may be employed for determining the capacity for all applications except where it is necessary to take into account the variations in so called 'constant losses'.

The equivalent torque method cannot be employed for applications where equivalent current method cannot be applied or in cases in which flux does not remain constant like DC series motors.

The equivalent power method cannot be used for motors whose speed varies considerably under load, especially when dealing with starting and braking conditions.

2.17 METHODS OF VENTILATION AND COOLING OF MACHINES

It is necessary to provide suitable ventilation and cooling for the machines so that the temperature rise at any part of the machine does not exceed the permissible limit governed the type of insulation employed in the construction. The cooling of electrical machine by means of an air stream is called ventilation of the machine.

The circulation of air in the machine can be arranged by the use of a built -in-fan or fan with a separate drive. Either a centrifugal type fan or an axial type fan is employed for rotating electrical machines. A centrifugal fan forces air from the centre to flow outwards and operates the same way irrespective of the direction of rotation of the machine. An axial or propeller type fan moves the air in the opposite direction when the direction of rotation of the machine reverses.

According to IS: 4722-1968, the cooling system for rotating electrical machines are classified into three types depending up the origin of cooling.

1. **Natural cooling:** The machine is cooled by air movements in the machine due to its rotation or due to temperature difference between inside parts and the outside temperature of the air. The machine thus is cooled without the use of a fan, by the movement of air and radiation. This method of cooling is employed only in small fractional kW machines with rating within several hundred watts, since the condition of cooling are comparatively favourable.

2. **Self cooling:** The machine is cooled by air, blown by fan integrally built with the motor mounted on the shaft.

3. **Separate cooling:** The machine is cooled either by a fan not driven by its shaft, or it is cooled by a cooling medium other than air put into motion by means not belonging to machine.

According to the manner of cooling the ventilation systems are classified as:

1. **Open-circuit ventilation:** The beat is given up directly to the cooling air flowing through the machine which is being continuously replaced. The open-circuit ventilation can be further divided into two types in accordance with how the air is brought into the machine.

    a. **Induced ventilation:** Induced ventilation method used in an electrical machine is illustrated in Fig. 2.39. In this arrangement, a fan produces a reduced pressure of air inside the machine and air from outside is sucked into the machine owing to the atmospheric pressure. The air is circulated through the machine and then pushed out by the fan into the atmosphere. The fan is normally arranged inside the machine, as illustrated in the figure. The fan can be provided externally also.

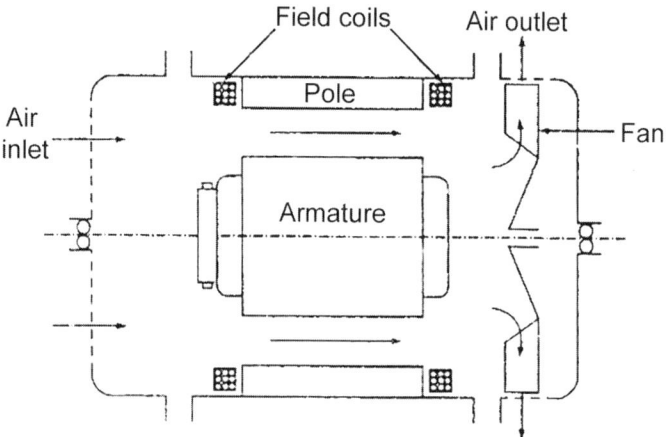

Fig. 2.39: Induced ventilation in electrical machine.

    b. **Forced ventilation:** In this method of ventilation, illustrated in Fig. 2.40. the fan sucks the air from outside and forces it into the machine and finally pushed out into the atmosphere. For forced ventilation, the pressure of air inside the machine is greater than atmospheric pressure outside. The fan may be arranged internally, as illustrated in figure or externally.

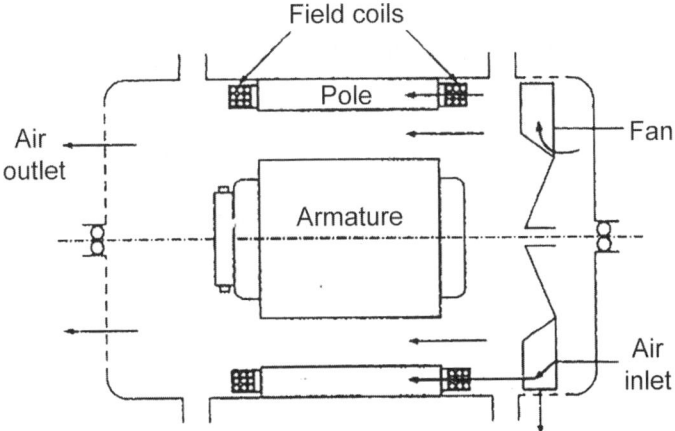

Fig. 2.40: Forced ventilation in electrical machine.

2. **Closed circuit ventilation:** In this system of cooling, heat is transferred to the cooling medium such as air or hydrogen through an intermediate cooling medium circulating in a closed circuit between the interior of the machine and the cooling device, e.g. a water cooler can be used to cool the cooling medium such as air.

   For increasing the surface of contact for cooler, stator, and rotor cores of electrical machines are provided with ducts. These ducts may be radial or axial type depending on the direction of air flow through them.

3. **Surface ventilation:** The heat is transferred from inside of the machine, by cooling medium, to the external surface of a totally enclosed machine. The external surface is being cooled by natural means or mainly by air blown by fan.

2.18 LOAD EQUALIZATION

If the load fluctuates between wide limits in a span of a few seconds, large peak demands of current is taken from the supply and a huge drop of voltage is experienced. A Large sized conductor is also required for this. The Process of stabilizing these fluctuating loads is commonly known as load equalization and involves storage of energy during light load periods which can be given out during the peak load period, so that demand from supply is almost kept constant. Tariff is also affected as it is based and dependent on MD (maximum demand). For example, in steel rolling mill, when the billet is in between the rolls it is a peak load period and when it comes out it is a light load period, when the motor has to supply only the friction and internal losses, as shown in Fig. 2.41.

2.19 USE OF FLYWHEELS

The method of Load Equalization most commonly employed by means of a flywheel. During peak load period, the flywheel decelerates and gives up its stored kinetic energy and as a result reducing the load demanded from the supply. During light load periods, energy is borrowed from supply to accelerate the flywheel, and replenish its stored energy ready for the next peak. Flywheel is mounted on the motor shaft near the motor. The motor must have drooping speed characteristics, that is, there should be a drop in speed as the load comes to enabling the flywheel to give up its stored energy. When the Ward Leonard system is used with a flywheel, then it is called Ward Leonard Ilgner control.

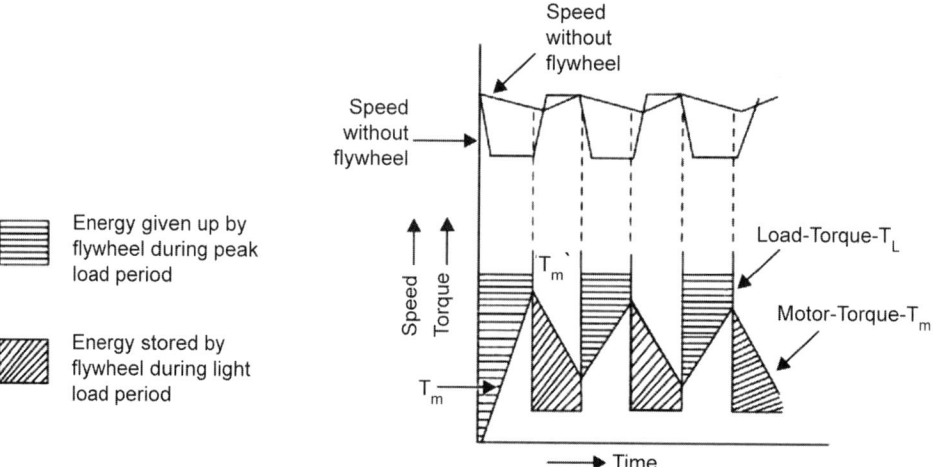

Fig. 2.41: Variation of speed, load torque and motor torque against time.

Flywheel Calculations

The behaviour of flywheel may be determined as follows:

☞ *Flywheel Decelerating (or Load Increasing)*

Let T_L → Load torque assumed constant during the time for which load is applied in kg.m.

T_f → Torque supplied by flywheel in kg.m.

T_0 → Torque required on no load to overcome friction internal losses, etc. in kg.m.

T_m → Torque supplied by the motor at any instant, in kg.m.

ω_0 → No Load speed of motor in rad/sec.

ω → Speed of motor at any instant in rad/sec.

s → motor slip speed $(\omega_0 - \omega)$ in rad/sec.

I → Moment of inertia of flywheel in kg.m².

g → Acceleration due to gravity in m/sec².

t → time in sec.

When the flywheel decelerates, it gives up its stored energy.

$$T_m = T_L - T_f \quad \text{or} \quad T_L = T_m + T_f \tag{2.9}$$

Energy stored by flywheel when running at speed 'ω' is $1/2\, I\omega^2/g$.

If speed is reduced from ω_0 to ω.
The energy given up by flywheel is

$$= \frac{1}{2}\frac{I}{g}\left(\omega_0^2 - \omega^2\right)$$

$$= \frac{1}{2}\frac{I}{g}\left(\omega_0 + \omega\right)\left(\omega_0 - \omega\right) \qquad (2.10)$$

$\left(\dfrac{\omega_0 + \omega}{2}\right)$ mean speed. Assuming speed drop of not more than 10%, this may be assumed equal to ω.

$\therefore \left(\dfrac{\omega_0 + \omega}{2}\right) \cong \omega$ Also $(\omega_0 - \omega) = S$

$\therefore$ From equation (2), energy given up $= \dfrac{I}{g}\omega S$

Power given up $= \dfrac{I}{g}\omega\dfrac{ds}{st}$

but torque $= \dfrac{\text{Power}}{}$

$\therefore$ Torque supplied by flywheel.

$$T_f = \frac{I}{g}\frac{ds}{dt}$$

$\therefore$ From equation (1),

$$T_m = T_L - \frac{I}{g}\frac{ds}{dt}$$

For values of slip speed up to 10% of No - load speed, slip is proportional to torque

Or $\qquad\qquad\qquad s = K T_m$

This equation is similar to the equation for heating of the motor $W - A\lambda\theta = G.S.\dfrac{d\theta}{dt}$

i.e. $\qquad\qquad (T_L - T_m) = \dfrac{I}{g}K\dfrac{dT_m}{dt}$

$$g\frac{dt}{IK} = \frac{dT_m}{(T_L - T_m)}$$

By integrating both sides.

$$-\mathrm{In}\left(T_L - T_m\right) = \frac{gt}{IK} + C_1 \qquad (2.11)$$

At $t = 0$, when load starts increasing from no load, i.e. $T_m = T_0$
Hence, at $t = 0$ $T_m = T_0$

$$\therefore \qquad C_1 = -\ln (T_L - T_0)$$

By substituting the value of C_1 above, in equation (2.11)

$$-\ln (T_L - T_m) = \frac{gt}{IK} - \ln (T_L - T_0)$$

$$\therefore \qquad \ln \left(\frac{T_L - T_m}{T_L - T_0} \right) = -\frac{gt}{IK}$$

$$\left(\frac{T_L - T_m}{T_L - T_0} \right) = e^{-\frac{gt}{IK}}$$

$$(T_L - T_m) = (T_L - T_0) e^{-\frac{gt}{IK}}$$

$$\therefore \qquad T_m = T_L - (T_L - T_0) e^{-\frac{gt}{IK}}$$

If the load torque falls to zero between each rolling period, then

$$T_m = T_L - \left(1 - e^{-\frac{gt}{IK}} \right) \quad (\because T_0 = 0)$$

☞ **Load Removed (Flywheel Accelerating)**

Slip speed is decreasing and therefore $\frac{ds}{dt}$ is negative

$$T_m = T_0 + T_f = T_0 - \frac{I}{g} \frac{ds}{dt}$$

$$T_0 - T_m = \frac{I}{g} K \frac{dT_m}{dt}$$

$$g \frac{dt}{IK} = \frac{dT_m}{(T_0 - T_m)}$$

After integrating both sides,

$-\ln (T_0 - T_m) = \frac{gt}{IK} + C$ At $t = 0$, $T_m = T'_m$ motor torque at the instant, when load is removed.

$\therefore C = -\ln (T_0 - T'_m)$ Putting this value of C in the above equation.

$$-\ln (T_0 - T_m) = \frac{gt}{IK} - \ln (T_0 - T'_m)$$

$$\therefore \quad \ln \left(\frac{T_0 - T_m}{T_0 - T'_m} \right) = \frac{gt}{IK}$$

$$T_0 - T_m = \left(T_0 - T_m'\right)e^{-\frac{gt}{IK}}$$

$$T_m = T_0 + \left(T_m' - T_0\right)e^{-\frac{gt}{IK}}$$

2.20 MECHANICAL FEATURES OF ELECTRIC MOTOR

The electric motor in this section will be discussed with reference to the following:

1. Types of enclosures.
2. Types of bearings.
3. Types of mountings.
4. Transmission of drive.
5. Noise emitted.

2.21 TYPES OF ENCLOSURES

The main function of an enclosure is to provide protection not only to the working personnel but also to the motor itself against the harmful ingress of dirt, abrasive dust, vapours and liquids and solid foreign bodies such as a spanner or screw driver, etc. At the same time, it should not affect the cooling system of the motor. Hence, different types of enclosures are used for different motors depending upon the environmental conditions. Some of the commonly used motor enclosures are as under:

1. **Open type:** In this case, the machine is open at both ends with its rotor being supported on pedestal bearings or end brackets. There is free ventilation system available since the stator and rotor ends are in free contact with the surrounding air. Such, machines are housed in a separate neat and clean room. This type of enclosure is used for large machines such as DC motors and generators.

2. **Screen protected type:** In this case, the enclosure has large openings for free ventilation. However, these openings are fitted with screen covers which safeguard against accidental contacts and rats entering the machine but afford no protection from dirt, dust and falling water. Screen protected type motors are installed where dry and neat conditions prevail without any gases or fumes.

3. **Drip proof type:** This enclosure is used in very damp conditions, i.e. for pumping sets. Since motor openings are protected by overhanging cowls, vertically falling water and dust are not able to enter the machine.

4. **Splash-proof type:** In such machines, the ventilating openings are so designed that liquid or dust particles at an angle between vertical and 100° from it cannot enter the machine. Such type of motors can be safely used in rain.

5. **Totally enclosed (TE) type:** In this case, the motor is completely enclosed and no opening are left for ventilation. All the heat generated due to losses is dissipated from the outer surface which is finned to increase the cooling area. Such motors are used for dusty atmosphere, i.e. sawmills, coal-handling plants and stone-crushing quarries, etc.

6. **Totally-enclosed fan-cooled (TEFC) type:** In this case, a fan is mounted on the shaft external to the totally enclosed casing and air is blown over the ribbed outer surfaces of the stator and end shields. Such motors are commonly used in flour mills, cement works and sawmills, etc. They require little maintenance

apart from lubrication and are capable of giving years of useful service without any interruption of production.

7. **Pipe-ventilated type:** Such an enclosure is used for very dusty surroundings. The motor is totally enclosed but is cooled by neat and clean air brought through a separate pipe from outside the dust-laden area. The extra cost of the piping is offset by the use of a smaller size motor on account of better cooling.

8. **Flame-proof (FLP) type:** Such motors are employed in atmospheres which contain inflammable gases and vapours, i.e. in coal mines and chemical plants. They are totally enclosed but their enclosures are so constructed that any explosion within the motor due to any spark does not ignite the gases outside. The maximum operating temperature at the surface of the motor is much less than the ignition temperature of the surrounding gases.

2.22 BEARINGS

These are used for supporting the rotating parts of the machines and are of two types:
1. Ball or roller bearings
2. Sleeve or bush bearings
a. **Ball bearings:** Up to about 75 kW motors, ball bearings are preferred to other bearings because of their following advantages :
   1. They have low friction loss.
   2. They occupy less space.
   3. They require less maintenance.
   4. Their use allows much smaller air-gap between the stator and rotor of an induction motor.
   5. Their life is long.
        Their main disadvantages are with regard to cost and noise particularly at high motor speeds.
b. **Sleeve bearings:** These are in the form of self-aligning pourous bronze bushes for fractional kW motors and in the form of journal bearings for larger motors. Since they run very silently, they are fitted on super-silent motors used for driving fans and lifts in offices or other applications where noise must be reduced to the absolute minimum.

2.23 TYPE OF MOUNTINGS

Motors are generally made in following mountings.
1. **Foot mounting type.** Where belt tension adjustment is required, slide rail, mounted motors are used, Fig. 2.42a.

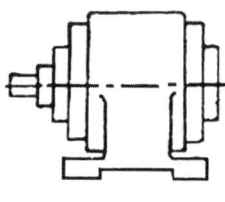

a

Fig. 2.42a: Foot mounting type.

2. **Ceiling mounting type.** Fig. 2.42b.

b

Fig. 2.42b: Ceiling mounting type.

3. **Wall mounting type.** With shaft horizontal or vertically down or vertically up Figs 2.42c to e respectively.

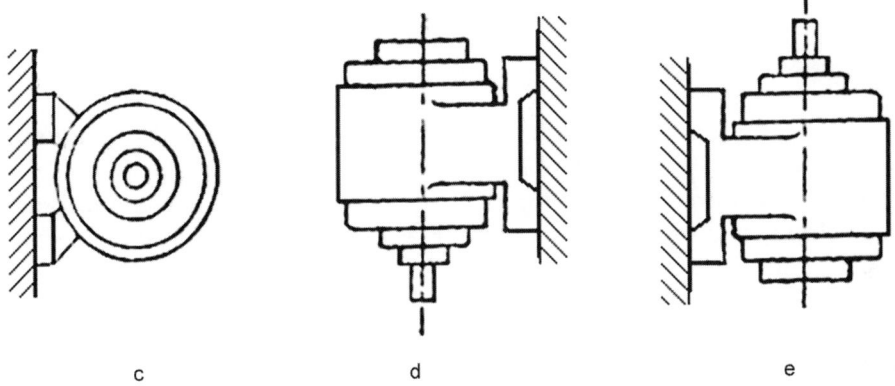

c   d   e

Figs 2.42c to e: Wall mounting type: (c) With shaft horizontal, (d) Vertically down and (e) Vertically up.

4. **Flange mounting type.** Motor with flange is fixed directly on the driven machine. Hollow shaft deep well turbine pump motor is an example of flange mounting Figs 2.42f and g.

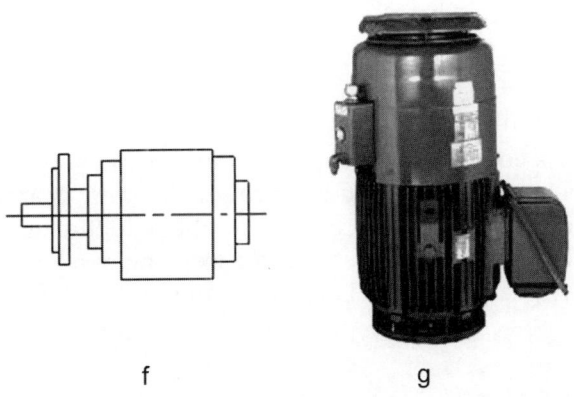

f   g

Figs 2.42f and g: (f) Schematic of flange mounting type, (g) Flange mounting motor.

5. **Cradle mounting type.** These motors are used as loom motors Fig. 2.42h.

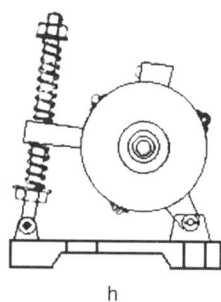

h

Fig. 2.42h: Cradle mounting type.

6. **Body turned mounting type.** For many of machine tools, these motors are filted into the body of the tool. This gives clean appearance.

2.24 TRANSMISSION OF DRIVE

There are many ways of transmitting mechanical power developed by a motor to the driven machine.

1. **Direct drive:** In this case, motor is coupled directly to the driven machine with the help of solid or flexible coupling. Flexible coupling helps in protecting the motor from sudden jerks. Direct drive is nearly 100% efficient and requires minimum space but is used only when speed of the driven machine equals the motor speed.

2. **Belt drive:** Flat belts are extensively used for line-shaft drives and can transmit a maximum power of about 250 kW. Where possible, the minimum distance between the pulley centres should be 4 times the diameter of the larger pulley with a maximum ratio between pulley diameters of 6 : 1. The power transmitted by a flat belt increases in proportion to its width and varies greatly with its quality and thickness. There is a slip of 3 to 4 percent in the belt drive.

3. **Rope drive:** In this drive, a number of ropes are run in V-grooves over the pulleys. It has negligible slip and is used when the power to be transmitted is beyond the scope of belt drive.

4. **Chain drive:** Though somewhat more expensive, it is more efficient and is capable of transmitting larger amounts of power. It is noiseless, slipless and smooth in operation.

5. **Gear drive:** It is used when a high-speed motor is to drive a low-speed machine. The coupling between the two is through a suitable ratio gearbox. In fact motors for low-speed drives are manufactured with the reduction gear incorporated in the unit itself. Fig. 2.43 shows such a unit consisting of a flange motor bolted to a high-efficiency gearbox which is usually equipped with feet, the motor being overhung.

2.25 NOISE

The noise produced by a motor could be magnetic noise, windage noise and mechanical noise. Noise level must be kept to the minimum in order to avoid fatigue to the workers in a workshop. Similarly, motors used for domestic and hospital appliances and in offices and theatres must be almost noiseless. Tranmission of noise from the building where the motor is installed to another building can be reduced if motor foundation is flexible, i.e. has rubber pads and springs.

Fig. 2.43: Flange motor bolted to a high-efficiency gearbox.

2.26 COST CONSIDERATION

Cost is one of the most important factors in the selection of a motor. From pure engineering considerations, discussed above, we might end up choosing a motor which is very costly. But in our analysis we have to balance the extra cost to be paid with special features obtained. Many a times, we need to sacrifice special features because they are not worth the additional investment. In most situations, we find equipment very cheap at first but it is operating efficiency becomes minimal as the time progresses. On one band we will be making some annual saving on the interest and depreciation if we go for cheap motor, on the other hand we will be incurring more annual expenditure on the energy bill. Thus rate of interest and duty cycle become the deciding factors in judging the ultimate economy of particular drive.

2.27 MOTORS FOR PARTICULAR SERVICE

Various factors are to be considered while selecting a motor described above. In this article, we will now suggest, possible choice of motor for particular application, keeping in mind the various requirements and duty cycle.

1. Rolling Mills

There are two types of mills and selection of the driving motor depends upon the type of mill. For instance in blooming and slabing mill, rolling operation requires production of heavy torque at low temp to effect heavy reduction of material from ingots to blooms or slabs. Blooming mill may have 2 rolls or 3 rolls. Two high blooming mill has arrangement of reversing the direction of rotation of the rolls in order to carry out rolling of material through successively reducing passes. In case of three high mill, direction of rotation need not be changed. By the use of tilt tables, material can be passed between lower and middle rolls or between middle and top rolls. Manipulators are not only used to guide the blooms from groove to groove and to 'turn the bloom through

900 but they also straighten the pieces during their entry into the passes. Blooms and slabs as obtained from blooming mill are further subjected to rolling to produce plates, structural; bars or billets. Sheet and tin bars are further rolled to sheets and tin plates. Billets when further rolled produce rounds, squares octagons, flats angles, etc. Rolling subsequent to blooming, in most of the cases, is carried out in continuous mills. Here rolling mill consists of a number of stands in series. Stock after coming out from first stand is turned through 90° by twist guides before being fed into the next stand. In order to minimize the formation of loops or tension in the material between stands, it becomes necessary to coordinate the speeds of rolls in various stands. If all the stands are driven by one motor, this coordination of speeds is achieved by giving the drive from, the shafts to various stands through bevel gears of different speed reduction ratio. If individual drive to each stand is adopted the speeds of different motors have to be synchronized. In order to achieve varied roiling programme say 6 to 25 mm diameter wire from 75×75 mm square billet, it requires reduction of cross section area from $\dfrac{1}{11}$ to $\dfrac{1}{200}$. Taking into account this varied reduction of cross section area and the initial and final speeds of stock being rolled, it becomes necessary to employ motors for continuous rolling mill having speed range in the ratio of 8. Depending upon the type of mill, selection of motor has to satisfy following condition also

1. Motor should have robust (construction to withstand severest duty).
2. It should be capable of having wide speed variation.
3. It should be capable of developing short-time torque to the extent of 2 to 2 times of rated torque.
4. It should be in a position to maintain preselected speed within close tolerances from no load to full-load. Therefore it should have quick response to change.

Having described the requirements in general, we will now discuss various motors in details.

a. **Induction motors:** We can change the speed of slip-ring motor by introducing resistance in the rotor circuit. This method is wasteful as the reduction in speed is proportional to copper losses taking place in rotor circuit. As such, slip-ring motors are used for small re-rolling mills where wide speed range is not required. Besides wastage of energy taking place in rotor, introduction of resistance in rotor circuit converts the speed torque characteristic of induction motor from shunt to series. Therefore speed cannot be maintained constant from no load to full-load. This will therefore cause the motor to fly off to no load speed as soon as stock passes out of the stand. This reflects very much on the quality of rolling. As such slip-ring motors are suitable for roughing stands of the rolling mill. Wastage of energy in the rotor of slip-ring motor which is proportional to the reduction of speed, can be avoided by employing cascaded induction motor. Rotor power at slip frequency is rectified by means of silicon rectifiers. In case of constant HP cascading, this rectified power is fed to a separately excited DC. Motor which is mechanically coupled to main induction motor. In case of constant torque cascading rectified power is fed back to the AC system by means of mercury arc rectifiers operating as inverters. In the first case speed of the main motor is adjusted by varying the excitation of the separately excited DC motor. In the second case, the amount of electrical power fed back to the system can be regulated by the grid control of the rectifier. Rating of the equipment required for cascading depends upon the range of the speed variation desired. Cascading as compared to DC motor becomes cheaper where speed variation desired is not more than 30%.

b. **DC motor:** DC motor is the only type of motor which can satisfy the stringent requirements of modern rolling mills. By varying the applied voltage to the armature and keeping field constant, it is possible to change the speed of motor at constant torque up to rated speed. In this way requirement of higher constant torque at low speed needed for effecting heavy reduction in first few passes is achieved.

On the other band low torque and higher speed is required for rolling operation in final passes when for the section or stock having become smaller, torque needed is low and elongated material requires higher speed to clear off in the shortest possible time. This is achieved by weakening the field. After rated speed it is achieved by the increase of applied voltage to the armature. In this way we get hick speed at reduced torque, i.e. HP remains same. For motors required specially for blooming and slabbing mill where heavy reduction in the stock has to be carried out, it becomes necessary to develop torque 4–5 times the rated value for short duration of time. This will therefore need as much armature current. In order to carry armature current peaks for short duration without causing sparking, it becomes necessary to employ compoles and compensating windings in the pole faces. In order that flux of the compoles follow the changes in armature current instantaneously, motors are provided with laminated yoke. In case of reversible mills, in order to ensure high rate of acceleration, it becomes necessary to reduce the moment of inertia of motors to minimum. This is achieved by increasing the length and reducing the diameter of armature.

DC motors of rolling mills can be supplied from three phase AC system by any of the following methods: (1) Grid controlled mercury arc rectifier. (2) Silicon controlled rectifier (3) Ward Leonard set. The main advantages of grid controlled rectifier are its overload capacity, stepless control of voltage from 0–100%, high efficiency and quick response to very rapid changes in operating variables. The only disadvantage of mercury arc rectifier is that the load peaks caused by the operating of the mill are transferred to the feeding system. Due to this reason mercury arc rectifier or silicon controlled rectifier which works only as a valve is recommended for supplying the rolling mills where AC supply network is sufficiently strong. The main advantage of silicon rectifier is its small size. It however suffers from the drawback that it does not have overload capacity. This fact coupled with low reverse voltage blocking capacity of silicon rectifiers, needs numbers of cells connected in series and parallel. This increases installation cost and voltage control becomes a complicated affair. By means of Ward Leonard set it is possible to have speed variation within ±100% of I cited speed. It is also possible to apply braking by working the mill motor as a generator. Main advantage of employing Ward Leonard set is that load peaks are not transferred to power supply. Specially in blooming mill where load peaks are severe, flywheel is employed with Ward Leonard set. By suitable speed control, KE of flywheel can be used to smoothen the load peaks to great extent. Ward Leonard system however suffers from having poor efficiency, longer response time and requires more maintenance. Uniform quality of the rolling can be ensured by keeping the rolling speed constant irrespective of fluctuation in the load or voltage. This requires power supply system for rolling mill drive which has high inherent response and which is capable of being operated on closed loop fed back system. In this respect grid controlled rectifier coupled with electronic speed control equipment is far superior to Ward Leonard system of power supply for rolling mill drive.

Special mention may here be made about the cooling of rolling mill motor. In case of induction motor, the problem of cooling is not accurate. Since this motor runs at constant speed which is near to the synchronous speed, proper cooling of the winding can be ensured by directing sufficient volume of air through the winding. This is done by fitting fan on the shaft of the motor. Problem of cooling is more accurate in case of DC motor which has to develop often many times full-load torque at low speed. Cooling air fan if fitted on the motor shaft will not be in a position to direct sufficient volume of air at low speed of rolling with a consequent danger to the insulation of motor. In $ order therefore to ensure proper cooling of motor at all speeds, closed circuit air cooling system is adopted. Here fan forces the air to the motor through water cooler in a closed circuit. Leakage air is made good by drawing in fresh air from atmosphere through dry filter since rolling mill motor is totally enclosed and cooling air is supplied and withdrawn from it by duct system, it can be safely installed on the floor of the mill itself and there is no need of having separate motor house.

c. **Synchronous motor:** Synchronous motors are used for driving constant speed continuous mills where load is maintained for relatively long periods. Flywheel cannot be used and consequently the motor carries the entire load. For slow speed, synchronous motors can be built more economically, than induction motors and we, at the same time, get power factor correction.

Besides main mill drive, motors are required for auxiliary drive such as for driving the rolls of roller table, adjusting the gap between rolls, driving the mainpulaters, for chopping the tail ends, cutting the material to the required length, winding the end product into coils, etc.

In most of auxiliary drives, a large part of the energy converted is used for acceleration and braking of the flywheel masses. It is therefore necessary that the inherent moment of inertia of motor is kept low. It is also desirable that these motors should be compact as in most of the cases space available for installation is restricted.

2. Textile Mills

Loom motors operate in atmosphere which is laden with wool-lint and moisture. Further, the motor should have high starting torque of about 2 to 2k times full-load torque. This is needed to ensure rapid acceleration of loom from rest to full speed as first pick has to be made at same velocity as subsequent picks if defective material is not required. Loom motors have to start and stop many times a day. This increases the motor heating during starting. Motors of low temperature rise under full-load conditions are, therefore, selected so that under working conditions of frequent starting and stopping, permissible temperature rise is not exceeded. Keeping above requirements in view, totally enclosed, fan cooled, high torque squirrel cage motors are selected. Where only wool lint is present and no moist atmosphere, screen protected motors are used. Splash proof motors are used in wet locations such as dye house. Brush shift adjustable speed motor is used on printing machines.

3. Cranes

Essential requirement of a hoist or crane motor is that it should develop high starting torque and should with stand great number of switching operations.

    a. DC is usually employed on cranes in steel mills where safety and flexibility of control are of first importance. Series motor is by far the most commonly type of motor which is used on account of its inherent characteristic such that it adjusts its speed to load. This characteristic prevents excessive power demand on heavy loads. During dynamic braking, series motor works as shunt generator.

    b. On cranes using AC slip-ring induction motors are used where speed control is obtained by rotor resistance. Speed range of slip-ring motor is less than that of series motor. Dynamic braking is applied by giving unbalanced supply to primary. This is done by giving single phase supply to the primary such that two phase windings are in parallel and this in turn is in series with third winding.

For outdoor service, totally enclosed motors are best. Hoist motors have special electromechanical brakes which are applied by springs to hold the load as soon as supply to motor goes off. As soon as supply to motor is resumed, solenoid connected across motor terminals is energized which releases the brakes. These motors are half or one hour rated developing 200% starting torque. They are robust to withstand severe strains to which they are exposed.

4. Mines

Motor used inside mines should have flame proof enclosure. For mine winders, it is essential to have both speed control as well as braking capacity. Ward Leonard Iligner system is most effective drive. Slip-ring induction motor with DC dynamic braking is also used. We may also use DC motors supplied from controlled rectifier.

5. Paper Mill Drive

Pulp stuff from storage tank, after passing through strainers, comes to first section of paper making machine which consists of moving wire mesh. Subsequently this sheet of paper goes in between serifs of heated press rolls, drying rolls, calenders, smoothing rolls and finally to reelers.

Following requirements of drive must be fulfilled. (i) To manufacture different types of paper, we should be able to vary the speed of entire series of rolls. (ii) Relative speeds of rolls should be constant otherwise tearing of paper may result. (iii) We should be able to adjust the speed of any one group of rolls relative to others in order to allow for the draw of the paper.

All above requirements are fulfilled by the use of schrage motors as shown in Fig. 2.44.

Master schrage motor drives a synchronous alternator from where variable frequency supply is obtained. This is supplied to the stators of all the sectional synchronous motors. Sectional synchronous motors are such that both stator and rotor

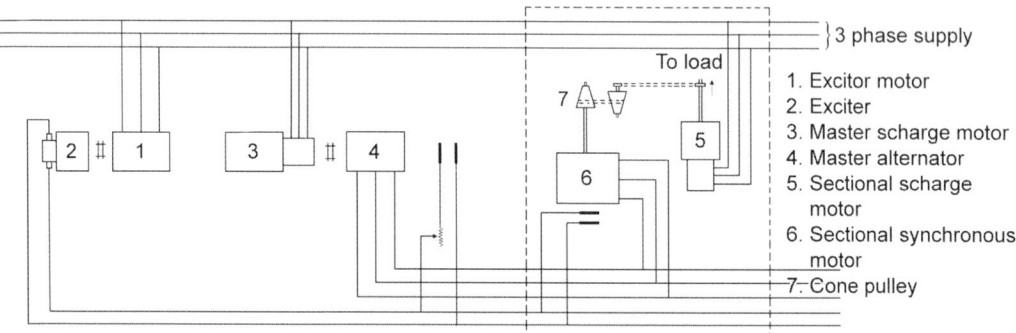

Fig. 2.44: Fulfilling the requirements of paper mill drive by the use of schrage motors.

are free to rotate. Stator of synchronous motor carries the brush gear of the sectional stage motor. Rotor of the synchronous motor is driven mechanically by cone pulley drive from the shaft of sectional schrage motor. Normally the speeds of rotor of synchronous motor and rotating magnetic field set up by stator currents are same. If due any reason speed of section motor tends to fall, this will produce torque on the stator of synchronous motor in such a manner as to increase the brush distance of sectional schrage motor, thereby condition (ii) is fulfilled. If we change the speed of master schrage motor, frequency of supply to stators of all synchronous motors will change. This will again tend to rotate the stators of synchronous motors. This results in the movement of brush gear of all sectional schrage motors. This fulfills requirement (0. Speed of any group of rolls can be adjusted by means of sectional cone pulley drive). This fulfills requirement (iii).

6. Marine Drive

Except for very small vessels, electric drive for propeller of ship is universal. It has following advantages.

1. Steam turbine or diesel engine has most economical spew of about 3000 rpm and 400 to 800 rpm respectively. 1/4 propeller speed on the other hand is in between 100 and 200 rpm. We can run propeller and prime mover independently. This enables prime mover to run at its most economical speed.
2. At the time of slow speeds, we can run only few prime mover sets to supply all the propellers. Low speed cruising is now no longer uneconomical.
3. Speed control and reversing in case of electric propulsion are easy and can be carried out from any convenient position instead of signalling to engine room.
4. Position of prime mover is not fixed by the propeller shaft. As such there is greater flexibility in the layout in case of very big ships, electric drive will be by three phase induction or synchronous motors. Prime mover will be steam turbine, driving three phase alternator. Speed control is done partly by variation of applied voltage and partly by frequency variation. Reversal is obtained by changing the phase sequence of the supply. Speed of induction motor can also be changed by pole changing method which gives two economical speeds. Disadvantage of induction motor is that is Power Factor (PF) at low speed is low between 0.7 and 0.75. Power at the ship is generated by turbo alternator at 2.2 to 6.6 KV.

5. Small ships can be driven by DC motors, which are controlled by means of Ward Leonard method. Power on the ship is generated by diesel electric sets, relatively at low voltage of 650 V.

   Instead of constant voltage system, we can use constant current system of supply to feed armatures of propeller motors, whose fields are supplied from constant voltage supply. This system of supply is inherently proof against overloads, provides maximum safe torque without any risk of interruption of current. Motor control is exercised through field control. This is very suitable for small vessels requiring maximum manoeuvrability in restricted waters such as tugs and ferries.

6. In case of fish trawlers, dredgers or refrigerated vessels, power for auxiliary services required is more when there is no demand for propulsion power! Changeover switch can switch power from propulsion motors to the motors of auxiliary equipment such as trawl winch, refrigeration plant, etc.

   In case of very small vessels, complications of using electrical generator and motor are not warranted. Propeller shafts are directly coupled to diesel engine through slip coupling. Speed of propeller shaft can be controlled by varying DC excitation of the slip coupling.

7. *Punches presses and shears.* Since heavy sudden loads are applied, DC compound motor or slip-ring motor with flywheel is used.

8. *Planing machine.* Cutting stroke is slow and return stroke is quick. Heavy mass of the job has to be accelerated and decelerated. It will require DC compound motor.

9. *Lathe, grinding and milling machine.* These machine tools require low starting torque and constant operating speed. Therefore DC shunt motors or squirrel cage motors are needed, case of drills, requiring different operating speeds for different metals, motors, having two sets of stator pole changing winding are used which give four economical speeds.

10. *Domestic use.* Univers mal series motor is used for house bold appliances, such as mixy, washing machine, sewing machine, vacuum cleaner, fans, etc.

11. *Lifts.* High smooth accelerating torque of 200 to 250% full-load torque at starting, high overload capacity and pull out torque and maximum degree of silence are essential requirements. Motors are normally one hour rated for duty cycle of 150 to 180 starts per hour. DC compound motors or slip-ring motors suit the above requirement.

12. *Wood working machinery.* When there are chances of motor being burried in sawdust, surface cooled motor is preferred, otherewise screen protected motor will do. Some wood working machines require speeds higher than 3000 rpm This can be obtained with induction motors only in conjunction with induction type frequency changer or by the use of commutator DC motors.

13. *Woollen mill.* Loom motors are totally enclosed type, capable of developing 300% torque at starting if direct driven or alternatively started light and controlled through clutch. Motors including switch gear in dyeing section should be totally enclosed, and should also be treated with acid proof paint as atmosphere is laden with fumes.

14. *Printing.* Screen protected slip-ring motors where speed control is required, otherwise squirrel cage motors are used.

15. *Quarrying.* These motors must be of robust construction and preferably, because of excessive dirt, of surface cooled type so that grit does not get into and spoil the winding.

16. *Pumps.* Drip proof or totally enclosed surface cooled motors are most frequently used, often mounted on a common bed plate and direct coupled to the pump. Some times motor is flange mounted. Were pump speed does not come within the range of the fixed motor speed, V belt drive is employed. Starting torque necessary for centrifugal pumps is about 40 to 45% of full-load torque. For reciprocating pumps, starting torque may be 100 to 200% of full-load torque. It is advisable to use slip-ring motors for reciprocating pumps and squirrel cage motors for centrifugal pumps.

17. *Refrigeration and air conditioning.* In vapour compression system of refrigeration, motor is used to drive the compressor. This motor is controlled by thermostat. When motor is restarted, it has to drive compressor against high head pressure. Motor should, therefore, be required to develop starting torque of 200 to 250% full-load torque.

    For small unit air conditioners, capacitor type single phise 230 V- motor with D.0.1, starter is most commonly selected. High torque squirrel cage induction motors or slip-ring induction motors are used for large installations. For very large plant size, synchronous motor, driving turbocompressor, may be found suitable specially when Power Factor (PF) correction is also required. Starting torque produced with star delta starter is only 66% of full-load torque, this type of starter can be only used when cylinder untoaders are used on the compressors.

18. *Breweries.* Motors have to work mainly in damp atmosphere. Therefore, totally enclosed surface cooled type motors are recommended.

19. *Excavators.* Due tom arduous and rugged work, totally enclosed motors arc used.

20. *Belt conveyors.* Belt conveyors are used for moving sand and gravel. Heavy loads have to accelerated. High torque squirrel cage or preferably slip-ring motors should be made our choice. Due to exposed working conditions and presence of grit and dust in atmosphere, totally enclosed surface cooled motors are used.

21. *Cement works.* Owing to dusty atmosphere, pipe ventilated or totally enclosed surface cooled motors are recommended.

22. *Flour mills.* Pipe ventilated or totally enclosed type motors should be used. Speed of motor varies from 600–1000 rpm.

23. Sugar centrifuge.

In sugar industry, a centrifuge is employed for separating out crystalized sugar from the syrup obtained from steam evaporator and for drying it out by the action of centrifugal force. The duty cycle of a centrifuge motor is illustrated in Fig. 2.45. The functions involved are charging, intermediate spinning, spinning, regenerative and reverse current braking and plugging for ploughing. All these functions are to be performed at different speeds, variation of which may be as much as 1:30. The motors used for this purpose are usually four-speed, pole-changing motors having two sets of stator windings that enable us to obtain synchronous speeds 1,500/750/214/107 or 1000/500/214/107 rpm. They not only are capable of providing the desired fixed speeds of operation, but also of returning a portion of energy back to the supply mains during regenerative braking accomplished by switching over to higher pole operation from a lower pole, one.

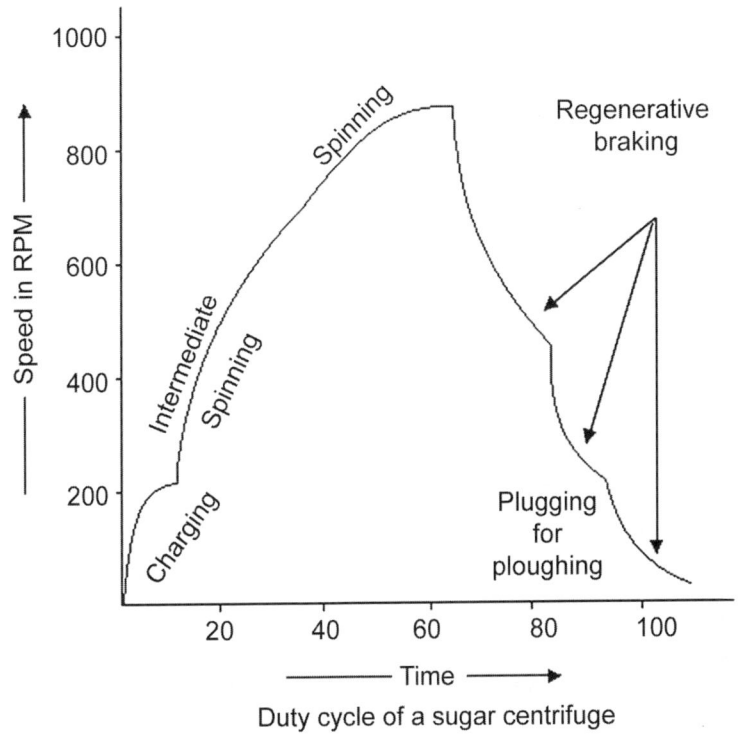

Duty cycle of a sugar centrifuge

Fig. 2.45: Duty cycle of a centrifuge motor.

In order to satisfy the duty cycle illustrated in fig. 2.45, first 28 pole winding is switched on to provide a speed of approximately 200 rpm. During charging supply is cut-off and then intermediate spin speed of about 450 rpm is obtained by energizing the 12-pole winding. The final spin speed of about 950 rpm is achieved by switching over to the 6-pole winding. After centrifuging operation is complete, super-synchronous regenerative braking is applied by connection the motor for 12, 28 and 56 poles successively which brings the speed down to 500, 214 and 10 rpm respectively. Ploughing speed of about 50 rpm is obtained by the application of reverse current braking. Automatic control of the entire duty cycle is achieved by feeding the control equipment from the output of a tachogenerator directly coupled to centrifuge motor.

Motors employed for driving centrifuges have their own special features in construction. They are vertically mounted so as to be coupled with the centrifuge shaft. The motors may be made with larger air gap so as to take care of the possible rotor oscillations about the vertical axis. Insulation used in motors must be humid proof so that it can operate successfully in a humid environment. For protection of motor against overheating, thermal elements are embedded in the windings. These elements, which are also called sensotherms by trade name, operate few degrees below the maximum permissible temperature of the winding. Operation of thermal elements will either directly trip the motor or give visible and audible warning signals so that the particular duty cycle may be completed. Following this, next cycle cannot be commenced until temperature of the motor has come down, to normal.

WORKED EXAMPLES

● **EXAMPLE 2.1**: A 230 V DC series motor has an armature resistance of 0.08 ohm and field resistance of 0.05 ohm. The magnetisation characteristic of the machine at 700 rpm is as follows:

Field current	30	60	90	120	150 A
EMF	70	125	180	210	230 V

The motor is run on a 230 V supply. An additional resistance of 1.5 ohm is connected in series with armature. Determine the torque and speed when the motor draws a current of 90 A.

Solution: Total resistance between motor terminals,

$R_m = R_a + R_{Se}$ 0.08 + 0.05 = 0.13 ohm

Field current, I_f = Line current I_1 = 90 A

Additional resistance inserted in armature circuit, R = 1.5 ohm

Back emf developed, $E_{bf} = V - I_1 (R_m + R) = 230 - 90 (0.13 + 1.5) = 83.3$V

Let speed be N_1,

From given datas for magnetisation curve when field current is 90 A

EMF developed, $E_2 = 180$ V

Speed, $N_2 = 700$ rpm

Since $N \propto \dfrac{E_b}{\phi}$

$\dfrac{N_1}{N_2} = \dfrac{E_{b1}}{E_2} \times \dfrac{\phi_2}{\phi_1} = \dfrac{E_{b1}}{E_2}$ $\phi_2 = \phi_1$,

or, $N_1 = N_2 \times \dfrac{E_{b1}}{E_2} = 700 \times \dfrac{83.3}{180} = 324$ rpm

Torque developed, $T = \dfrac{9.55 \times E_b I_a}{N} = \dfrac{9.55 \times 83.3 \times 90}{324} = 221$ Nm

● **EXAMPLE 2.2**: A 230 V de shunt motor with constant field drives a load whose torque is proportional to its speed. When running at 750 rpm it takes 30A. Find the speed at which it will run if a 10 ohm resistance is connected in series with its armature. The resistance of armature may be neglected.

Solution: Armature current, I_a = Line current I_L = 30 A neglecting shunt field current.

Speed, $N_1 = 750$ rpm.

Let the speed be reduced to N_2 rpm where $N_2 = KN_1$ after inserting additional resistance of 10 ohm in the armature circuit.

Since with constant field excitation, torque developed is proportional to armature current i.e.

$$T \propto I_a \qquad (i)$$

Load torque, $T \propto I$ $\qquad (ii)$

From equations (i) and (ii), we have

$$I_a \propto T \propto N$$

or, $I_{a2} = I_{a1} \times \left(\dfrac{N_2}{N_1}\right) = KI_{a1} = 30\,K$ $\qquad (iii)$

Since $N \propto [V - I_a(R_a + R)]$ flux ϕ remaining the same

Or $\dfrac{N_2}{N_1} = \dfrac{V - I_{a2}R}{V}$ armature resistance is negligible

Substituting $\dfrac{N_2}{N_1} = K$, $I_{a2} = 30\,K$ from equation (iii), and $R = 10$ ohm in above

equation we have

$$K = \dfrac{230 - 30K \times 10}{230}$$

or

$$K = \dfrac{230}{230 + 300} = 0.434$$

New Speed, $N_2 = KN_1 = 0.434 \times 750 = 325.5$ rpm.

◉ **EXAMPLE 2.3**: A 50 kW motor with a heating time constant of 100 minutes has a final temperature rise of 50°C on continuous rating. Find the half-hour rating of the motor for this temperature rise assuming that it cools down completely, between each load period. The motor has maximun efficiency of 80% at its full-load.

Solution: Heating times constant, $\tau = 100$ minute

Final temperature, rise, $\phi_F = 50°C$ in case of continuous rating

Let P kW be the half-hour (30 minutes) rating, then final temperature rise,

$$\phi'_F = \dfrac{\phi}{1 - e^{-\frac{t}{\tau}}} = \dfrac{50}{1 - e^{-0.03}} = 193°C$$

Since maximum efficiency occurs at 80% of full-load, i.e. at a load of 40 kW, therefore, at 40 kW load copper loss is equal to constant losses, let each be equal to W watts.

$$\text{Losses at 50 kW} = W + \left(\dfrac{50}{40}\right)^2 W = 2.5625\,W$$

$$\text{Losses at P kW} = W + \left(\dfrac{p}{40}\right)^2 W = W + \dfrac{p^2}{1600}W = W\left(1 + \dfrac{p^2}{1600}\right)$$

(Since copper loss varies as the square of the load)

$$\dfrac{\text{Total losses at P kW}}{\text{Total losses at 50 kW}} = \dfrac{\phi_F}{\phi_F}$$

or
$$\frac{W\left(1+\dfrac{P^2}{1600}\right)}{2.5625\,W} = \frac{193}{50}$$

or
$$P = 119.24 \text{ kW}$$

☺ **EXAMPLE 2.4**: A motor has a thermal time constant of 46 minute. When the motor runs continuously on full-load, its final temperature rise is 80°C. (i) What is the temperature rise after 1 hour if the motor runs continuously on full-load? (ii) If the temperature on one-hour rating is 80°C, find the maximum steady-state temperature at this rating. (iii) How much time does the motor take for its temperature rise from 50 to 80°C if it is working at its 1 hour rating?

   Solution: (i) From given datas:
   Final temperature rise, $\phi_F = 80°C$
   Thermal time constant, $\tau = 45$ minutes $= 0.75$ hour
   Temperature rise after 1 hour (i.e. when $t = 1$) can be had by substituting $\phi_F = 80°C$; $\tau = 0.75$ hour and $t = 1$ hour

   i.e. $\phi = \phi_F\left(1-e^{-\frac{t}{\tau}}\right) = 80\left(1-e^{-\frac{1}{0.75}}\right) = 58.9°C$

   (ii) Let the maximum steady-temperature rise at one hour rating be ϕ_F
   $\tau = 0.75$ hour, $\phi = 80°C$ when $t = 1$ hour.
   Now from given datas

   ∴ $\phi_F' = \dfrac{\phi_F}{\left(1-e^{-\frac{t}{\tau}}\right)} = \dfrac{80}{\left(1-e^{-\frac{1}{0.75}}\right)} = 108.64°C$

   (iii) Time taken to attain temperature of 80°C (assuming initial temperature as 0°C) in one hour or 60 minutes,
   Final temperature rise, $\phi_F' = 108.64°C$
   Now let the time taken to attain temperature of 50°C from 0° be t hours then
   $$50 = 108.64\,(1-e^{-1/0.750})$$
   Or $t = 0.4625$ hour or 27.75 minutes
   Hence time taken to increase temperature from 50 to 80°C
   $= 60 - 27.75 = 32.25$ minutes

☺ **EXAMPLE 2.5**: The heating and cooling time constants of 100 kW motor are 90 and 120 minutes respectively. Find the rating of motor when subjected to a duty cycle of 18 minutes on certain load and 30 minutes on no load. Assume that the losses are proportional to the square of load.

   Solution: Heating time constant, $\tau = 90$ minutes
   Cooling time constant, $\tau' = 120$ minutes
   On-load period, $t_1 = 18$ minutes
   Off-load period, $t_2 = 30$ minutes

$$\frac{\phi'_F}{\phi_F} = \frac{1-e^{-\left(\frac{t_1}{\tau}+\frac{t_2}{\tau}\right)}}{1-e^{\frac{t_1}{\tau}}} = \frac{1-e^{-\left(\frac{18}{90}+\frac{30}{120}\right)}}{1-e^{-18/90}} = \frac{1-0.6376}{1-0.8187} = 2$$

Since losses are proportional to the square of load

$$\frac{\text{Total losses at } P \text{ kW output}}{\text{Total losses at } 100 \text{ kW output}} = \left(\frac{P}{100}\right)^2 = \frac{\phi'_F}{\phi_F} = 2$$

where P is the rating of motor when subjected to a duty cycle of 18 minutes on certain load followed by 30 minutes on no-load

or $P = 100\ \sqrt{2} = 141.4$ kW   **Ans.**

☺ **EXAMPLE 2.6**: An induction motor has a temperature rise of 25°C after one hour and 42°C after two hours while operating at full-load output of 7.5 kW. The constant losses of the motor are 80% of full-load copper losses. Determine how long the motor can be operated at twice the rated output without overheating.

Solution : From equation $\phi = \phi_F\left(1-e^{-\frac{t}{\tau}}\right)$ (i)

Now from given datas when $t = 1$ hour, $\phi = 25°C$ and when $t_2 = 2$ hours, $\phi = 40°C$. Substituting these values for t and ϕ in above equation (i), we have

$$25 = \phi_F\left(1-e^{-\frac{1}{\tau}}\right)$$ (ii)

$$40 = \phi_F\left(1-e^{-\frac{2}{\tau}}\right)$$ (iii)

Solving above equations (ii) and (iii), we have

Heating time constant, $\tau = \dfrac{-1}{\log_e\left(\dfrac{40}{25}-1\right)} = \dfrac{-1}{\log_e 0.6} = 1.958$ hours

Substituting $\tau = 1.958$ in equation (ii), we have
Final steady-temperature rise,

$$\phi_F = \frac{25}{1-e^{-1/1.958}} = 62.5°C$$

Let the full-load copper losses be W watts.

Constant losses = 80 percent of full-load copper losses = 0.8 W
Total losses at full-load = 0.8 W + W = 1.8 W
Total losses at twice full-load = 0.8 W + 22 W = 4.8 W
Let the new maximum temperature rise be ϕ'_F, then

$$\frac{\phi'_F}{\phi_F} = \frac{4.8\ W}{1.8\ W} = 2.6667$$

Or $\phi = 2.6667\ \phi_F = 2.6667 \times 62.5 = 166.67°C$

Let the motor be run for t hours at twice the continuously rated output without overheating (i.e. attaining temperature of 62.5°)

Since $\qquad \phi = \phi_F\left(1 - e^{-\frac{t}{\tau}}\right)$

$$62.5 = 166.67°\left(1 - e^{-\frac{t}{1.958}}\right)$$

Or $t = -1.958 \log_e (1 - 0.375) = 0.92$ hour or 55 minutes. **Ans.**

☺ **EXAMPLE 2.7**: Calculate the maximum overload that can be carried by a 25 kW motor if the temperature rise is not to exceed 50°C after one hour on overload. The temperature rise on full-load after one hour is 30°C and after 2 hours is 40°C. The losses vary as square of the load.

Solution: The expression for temperature rise is given as

$$\phi = \phi_F\left(1 - e^{-\frac{t}{\tau}}\right) \tag{i}$$

From given datas when $t = 1$ hour, $\phi = 30°C$ and when $t = 2$ hours, $\phi = 40°C$. Substituting these values of t and ϕ above expression (i), we have

$$30 = \phi_F\left(1 - e^{-\frac{1}{\tau}}\right) \tag{ii}$$

$$40 = \phi_F\left(1 - e^{-\frac{2}{\tau}}\right) \tag{iii}$$

Solving above equations (ii) and (iii), we have

$$e^{-\frac{1}{\tau}} = \frac{40}{30} - 1 = \frac{1}{3}$$

Substituting $e^{-\frac{1}{\tau}} = \frac{1}{3}$ in equation (ii), we have

$$\phi_F = \frac{30}{1 - \frac{1}{3}} = 45°C$$

If temperature rise after 1 hour is not to exceed 50°, final temperature rise, ϕ_F will be given as

$$\phi_F' = \frac{50}{\left(1 - e^{-\frac{1}{\tau}}\right)} = \frac{50}{1 - \frac{1}{3}} = 75°C \qquad t = 1 \text{ and } e^{-\frac{1}{\tau}} = \frac{1}{3}$$

Let the maximum overload that can be carried by a 25 kW motor be P kW. Since losses vary as the square of load,

$$\frac{\phi'_F}{\phi_F} = \frac{\text{Losses at P kW}}{\text{Original losses}}$$

$$\frac{75}{45} = \left(\frac{P}{25}\right)^2$$

$$P = 25\sqrt{75/45} = 32.3 \text{ kW}$$

⊕ **EXAMPLE 2.8**: An electric motor has load variation as given below: Torque 240 Nm for 20 minutes, 140 Nm for 10 minutes, 300 Nm for 10 minutes, 200 Nm for 20 minutes. If the speed of the motor is 720 rpm, find power rating of motor.

Solution: Equivalent load torque, from equation (2.7),

$$T_{eq} = \sqrt{\frac{T_1^2 t_1 + T_2^2 t_2 + T_3^2 t_3 + T_n^2 t_n}{t_1 + t_2 + t_3 + t_n}}$$

$$= \sqrt{\frac{240^2 \times 20 + 140^2 \times 10 + 200^2 \times 20}{20 + 10 + 10 + 20}}$$

$$= 225.39$$

Speed of motor N =720 rpm
Power rating of motor,

$$I_{sh1} = \frac{V}{R_{sh}} = \frac{250}{250} = 1A$$

⊕ **EXAMPLE 2.9**: A 250 V DC shunt motor has armature resistance of 0.5 ohm and a field resistance of 250 ohm. When driving a constant torque load at 600 rpm the motor draws 20 A. What will be the new speed of the motor if an additional 250 ohm resistance is inserted in the field circuit?

Solution. Given: V = 250 volts ; R_a = 0.5 ohm, R_{sh} = 250 ohm, N_1 = 600 rpm, I = 21A
New speed, N_2:
Shunt field current, $I_{sh1} = \frac{V}{R_{sh}} = \frac{250}{250} = 1A$

Armature current, $I_{a11} = 21 - 1 = 20A$

Back emf, $E_{b1} = V - I_a R_a = 250 - 20 \times 0.5 = 240 V$

Shunt field current when an additional 250 ohm resistance is inserted in the field circuit,

$$I_{sh2} = \frac{250}{250 + 250} = 0.5 A$$

Neglecting magnetic saturation,

$$\varphi_1 \propto I_{sh1}$$

$$\frac{\phi_1}{\phi_2} = \frac{I_{sh1}}{I_{sh2}} \text{ and } \phi_2 \propto I_{sh2}$$

For constant load torque, T_1 and T_2

$$\phi_1 I_{a1} = \phi_2 I_{a2}$$

$$I_{a2} = I_{a1} \times \frac{\phi_1}{\phi_2} = I_{a1} \times \frac{I_{sh1}}{I_{sh2}} = 20 \times \frac{1}{0.5} = 40 \text{ A}$$

Back emf $E_{b2} = 250 - 40 \times 0.5 = 230 \text{ V}$

also, $\dfrac{N_2}{N_1} = \dfrac{E_{b2}}{E_{b1}} \times \dfrac{\phi_1}{\phi_2}$

$$\frac{N_2}{600} = \frac{230}{240} \times \frac{I_{sh1}}{I_{sh2}} = \frac{230}{240} \times \frac{1}{0.5}$$

$$N_2 = 600 \times \frac{230}{240} \times \frac{1}{0.5} = 1150 \text{ rpm}$$

☺ **EXAMPLE 2.10**: A 15 kW, 230 V, 1150 rpm, 4-pole, DC shunt motor has a total of 882 armature conductors arranged in four parallel paths and yielding an armature circuit resistance of 0.2 ohm. When it delivers rated power at rated speed, the motor draws an armature current of 73 A at a field current of 1.6 A. Calculate the developed torque. Also find new operating speed if the field flux is reduced by 80% of its original value for the same developed torque.

Solution. Given: Power output = 15 kW ; V = 230 volts ; N = 1150 rpm, Z = 882, R_a = 0.2 ohm, a = 4 ; I_a = 73 A ; I_{sh} = 1.6 A ; $\phi_2 = 0.8 \, \phi_1$
Torque developed, T_a :
We know that, $T_a = 0.159 \, Z\phi p.\dfrac{I_a}{a} N - m$

$$E_b = \frac{p\phi ZN}{60a} = V - I_a R_a$$

$$\phi = \frac{(V - I_a R_a) \times 60a}{pZN}$$

Putting the given values, we have

$$\phi = \frac{(230 - 73 \times 0.2) \times 60 \times 4}{4 \times 882 \times 1150} = 0.01274 \text{ Wb}$$

$$T_a = 0.159 \times 882 \times 0.01274 \times 4 \times \frac{73}{4}$$

$$= 130.4 \text{ N–m}$$

Also, $T_a \propto \phi I_a$

$$\frac{T_1}{T_2} = \frac{\phi_1 I_{a1}}{\phi_2 I_{a2}}$$

Again $\quad N \propto \dfrac{E_b}{\phi}$ or $\phi \propto \dfrac{E_b}{N}$

$$\frac{T_1}{T_2} = \frac{E_{b1}}{N_1} \times \frac{N_2}{E_{b2}} \times \frac{I_{a1}}{I_{a2}} \qquad\qquad (i)$$

When $T_1 = T_2$ (given), we have

$$\phi_1 I_{a1} = \phi_2 I_{a2}$$

$$I_{a2} = I_{a1} \times \frac{\phi_1}{\phi_2} = 73 \times \frac{1}{0.8} = 91.25 A$$

Now $E_{b1} = V - I_{a1}R_a = 230 - 73 \times 0.2 = 215.4$ V

$\qquad E_{b2} = V - I_{a2}R_a = 230 - 91.25 \times 0.2 = 211.75$ V

Substitute the values in (i), we have

$$1 = \frac{215.4}{1150} \times \frac{N_2}{211.75} \times \frac{73}{91.25}$$

$$N_2 = \frac{1150 \times 211.75 \times 91.25}{215.4 \times 73} = 1413 \text{ rpm}$$

☻ **EXAMPLE 2.11**: A 230 V, 1000 rpm DC shunt motor has field resistance of 115 ohm and armature circuit resistance of 0.5 ohm. At no load, the motor runs at 1000 rpm with armature current of 4 A and with full field flux.

   i. For a developed torque of 80 Nm. compute armature current and speed of the motor.

  ii. If it is desired that motor develops 8 kW at 1250 rpm, determine the value of external resistance that must be inserted in series with the field winding. Saturation and armature reaction are neglected.

Solution: Given: $V = 230$ volts; $N_1 = (N_0)$ 1000 rpm, $I_{a0} = 4$ A ;
$R_{sh} = 115$ ohm, $R_a = 0.5$ ohm, $T_a = 80$ Nm
(i) I_a; N_2

$$I_{sh} = \frac{V}{R_{sh}} = \frac{230}{115} = 2A$$

At no load:

$$E_{b0} = V - I_{a0}R_a$$
$$= 230 - 4 \times 0.5 = 228 \text{ V}$$

Also, $\qquad\qquad E_{b0} = \dfrac{p\phi ZN}{60a}$

or $\qquad\qquad 228 = \left(\dfrac{p\phi ZN}{a}\right) \times \dfrac{1000}{60}$

$$\frac{p\phi ZN}{a} = \frac{228 \times 60}{1000} = 13.68$$

Now torque in Nm,

$$T_a \quad \frac{p\,Z}{}I_a$$

$$80 = \frac{13.68}{} \times$$

$$I = \frac{80 \times 2\pi}{13068} = 36.74\ A$$

☻ **EXAMPLE 2.12**: A 50 HP, 440 V DC shunt motor is braked by plugging. Calculate the value of resistance to be placed in series with the armature circuit to limit the initial braking current to 150 A. Calculate the braking torque so obtained. Assume armature resistance as 0.1 ohm; full-load armature current =100 A, full-load speed = 600 rpm.

Solution. Given : Power, $P = 50$ HP; $V = 440$ volts ; $I_b= 150$ A;
$R_a = 0.1$ ohm; $I_f = 100$ A; $N = 600$ rpm.
External resistance to be connected in series, R_{ext} :
Induced emf in the motor, $E = V - I_{fRa}$
$$= 440 - 100 \times 0.1 = 430\ V$$
Voltage across the armature at the instant of braking
$$= V + E = 440 + 430 = 870\ V$$
Resistance required in the armature circuit to limit the current to 150 A,

$$R = \frac{870}{150} = 5.8\ \text{ohm}$$

External resistance required in the armature circuit, $R_{ext} = R - R_a = 5.8 - 0.1$ = 5.7 ohm (Ans.)
Braking torque, T_b:
Full-load torque,

$$T_f = \frac{P \times 735.5}{(2\pi N / 60)}\,Nm = \frac{50 \times 735.5}{\left(2\pi \times 600/60\right)} = 585.3\ Nm$$

As in case of a DC shunt motor the flux remains unchanged, therefore, torque is proportional to armature current.
∴ Initial braking torque,

$$T_b = T_f \times -$$

$$= 585.3 \times \frac{150}{100} = 877.95\ Nm$$

☻ **EXAMPLE 2.13**: A 500 V, 45 kW, 600 rpm DC shunt motor has a full-load efficiency of 90%. The field resistance is 200 Ω and armature resistance is 0.2 Ω. Find the speed under each of the following conditions at which will develop an electromagnetic torque equal to rated value:

i. Regenerative braking: no limiting resistance.
ii. Plugging; external limiting resistance of 5.5 Ω inserted.
iii. Dynamic braking external limiting resistance of 2.6 Ω inserted.

The field current is maintained constant and armature reaction and the brush drop may be neglected. (AMIE Elec drives and their control)

Solution: $V = 500$ volts; $P = 45$ kW; $N_f = 600$ rpm; $\eta = 90\%$; $R_{sh} = 200$ Ω; $R_a = 0.2$ Ω

$$\text{Rated line current, } I_{Lf} = \frac{\text{Output in kW} \times 1000}{\text{Supply voltage} \times \text{full-load efficiency}}$$

$$= \frac{45 \times 1000}{500 \times 0.9} = 100\text{A}$$

$$\text{Shunt field current, } I_{sh} = \frac{V}{R_{sh}} = \frac{500}{200} = 2.5\text{A}$$

Armature current on full-load, $I_{af} = I_{Lf} - I_{sh} = 100 - 2.5 = 97.5$ A
Induced emf on full-load, $E_f = V - I_{af} R_a = 500 - 97.5 \times 0.2 = 480.5$ V

Since field current remains constant and the armature reaction and brush drop are negligible so electromagnetic torque developed is directly proportional to armature current and speed is directly proportional to induced emf.

As torque to be developed in each case is equal to rated torque so armature current in each case is equal to armature current at full-load, i.e. 97.5 A

Speeds under various conditions, N_1, N_2, N_3:

i. Induced emf to give armature current of 97.5 A for regenerative braking.
$E_1 = V + I_{af} R = 500 + 97.5 \times 0.2 = 519.5$ V

$$\text{And speed,} \quad N_1 = N_f \times \frac{E_1}{E_f} = 600 \times \frac{519.5}{480.5} = 648.7 \text{ rpm}$$

ii. Induced emf to give armature current of 97.5 A for plugging with external limiting resistance of 5.5 Ω,
$E_2 = I_{af} (R + R_a) - V = 97.5 (5.5 + 0.2) - 500 = 55.75$ V
And speed,

iii. Induced emf to give armature current of 97.5 A for dynamic braking with external limiting resistance of 2.6 Ω,
$E_3 = I_{af} (R + R_a) = 97.5 (2.6 + 0.2) = 273$ V

$$\text{And speed, } N_2 = N_f \times \frac{E_2}{E_f} = 600 \times \frac{55.75}{480.5} = 69.6 \text{ rpm}$$

◉ **EXAMPLE 2.14:** An induction motor runs at a slip frequency of 2 Hz when supplied from a three-phase 400 V, 50 Hz supply. For the same developed torque, find the slip frequency at which motor will run when supplied from a three-phase 340 V, 40 Hz system. Slip at which machine develops maximum torque using 50 Hz supply is 0.1. Neglect the stator impedance and assume linear torque-slip characteristics between zero torque and maximum torque in the working region.

Solution: Rating of induction motor = 400 V, 50 Hz

Slip frequency = 2 Hz

Slip at maximum torque, $S_{mt} = 0.1$

When the slip frequency is 2 Hz, the slip at this frequency $S_1 = \dfrac{2}{50} = 0.04$

Let the slip at 340 V, 40 Hz be s_2

$$\text{Torque } T \propto \frac{sE_2^2}{R_2^2 + (sX_2)^2} \qquad ....(\text{i})$$

Here stator impedance is neglected and as such V = E2

Also slip at maximum torque, $S_{mT} = \dfrac{R_2}{X_2} = 0.1$

$$R_2 = 0.1X_2$$

Substituting this value in (i), we get

$$T \propto \frac{sV^2}{\left(0.01 + s^2\right)X_2^2}$$

Since the developed torque for both the cases is same, therefore,

$$\frac{s_1 V_1^2}{\left(0.01 + s_1^2\right)X_2^2} = \frac{s_2 V_2^2}{\left(0.01 + s_1^2\right)X_2^2}$$

$$\frac{0.04 \times (400)^2}{0.01 + (0.04)^2} = \frac{s_2 \times (340)^2}{0.01 + (s_2)^2}$$

$$\frac{6400}{0.0116} \propto \frac{s_2 \times 115600}{0.01 + s_2^2}$$

$$6400\,(0.01 + s_2^2) = 0.0116 \times s_2 \times 115600$$

$$64 + 6400s_2^2 = 1340.96\,s_2$$

$$6400s_2^2 - 1340.96s_2 + 64 = 0$$

As the slip cannot be high, thus select the value of slip as 0.0735.

Slip at 40 Hz = 0.0735

Hence, slip frequency = $0.0735 \times 40 = 2.94$ Hz. (Ans.)

◑ **EXAMPLE 2.15**: A 3-phase star-connected 6.6 kV, 20 pole 50 Hz induction motor has rotor resistance of 0.12 ohm and standstill reactance of 1.12 ohm.

The motor has speed of 292.5 rpm at full-load.

Calculate slip at maximum torque and ratio of maximum torque to full-load torque.

Solution. Given: $p = 20; f = 50$ Hz ; $R_2 = 0.12$ ohm; $X_2 = 1.12$ ohm; $N = 292.5$ rpm

$$S_{mT} ; \frac{T_m}{T_f}$$

Slip corresponding to maximum torque is given by

$$S_{mT} = \frac{R_2}{X_2} = \frac{0.12}{1.12} = 0.107 \text{ or } 10.7\%$$

We know that,

$$\frac{T_m}{T_f} = \frac{2S_f S_{mT}}{S_f^2 + s_{mT}^2}$$

Where, $S_{mT} = \dfrac{R_2}{X_2}$, and S_f = full-load slip of motor,

Now,

$$S_{mT} = \frac{0.12}{1.12} = 0.107$$

$$N_s = \frac{120\, f}{p} = \frac{120 \times 50}{20} = 300 \text{ rpm}$$

$$S_f = \frac{N_s - N}{N_s} = \frac{300 - 292.5}{300} = 0.025$$

Substituting the values in the above equation, we get

$$\frac{T_f}{T_m} = \frac{2 \times 0.107 \times 0.025}{(0.107)^2 + (0.025)^2} = 0.443$$

Hence,

$$\frac{T_m}{T_f} = \frac{1}{0.443} = 2.257$$

◉ **EXAMPLE 2.16**: A 3-phase induction motor has starting torque of 100% and a maximum torque of 200% of the full-load torque. Find slip at maximum torque.

Solution: Given: Starting torque, $T_{st} = 100\%$ of T_f or $= T_f$
Maximum torque, $T_m = 200\%$ of T_f or $= 2T_f$
Slip at maximum torque, S_{mt} :
We know that,

$$\frac{T_{st}}{T_m} = \frac{2S_{mT}}{S_{mT^2} + 1}$$

$$\frac{T_f}{2T_f} = \frac{2S_{mT}}{S_{mT}^2 + 1}$$

$$1 = \frac{2s_{mT}}{S_{mT}^2 + 1}$$

or

$$s_{mT}^2 - 4\, s_{mT} + 1 = 0$$

☻ **EXAMPLE 2.17**: A 50 kVA, 400 V 3-phase, 50 Hz squirrel cage induction motor has full-load slip of 5%. Its standstill impedance is 0.866 ohm/phase. It is started using a tapped autotransformer. If the maximum allowable supply current at the time of starting is 100 A, calculate the tap position and the ratio of starting torque to full-load torque.

Solution. Given: Rating of induction motor = 50 kVA, 400 V; S_f = 5% or 0.05; standstill impedance = 0.866 ohm per phase, I_{st} = 100 A.

Tap position $\dfrac{T_{st}}{T_f}$;

Full-load current,

$$= \frac{\text{output in kVA} \quad 1000}{\sqrt{3} \times \text{line voltage}}, \text{neglecting losses}$$

$$= \frac{50 \times 1000}{\sqrt{3} \times 400} = 72.2 \ A$$

Short-circuit current, $I_{sc} = \dfrac{\left(400\sqrt{3}\right)}{0.886} = 266.7\,A$

Tap position of the transformer,

$$K = \sqrt{\frac{I_{st}}{I_{sc}}} = \sqrt{\frac{100}{266.7}} = 0.6123 \text{ or } 61.23\%$$

Now,

$$\frac{T_{st}}{T_f} = K^2 \left(\frac{I_{sc}}{I_f}\right)^2 s_f = (0.6123)^2 \times \left(\frac{266.7}{72.2}\right)^2 \times 0.05 = 0.256$$

☻ **EXAMPLE 2.18**: A 30 kW rated output, 400 V, 3-phase delta-connected, 4-pole, 50 Hz induction motor has full-load slip of 5%. If the ratio of standstill reactance to resistance per rotor phase is 4, estimate the plugging torque at full-load speed. Ignore stator leakage impedance and magnetising reactance.

Solution. Given: Rated output = 30 kW, supply voltage = 400 V;

$p = 4$; $f = 50$ Hz; $s_f = 5\%$ or 0.05; $\dfrac{x_2}{R_2} = 4$

Plugging torque at full-load speed:
Synchronous speed,

$$N_s = \frac{120f}{p} = \frac{120 \times 50}{4} = 1500 \ rpm$$

Full-load speed,
$N_f = N_s(1 - s_f) = 1500 \ (1 - 0.05) = 1425 \ rpm$
Full-load torque,

$$T_f = \frac{\text{Rated output in kW} \times 1000}{(2\pi N/60)}$$

$$= \frac{30 \times 1000 \times 60}{2\pi \times 1425} = 201.04 \ Nm$$

Now,

$$T \propto \frac{sR_2 E_2^2}{R_2^2 + (sX_2)^2}$$

$$\frac{T_p}{T_f} = \frac{\left(s_p R_2 E_2^2\right)/\left(R_2^2 + s_p^2 X_2^2\right)}{\left(s_f R_2 E_2^2\right)/\left(R_2^2 + s_f^2 X_2^2\right)} = \frac{s_p\left(R_2^2 + s_f^2 X_2^2\right)}{s_f\left(R_2^2 + s_p^2 X_2^2\right)}$$

$$\frac{T_p}{T_f} = \frac{s_p}{s_f} \frac{\left[1 + s_p^2\left(\dfrac{X_2}{R_2}\right)^2\right]}{\left[1 + s_f^2\left(\dfrac{X_2}{R_2}\right)^2\right]} = \frac{s_p}{s_f}\left[\frac{1 + 16s_f^2}{1 + 16s_p^2}\right] \qquad \left(\frac{X_2}{R_2} = 4\right)$$

Putting $s_f = 0.05$ and s_p = slip at plugging = $2 - s_1 = 2 - 0.05 = 1.95$, we get
Putting torque,

$$T_p = \frac{1.95}{0.05}\left[\frac{1 + 16 \times (0.05)^2}{1 + 16 \times (1.95)^2}\right]T_f$$

$$= 39 \times \frac{1.04}{61.84} \times 201.04 = 131.86 \ Nm$$

⊘ **EXAMPLE 2.19**: The initial to temperature of a machine is 45°C. Calculate the temperature of the machine after 1.2 hour if its final steady temperature rise is 85°C and the heating time constant is 2.4 hours. The ambient temperature is 25°C.

Solution. Initial temperature of a machine = 45°C
Ambient temperature = 25°C
Final steady temperature rise, ϕ_m = 85°C
Heating time constant, T_h = 2.4 hours
Temperature of the machine after 1.2 hour, ϕ:
Initial temperature rise, ϕ_i = Initial temperature of the machine – ambient temperature

$$= 45 - 25 = 20°C$$

Using the equation, $\phi = \phi_m\left(1 - e^{-\frac{t}{T_h}}\right) + \phi_i e^{-\frac{t}{T_h}}$

$$= 85\left(1 - e^{-\frac{1.2}{2.4}}\right) + 20e^{-\frac{1.2}{2.4}}$$

$$= 85\left(1 - e^{-\frac{1}{2}}\right) + 20e^{-\frac{1}{2}}$$

$$= 85(1 - 0.607) + 20 \times 0.607 = 33.4 + 12.14 = 45.54°C$$

Hence, the temperature of the machine after 1.2 hour

$$= 45.54 + 25 = 70.54°C$$

QUESTIONS

1. What are the advantages of Electrical Drive over other Drives?
2. Discuss the various factors that govern the choice of a motor for a given service.
3. What are the relative advantages and disadvantages of DC and AC electric drive?
4. Explain what you mean by "Individual drive" and "Group drive". Discuss their relative merits and demerits.
5. For what type of speed torque characteristic, would you recommend shunt motor?
6. What are advantages of speed control of shunt motor by Ward Leonard method?
7. Explain the thyristor control on AC side and DC side.
8. What do you understand by closed loop system of control of a motor?
9. Explain which properties make series motor more suitable for heavy torque applications.
10. Explain various methods of speed control of series motor. Which method is more effective in reducing the speed for a given torque?
11. What are the advantages and disadvantages of electric braking?
12. Explain regenerative braking of induction motor.
13. Explain the method of rheostatic braking.
14. Explain dynamic braking and regenerative braking.
15. What are braking systems applicable to a DC shunt motor?
16. What for Series motor Regenerative Braking is not suited?
17. What are the important stages in controlling an electrical drive?
18. Explain rheostatic braking of DC motors.
19. Define the following terms regarding the ratings of motor:
    (i) Continuous rating (ii) short-time rating (iii) Intermittent rating.
20. With the help of heating and cooling curves define and explain the terms:
    (i) Heating time constant (ii) Cooling time constant.
21. Derive an expression for the temperature rise of an equipment in terms of the heating time constant.
22. Explain what you mean by Lord Equalization and how it is accomplished?
23. What do you mean by 'load-equalisation' it is possible to apply this scheme for reversible drive? Why?
24. On that factors does the selection of type enclosure depend?
25. Suggest, with reasons the electric drive used for the following applications. (i) Rolling mills (ii)Textile mills (iii) Cement mills (iv) Paper mills (v) Coal mining (vi) Lift, cranes, lathes and pumps.
26. What are essential requirements of paper mill drive, rolling mill drive and sugar centrifuge?

ILLUMINATION

3.1 INTRODUCTION

Mankind in primitive era used to pass most of their time out of doors and their lighting needs were provided by nature. Today, however, they spend most of their times indoors specifically in buildings where artificial lighting plays an important role. The aim of artificial lighting is to supplement the sunlight or even to replace it completely. Artificial lighting produced electrically, on account of its cleanliness, ease of control, reliability, steady output, as well as its low cost, is playing an important part in modern civilization. It has been proven that good illumination ensures increased production, reduces worker's fatigue, protects his health, eyes and nervous system and reduces accidents. Thus economy of lighting installation cannot be judged only on the basis of its own installation and running cost but we have to consider its effect on the production also. The science of illumination engineering is, therefore, becoming of major importance.

3.2 NATURE OF LIGHT

For better understanding of the subject—illumination and for following the latest developments in lighting, it is mandatory to know the nature of light and the basic principle specifying its production. Visible light is composed of enormous number of trains of transverse waves arising out of electromagnetic oscillations. As a result of definite stimuli to atom, light radiations called are light emitted from within the atom in specific quantity in succession or photons and travel through the medium of ether. In fact Maxwell had shown light to form a small portion on the electro-spectrum. Magnetic waves have same speed of 3×10^8 metres/sec. in free space. They, however, differ in wavelength and, therefore, in frequency. If λ be the wavelength, if the frequency and v the velocity of propagation of the wave, then

$$v = \lambda f \tag{3.1}$$

Depending upon the method of stimulus to the atom, we get electromagnetic radiations of different wavelength and frequency. For instance stimulus may be thermal which is called incandescent radiations or nonthermal such as luminescent radiations. While incandescent radiations give rise to continuous spectrum, luminescent radiations produce bands or lines on the spectrum characteristic to the gas whose atoms are excited.

It is convenient to measure the wavelength of high frequency electromagnetic radiations in much smaller unit than centimeter, either in micron or angstrom where,

$$1 \text{ micron } (\mu) = 10^{-6} \text{ metre}$$

And

$$1 \text{ angstrom } (A) = 10^{-10} \text{ metre}$$

A look on the classification of electromagnetic waves Fig. 3.1 shows that visible portion of spectrum constitutes only a small fraction. Visible spectrum on a more magnified scale is shown in Fig. 3.2. It will be observed there from that radiations of different wavelengths produce different color sensation on the eyes. Also relative sensitivity of human eye of standard observer varies very widely, being maximum at about 5550 AU wavelength corresponding to yellow green color. The limit of visible spectrum is not well-defined as the eye sensitivity curve becomes asymptotic at both ends of spectrum. In order, therefore, to specify the limits, visible spectrum is said to extend over wavelengths where sensitivity of human eye falls to 1% of its maximum value. With this criterion, visible spectrum lies in between wavelengths of 4000 AU to 7000 AU Infrared radiation has longer wavelength than visible light and ultraviolet has shorter. Part of ultraviolet radiations can however by converted into visible light by fluorescent substances.

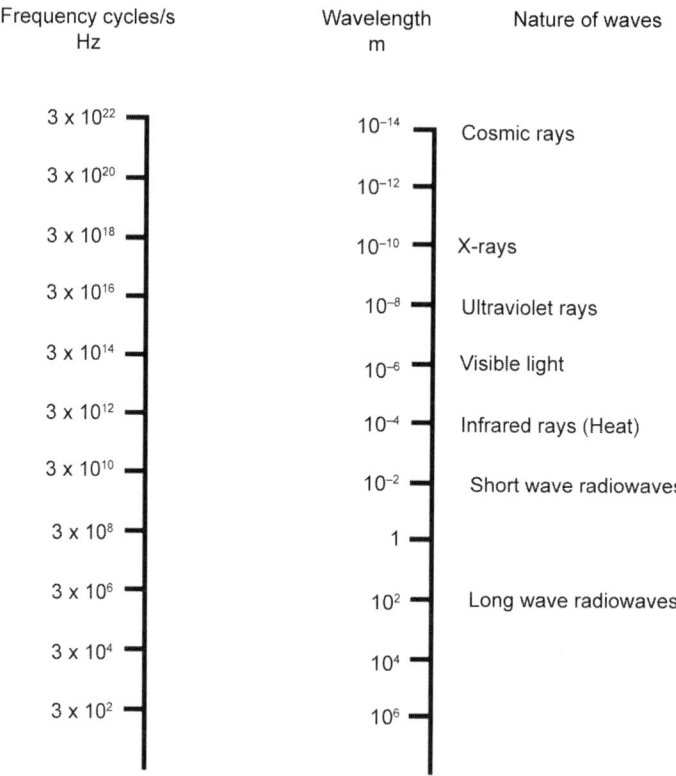

Fig. 3.1: Spectrum of electromagnetic waves.

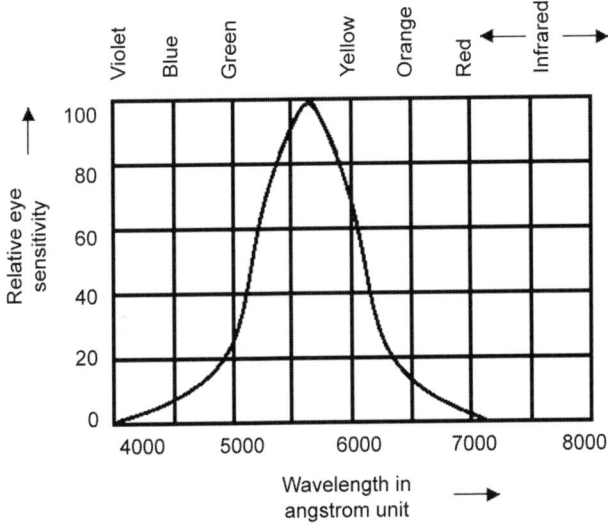

Fig. 3.2: Eye sensitivity curve.

3.3   DEFINITIONS

Before any mathematical treatment of the subject is attempted, it becomes imperative to know the exact definitions of various quantities as given below.

3.3.1 Plane Angle and Solid Angle

Plane angle is subtended at a point in a plane by two converging straight lines, Fig. 3.3a. Magnitude of the angle is given by $\theta = \dfrac{arc}{radius}$ radians. Solid angle (Fig. 3.3b) on the other hand is the angle generated by the line passing through the point in space and the periphery of the area. It is measured in steradians and is denoted by ω. its magnitude is given by $\omega = \dfrac{area}{(radius)^2}$ steradians. Thus if we take an area of R^2 metres on the surface of a sphere of radius K and if radius of this sphere moves along the boundry of this area which for convenience sake is assumed to be circulate, then radius of the sphere describes a cone which encloses 1 steradian.

Total solid angle subtended at a point in space is obtained by considering the point to be situated at the centre of the sphere.

$$\omega = \frac{area\ of\ sphere}{(radius)^2} = \frac{4\pi r^2}{r^2} = 4\pi \text{ steradians}$$

Relationship between plane and solid angle can be obtained as follows:

We will consider a segment of a sphere Fig. 3.3c. Curved surface of spherical segment = $2\pi rH$

The section of the spherical segment is shown in Fig. 3.3d from where we find

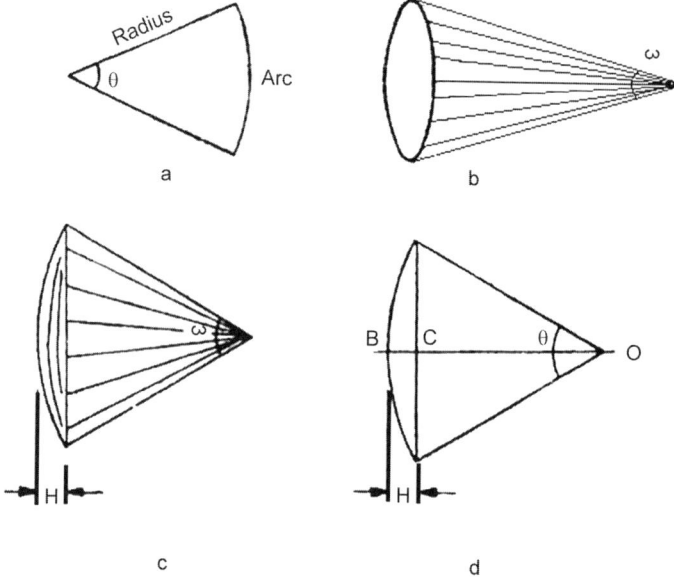

Figs 3.3a to d: (a) Plain angle, (b) Solid angle, (c) Segment of sphere, (d) Relationship between plane and solid angle.

$$H = OB - OC$$

$$= r - r\cos\frac{\theta}{2}$$

$$= r\left(1 - \cos\frac{\theta}{2}\right)$$

$$\therefore \quad \omega = \frac{\text{area of sphere}}{(\text{radius})^2} = \frac{2\pi r.r\left(1-\cos\frac{\theta}{2}\right)}{(r^2)}$$

$$= 2\pi\left(1 - \cos\frac{\theta}{2}\right) \tag{3.2}$$

3.3.2 Light Energy

It is the energy contained in visual radiations in a given time and is expressed in lumen hour—analogous to kWh. It is usually denoted by Q. Its importance can be realised when we want to ascertain the quantity of electrical energy used.

3.3.3 Luminous Flux

It is the rate of flow of luminous energy and is denoted by ϕ $\therefore$ $\phi = \dfrac{Q}{t}$. Its unit is lumen

(lm). The conception of luminous flux helps us to specify the output and efficiency of a given light source. For instance output of a bicycle lamp is 10 lm. And that of 150 W incandescent lamp is 1940 lm.

3.3.4 Luminous Intensity

It is the measure of luminous flux in lumens emitted per unit solid angle by a point source and is denoted by I $\therefore$ I $= \dfrac{\phi}{\omega}$ unit is candela (cd) or lumen/steradian.

It is the ratio of the brightness of a source of light to that of standard candle. One candle gives out luminous flux of 4π lumens in space. Thus lumens emitted by one candela source of light is one lumen/steradian. Actual measurement of candela of a given source of light in any given direction is found out with the aid of photometers described in article 3.15.

3.3.5 Candela

In scientific terms candela is defined as the luminous intensity in the perpendicular direction of a surface of $\dfrac{1}{600000}$ sq. metre of a full radiator at the temperature of freezing platinum under a pressure of 101325 N/sq. metre.

3.3.6 Illumination

It is defined as luminous flux density per unit area and is denoted by E.

Illumination of one lumen per sq. metre is called lux and one lumen per sq. ft. is called ft. candle.

$\therefore$

$$E = \frac{\phi}{A}$$

$$\text{Lux} = \frac{\text{Lumen}}{(\text{Metre})^2}$$

$$= \frac{\text{Lumen}}{(3.28 \text{ ft.})^2}$$

$$= \frac{1}{(3.28)^2} \frac{\text{Lumen}}{\text{ft.}^2}$$

$$= \frac{1}{10.76} \text{ ft. candle}$$

$$= 0.093 \text{ ft. candle}$$

$\therefore$ 1 ft. candle $= 10.76$ flux.

If a point source of light of one candela is placed at the centre of a sphere of radius one metre, illumination at surface

$$E = \frac{4\pi \times 1}{4\pi \times 1^3} = 1 \text{ lux}$$

Unit of illumination, instead of lux, is sometimes called metre candle also. Still bigger unit of illumination is Phot. It is lumen per sq. cm. and is 10,000 times lux.

The idea of illumination will be had from the following illumination levels obtained under different situations.

1. During sunset and sunrise   500 Lux
2. During summer midday   100,000 Lux
3. During winter midday   100,00 Lux
4. With full moon   0.25 Lux

3.3.7 Luminance (Brightness)

It is defined as luminous intensity per unit projected or apparent area of either a surface source of light or illuminated surface and is denoted by L

$$L = \frac{I}{A_a}$$

where A_a is apparent area

$$L = \frac{I}{A \cos \theta} \tag{3.3}$$

where θ is the angle between the normal to the surface and the direction of vision Fig. 3.4. We may distinguish between illumination and luminance by an example of an open book lying on a table. Even though illumination, i.e. light flux falling on both is same, yet the luminance of book will be more than that of table because the former has more reflection factor.

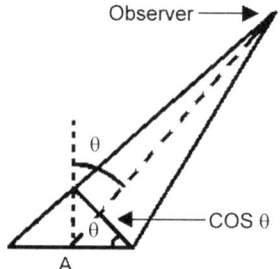

Fig. 3.4: Change in luminance with a change in angle.

☞ **Relationship between I, E and L**

We will first consider a self-source of light having radiating sphere of R metres and possessing luminous intensity of 1 cd.

Then
$$L = \frac{I}{\pi R^2}$$

And
$$E = \frac{4\pi I}{4\pi R^2} = \frac{I}{R^2}$$

∴
$$L = \frac{1}{\pi} \frac{I}{R^2}$$

$$= \frac{E}{\pi} \tag{3.4}$$

In the case of secondary source of light, light radiated is $E \times \gamma_r$ lm/m² where γ_r reflection factor.

$$L = \frac{E}{\pi} \gamma_r \qquad (3.5)$$

We thus see from equation 3.5 that luminance depends upon the illumination and reflection factor of the object.

3.3.8 Mean Horizontal Candle Power (MHCP)

It is the mean of the candle powers in all directions in horizontal plane passing through the source of light.

3.3.9 Mean Spherical Candle Power (MSCP)

It is the mean of candle powers in all directions in all planes.

3.3.10 Mean Hemispherical Candle Power (MHSCP)

It is mean of candle powers below horizontal plane passing through the light source.

3.3.11 Reduction Factor

The ratio of MSCP to MHSCP of source of light is called its reduction factor.

3.3.12 Lamp Efficiency

It is the ratio of lumens emitted per one watt of electric power. Sodium lamp produces 100 lm/W, mercury lamp 40–50 lm/W, fluorescent-lamp 75 lm/W and Incandescent lamp 12 to 20 1m/W.

3.3.13 Specific Consumption

It is the ratio of power input to the source of light to its luminous intensity. Its unit is watts per candela.

3.3.14 Space Height Ratio

In order to achieve even illumination and do away with dark areas, it is necessary that area illuminated by one lamp should overlap that illuminated by adjacent space lamps,

$$\text{Space height ratio} = \left(\frac{\text{space between lamps}}{\text{height of lamps above working plane}} \right)$$

Has special significance in this connection. Space height ratio employed with dispersive type of reflectors, fluorescent lamps and other direct fittings is 1.5 and with concentrating reflectors is 1.

3.3.15 Utilization Factor (UF)

The ratio of total lumens reaching the working plane to total lumens given out by the lamp is called utilization factor.

3.3.16 Maintenance Factor (MF)

It is the ratio of illumination under normal working condition to the illumination when the things are perfectly clean.

3.3.17 Depreciation Factor

It is defined as the ratio of initial metre-candles to the ultimate maintained metre-candles on the working plane. Its value is more than unity.

3.3.18 Waste Light Factor

In flood lighting scheme, a number of projector lamps is employed so that light of each projector not only overlaps that of its adjacent projector but at the edges of the area, some light falls outside. In order, therefore, to compute total lumens required, theoretical lumens required are multiplied with waste light factor. Value of this factor is about 1.2 for rectangular areas and 1.5 for irregular objects such as statues, monuments, etc.

3.3.19 Absorption Factor

In case of foundries or places where atmosphere is full of smoke fumes, there is possibility of absorption of light energy.

$$\text{Absorption factor} = \frac{\text{Net lumens available after absorption}}{\text{Total lumens emitted by source of light}}$$

Absorption factor varies from 1 for clean atmosphere to $0'5$ for casting floors of steel works and foundries.

3.3.20 Luminous Efficiency

If whole of the electric energy input to the filament lamp were to be produced at 5550 AU wavelength, it will give out 680 lumens/watt. If whole of electric energy input to filament lamp would produce only light spectrum similar to sunlight, it would give out 220 lumens/watt. This is maximum efficiency of the filament lamp which would have been possible, could the infrared and ultraviolet radiations be suppressed. However after taking infrared and ultraviolet radiations into account, filament lamp efficiency comes out to be in the range of 10 to 20 lumens/watt.

If E = Energy radiated at wavelength λ

λ = The relative sensitivity of eye at wavelength λ

K = Maximum possible efficiency if whole of the electrical input were transformed into radiating energy at 5550 AU

= 620 lumens/watt.

Efficiency at wavelength $\lambda = \eta K$

$\therefore$ Total energy converted into visual effect = $K \int_{\lambda_1}^{\lambda_2} d\lambda$

Total energy radiated on all wavelengths = $\int_0^\infty E \, d\lambda$

$$\text{Lu minous efficiency} = \frac{K\int_{\lambda_1}^{\lambda_2} E_\eta d\lambda}{\int_0^\infty E d\lambda}$$

3.3.21 Beam Factor

This term is used in connection with projector lamp. It is the ratio of lumens in the beam of a projector to lamp lumens. This factor takes into account the absorption of light by reflector and front glass of the projector lamp. Its value lies between 0.3 and 0.6.

3.3.22 Reflection Factor

All the light incident on a reflecting surface is not reflected. Some portion of the same is absorbed by the surface.

$$\text{Reflection factor} = \frac{\text{Reflected light}}{\text{Incident light}}$$

Reflection factor is always less than unity.

3.3.23 Coefficient of Utilisation

Not all the light emitted by the lamp reaches the surface to be illuminated. As shown in Fig. 3.5, some portion of light falling within angle (1) will reach the working plane directly while other portion of light will go to walls [falling within angle (2)], and to ceiling falling within angle (3)]. Out of this portion of light, some will be reflected to the illuminated surface.

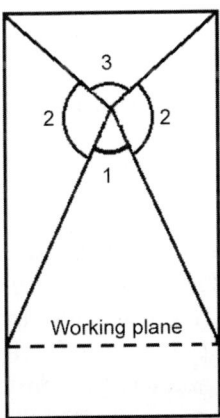

Fig. 3.5: Distribution of light across the surface.

$$\text{Coefficient of utilisation} = \frac{\text{Lumens reaching the working plane}}{\text{Total lumens given out by lamp}}$$

Coefficient of utilisation is high for large rooms having light coloured walls and low for small rooms with dark colours.

3.4   POLAR CURVE

For determining illumination of a point, the formula $E = \dfrac{1}{R^2} \cos \alpha$ is required and thus it becomes essential to know the value of luminous intensity of the source of light in that direction.

Due to several reasons luminous intensity of a source of light is not same in all directions. Distance of any point on the curve from origin gives to some scale the magnitude of luminous intensity in candela of source of light in that direction. If we determine luminous intensity in horizontal plane passing through the source of light and plotted for various angles, we get a horizontal polar curve. Similarly vertical polar curve gives the relationship of candle power in vertical plane passing through the lamp at various angles. Typical polar curves of filament lamp are shown in Figs 3.6a and b. Depression at 180° in vertical polar curve is due to the lamp holder. Slight depression at 0° in the horizontal polar curve is because coil filament occupies an arc which subtends an angle less than 360°.

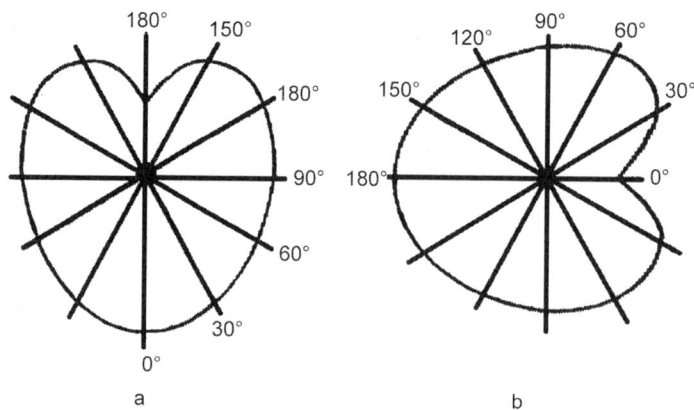

Figs 3.6a and b: (a) Polar curve for vertical plane, (b) Polar curve for horizontal plane.

3.5   ROUSSEAU DIAGRAM

Mean spherical candle power of a symmetrical source of light can be measured from the polar curve with the help of Rousseau diagram. In Fig. 3.7 the light distribution diagram or polar curve is shown, which is symmetrical about vertical axis of a filament lamp burning in the normal 'cap up' position. P1 represents in magnitude the candela of the lamp vertically downwards and P11 the candela of the lamp vertically upwards. With centre as P, a semicircle of radius greater than the greatest radius of the polar curve is drawn. The radii of the polar curve are drawn to cut the circumference of the circle as shown. These points of intersection are projected on the vertical line AB = diameter of the circle. On AB as base, abscesae B1 = P1, etc. are drawn, the last ordinate being A11 = P11. The resulting curve 1, 2, 3, 11 is the Rousseau diagram.

$$\text{Average candela of the source of light} = \frac{\text{Area of Rousseau diagram}}{\text{AB}}$$

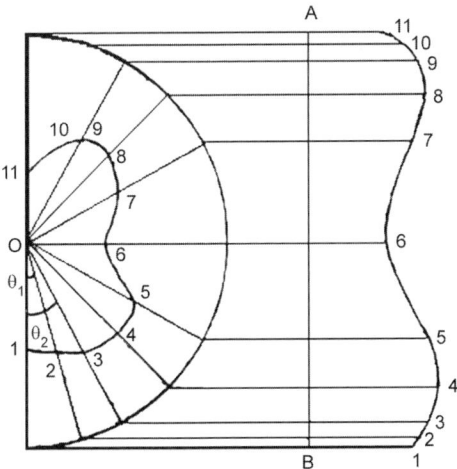

Fig. 3.7: Rousseau diagram.

The process is further simplified by the use of Russell angles. Angles θ_1, θ_2 are chosen in such a manner that it divides the diameter AB in equal number of parts (say 10). Then average candela of source of light will be the arithmetical average of the candela obtained at those angles. If R be the radius of the circle and θ_1, θ_2 etc. be the angles from the vertical so as to divide $\left(\dfrac{AB}{2}\right)$ in n equal parts

Then $R - R\cos\theta_1 = \left(\dfrac{R}{n}\right)$

And $R - R\cos\theta_2 = 2\left(\dfrac{R}{n}\right)$, etc.

$\therefore \cos\theta_1 = 1 - \dfrac{1}{n}$

And $\cos\theta_2 = 1 - \dfrac{2}{2n}$, etc.

It is however better to use mid ordinate method in which case

$$\cos\theta_1 = 1 - \dfrac{1}{2n}$$

$$\cos\theta_2 = 1 - \dfrac{3}{2n}$$

The values of Russell angles from the downward vertical corresponding to mid ordinates for $n = 5$ or total ten angles are 25.8, 45.5, 60, 72.5, 84.2, 95.8, 107.5, 120, 134.5, and 154.2 degrees.

3.6   LAWS OF ILLUMINATION

There are two laws of illumination. (i) Inverse Square Law (ii) Lambert's Cosine Law

i. **Inverse Square Law:** The illumination of a surface is inversely proportional to the square of the distance of the surface from the source. In other words, $E \propto \dfrac{1}{r^2}$.

Proof:

Consider surface areas A_1 and A_2 at distances r_1 and r_2 respectively from the point of source S of luminous intensity I and normal to the rays as shown in Fig. 3.8. Let the solid angle subtended be ω steradians.

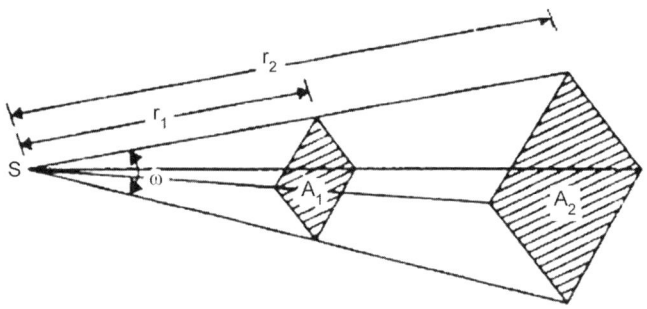

Fig. 3.8: Inverse square law.

Luminous flux radiated per steradians = I

Total luminous flux radiated = $I\omega$ lumens.

Illumination of the surface of the area A_1

$$= \frac{I\omega}{A_1} \text{ lumens/unit area}$$

And area $A_1 = \omega r_1^2$

Illumination on the surface of area $A_{1'}$

$$E_1 = \frac{I\omega}{\omega_1^2} = \frac{I}{r_2^2} = \text{lumens/unit area}$$

Similarly illumination on the surface of area,

$$E_2 = \frac{I\omega}{A_2} = \frac{I\omega}{\omega_2^2} = \frac{I}{r_2^2} = \text{lumens/unit area}$$

Hence the illumination of a surface is inversely proportional to the square of the distance between the surface the light source provided that the distance between the surface and the source is sufficiently large so that source can be regarded as appoint source.

ii. **Lambert's Cosine Law:** According to this law, E is directly proportional to the cosine of the angle made by the normal to the illuminated surface with the direction of the incident flux.

Proof:

As shown in Fig. 3.9, let ϕ be the flux incident on the surface of area A when in position 1. When this surface is turned back through an angle θ, then the flux incident on it is $\phi \cos \theta$. Hence, illumination of the surface when in position 1 is $E_1 = \phi/A$. But when in position 2.

$$E_2 = \frac{\phi \cos\theta}{A}$$

∴ $$E_2 = E_1 \cos\theta$$

Combining all these factors together, we get $E_2 = I \cos\theta / r^2$. Then unit is lm/m^2.

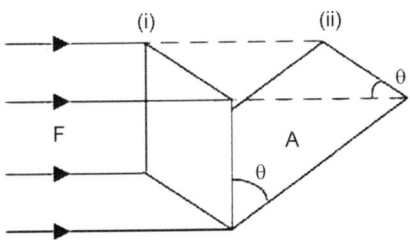

Fig. 3.9: Lambert's cosine law.

The above expression makes the determination of illumination possible at a given point provided the position and the luminous intensity or candle power (in the given direction) of the source (or sources) by which it is illuminated, are known as illustrated by the following examples. Consider a lamp of uniform luminous intensity suspended at a height h above the working plane as shown in Fig. 3.10. Let us consider the value of illumination at point A immediately below the lamp and at other points B,C,D, etc., lying in the working plane at different distances from A.

$$E_A = \frac{1}{h^2} \quad \text{since } \theta = 0 \quad \cos\theta = 1$$

$$E_B = \frac{1}{LB^2} \times \cos\theta_1 \quad \text{since } \cos\theta_1 = \frac{h}{LB}$$

$$E_B = \frac{I}{LB^2} \times \frac{h}{LB} = I \times \frac{h}{LB^2} = \frac{1}{h^2} \times \left(\frac{h}{LB}\right)^3$$

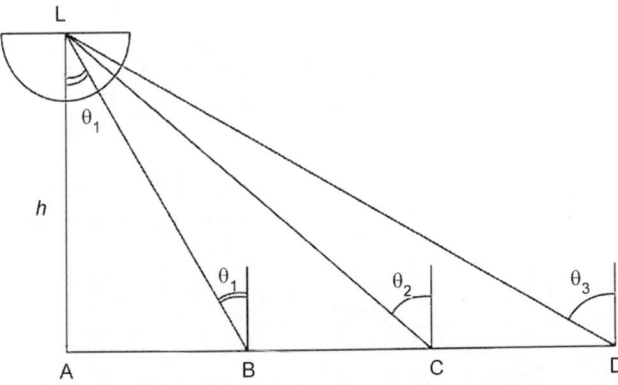

Fig. 3.10: Lamp of uniform luminous intensity suspended at a height above the working plane.

Now $\dfrac{1}{h^2} = E_A$ and $\left(\dfrac{h}{LB}\right)^3 = \cos^3 \theta_1$

$$E_B = E_A \cos^3 \theta_1$$

Similarly,

$E_C = E_A \cos \theta^3 \theta_2$ and $E_D = E_A \cos^3 \theta_3$ and so on

3.7 SOURCES OF LIGHT

According to principle of operation the light sources may be grouped as follows:

1. **Arc lamps:** Electric discharge through air provides intense light. This principle is utilised in are lamps.
2. **High temperature lamp:** Oil and gas lamps and incandescent filament type lamps, which emit light when heated to high temperature.
3. **Gaseous discharge lamps:** Under certain conditions, it is possible to pass electric current through a gas or metal vapour, which is accompanied by visible radiations. Sodium and mercury vapour lamps operate on this principle.
4. **Fluorescent type lamps:** Certain materials, when exposed to ultraviolet rays, transform the absorbed energy into radiations of longer wavelength lying within the visible range. This principle is employed in fluorescent lamps.

3.7.1 Arc Lamps

In an arc lamp, electric current is allowed to pass through two electrodes, which are in contact with each other. The result is an arc being struck. The arc maintains the current, and is a very efficient source of light. There are various forms of arc lamps such as carbon-arc, flame-arc or magnetic-arc lamps.

3.7.1.1 Carbon-Arc Lamp

This is one of the oldest types and is still used in projectors used in theatres and search lights. Fig. 3.11 shows a carbon arc lamp. When two hard carbon rods are placed end to end and connected to the terminals of DC supply mains of not less than 45 volts, the current flows through them and the ends of carbon rods soon become incandescence due to high resistance. If they are slightly pulled (say about 2 or 3 mm) an arc will be formed between two carbon rods and white light will be produced. The arc is maintained by transfer of carbon particles from one rod to another one. It is found that these particles travel from the positive carbon rod to negative one. That is why, the

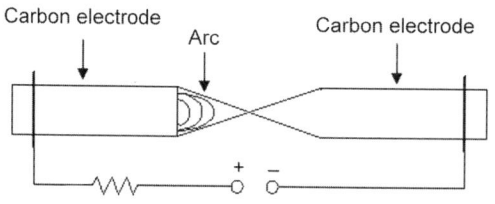

Fig. 3.11: Carbon arc lamp.

positive rod after sometime of use becomes hollow and negative becomes a pointed pencil.

Due to transfer of particles and oxidation both rods burn away by and by. The rate of burning of positive rod is twice of that of negative rod, that is why, the x-section of the positive rod is made twice of that of negative one. In case of AC supply the rate of burning of both rods is same and, therefore, they are made of equal X-section. Since the carbon is consumed during operation, therefore, requires adjustment of distance between two rods now and then. Except in small lamps, an automatic device is required for feeding the carbon into the arc as they burn away. The voltage required for maintaining the arc is given approximately by V = (39 + 2.8 l) where l is the length of the arc in mm.

The carbon arc is unstable, since as the current increases the amount of vaporised carbon increases, and the resistance decreases so much that the product of current and resistance, i.e., the voltage drop across the arc falls. A series resistance is, therefore, used for stabilising the arc, and this leads to a low overall luminous efficiency. The voltage drop across the arc is about 45 to 60 volts and the supply voltage is about 70 to 100 volts, in case of AC supply an inductor is used in place of series resistance for stabilising the arc, in a carbon arc lamp about 85% of light is given out by the positive carbon, 10% by the negative carbon and 5% by the air. The temperature of positive carbon is between 3,500 and 4,000°C and that of negative one is about 2,500°C. The luminous efficiency of such lamp is 12 lumens per watt.

3.7.1.2 Flame-Arc Lamp

The principle of operation is similar to that of carbon-arc lamp. Schematic diagram of a flame arc lamp is shown in Fig. 3.12. The electrodes of such a lamp has 5 to 15% fluoride (called the flame material) and 85 to 95% carbon. Generally core type carbon electrodes are used and the cavities are filled with fluoride. The fluoride has a characteristic which radiate light energy efficiently from a very high heated arc stream. Fluoride turns into vapour along with the carbon and these fluoride vapours cause a very high luminous intensities. In addition to fluoride, there are also other some flame materials. Different flame materials will produce different colours. Most of these colours do not appeal to eyes and at the same time they produce strain on them. The arc can be drawn

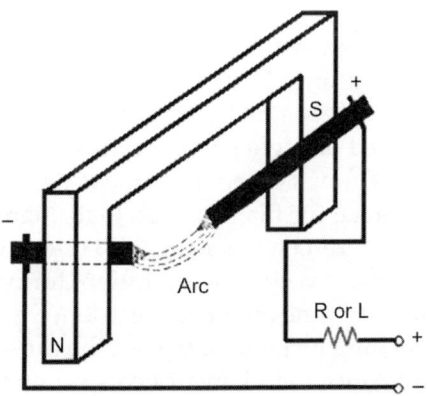

Fig. 3.12: Flame arc lamp.

out to one side with the help of a magnet. Though the arc is very efficient but owing to objection to its colours it has now been superseded by the electric discharge lamps. The luminous efficiency of such a lamp is 8 lumens per watt.

3.7.1.3 Magnetic-Arc Lamp

In such a lamp positive electrode made of copper and negative electrode made of magnetic oxide of iron are used. The arc is struck in the similar way as in case of carbon-arc lamp. Such lamps are rarely used.

3.7.2 Incandescent Lamps

Earlier the arc lamps were in use for general lighting purpose for a short-time but as the carbon arc lamps were complicated so these have been superseded by filament lamps for general lighting.

When an electric current is passed through a fine metallic wire, heat is produced and the temperature of the wire increases. At low temperatures the wire radiates heat energy as the temperature of the wire increases due to heating it radiates heat as well as light energy. The higher the temperature of the wire, higher is the amount of light energy radiated. A black body when heated to 6,250°C emits the maximum energy in the visible spectrum range.

The incandescent or filament type lamp consists of a glass globe completely evacuated and a fine wire, known as filament, within it. The glass globe is evacuated to prevent the oxidization and convection currents of the filament and also to prevent the temperature being lowered by radiation.

The material, which can be used for the filaments of incandescent lamps. must possess the properties of high melting point, low vapour pressure, high resistivity, low temperature coefficient, ductility and sufficient mechanical strength to withstand vibrations during use.

The materials which can be used for the filaments in incandescent lamps are carbon, osmium, tantalum and tungsten. Though the, carbon has a melting point of 3,500°C but its main drawback is that it starts vaporizing at very fast rate if it is operated beyond 1,800°C moreover its temperature coefficient of resistance is negative, i.e., its resistance decreases with the rise in temperature and, therefore, it takes more current from mains. Because of its low operating temperature its efficiency is quite low and is of the order of 3.5 lumens per watt. Osmium is very rare and expensive metal. The melting point of osmium is 2,600°C. Average efficiency of osmium lamp is of the order of 5 lumens per watt. The melting point of tantalum is only 2,800°C and, therefore, it cannot be operated at higher temperature to give more efficiency. The efficiency of tantalum filament lamps is about 5 lumens per watt. Nowadays tungsten is the most commonly used metal for filament due to its high melting point (3400°C), high resistivity, low temperature coefficient (0.0051), low vapour pressure, being ductile and mechanically strong to withstand vibration during use. The hot resistance of tungsten filament is about 15 times the cold resistance and, therefore there is an inrush of current at the switching instant. However, the inrush current is not beyond 15 times the normal current. Since the filament heats up immediately and resistance increases almost instantaneously, so the inrush current attains a maximum value in 0.003 second. The average efficiency of tungsten filament lamp is about 10 lumens per watt. The lights spectrum of an incandescent lamp is continuous, i.e. it contains all the colours, but contains relatively excess of red and yellow radiations and less of blue and violet radiations.

☞ Aging Effects

The light output of an incandescent lamp decreases gradually. As a tungsten lamp ages, its light output decreases for two reasons. Evaporation of the filament tends to cause the bulb to blacken. Also, evaporation makes the filament slowly decrease in diameter, which means that the resistance of the filament increases. Therefore, an old filament draws less current and operates at a lower temperature, which reduces its light output. In turn the efficiency of the lamp (lumens output/watt input) also decreases with the passage of time. The current drawn and the power consumed by the filament decreases at the same rate as the lamp ages. However, the efficiency decreases about four times as fast, and the light output decreases approximately five times as fast. The total depreciation of light output is roughly 15% over the useful life range.

☞ Effects of Voltage Variations

The operating characteristics of an incandescent lamp are materially affected by departure from its normal operating voltage. An increase of 5% in operating voltage increases the lumens output by 20% but shortens the life of the lamp by 50%, on the other hand, reduction of 5%, in operating voltage causes a reduction of approximately 20% in lumens output but doubles the life of the lamp. The efficiency of lamp (lumens/watt) increases with the increase in voltage owing to increase in temperature and is proportional to the square of the voltage. 1% change no applied voltage, modifies lamp wattage by 1.5%, efficiency by 2%, lumens output by 3.5%. The normal life of a filament lamp is about 1,000 working hours. The variations in power consumption, lumens output, efficiency and life of incandescent lamps with the variation in voltage are shown in Fig. 3.13.

Analytically these relationships at are given as below:

Lumens output $\propto (V)^{3.55}$

Power consumption $\propto (V)^{1.55}$

Luminous efficiency $\propto (V)^2$

Life $\propto (V)^{-13}$ for vacuum lamps.

$\propto (V)^{-13}$ for gas-filled lamps.

☞ Advantages

1. Operating power factor unity.
2. Direct operation on standard distribution voltage.
3. Availability in various shapes and shades.
4. Good radiation characteristic in the luminous range.
5. No effect of surrounding air temperature.

☞ Filament Dimensions

Depending upon the voltage and wattage the diameter of a tungsten lamp filament may be as small as 10 microns (about one-sixth of diameter of human hair).

The influence of the diameter of the filament on the heating is found easily if it is assumed that the heat lost by convection is negligible compared with that lost by radiation.

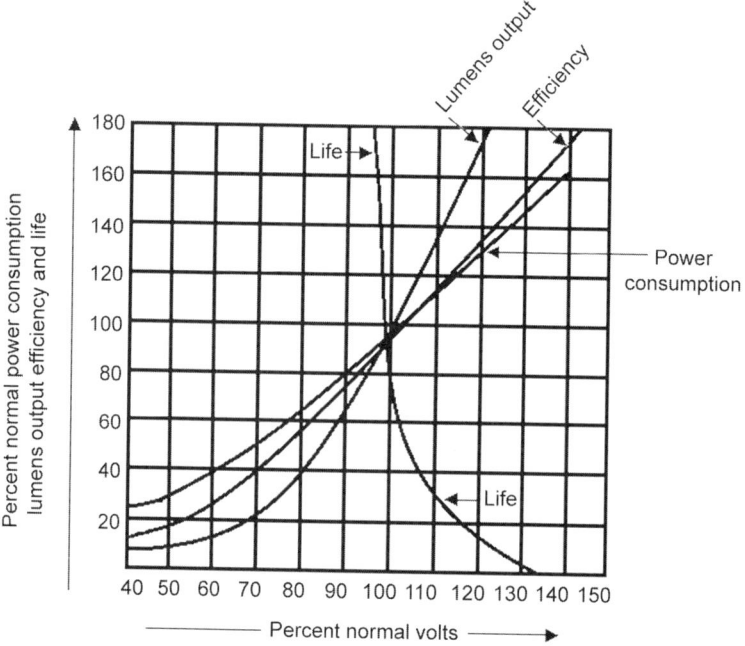

Fig. 3.13: Variations in power consumption, lumens output, efficiency and life of incandescent lamps with the variation in voltage.

The power absorbed by the filament $= I^2R = I^2R = I^2\rho\dfrac{l}{a} = I^2\rho\dfrac{l}{\dfrac{\pi}{4}d^2} = \dfrac{4\rho l}{\pi d^2}I^2$

where I is filament current in ampere. ρ is the resistivity of filament material at the operating temperature, l is length of the filament and d is diameter of the filament.

The power radiated by the filament $eK \times$ surface area $\times \left(T_1^4 - T_0^4\right)$

$$= eK \times \text{surface area } dl \times \left(T_1^4 - T_0^4\right)$$

where e is the emissivity of the surface, K is a constant, T_1 is the temperature of filament and T_0 is the temperature of the surrounding medium.

Since the power absorbed = Power radiated

or $\qquad\qquad\qquad I = d^{3/2}\sqrt{\dfrac{eK\pi^2}{4\rho}\left(T_1^4 - T_0^4\right)}$

For two filaments of same material operating at the same temperature the diameter d is proportional to $I^{2/3}$

3.7.2.1 Gas-Filled Lamps

Construction of a filament lamp is shown in Fig. 3.14. A metal filament can work in an evacuated bulb up to 2,000°C without oxidation and if it is worked beyond this temperature it vaporizes quickly and blackens the lamp. For higher efficiency it is necessary to use working temperature more than 2,000°C keeping down the evaporation,

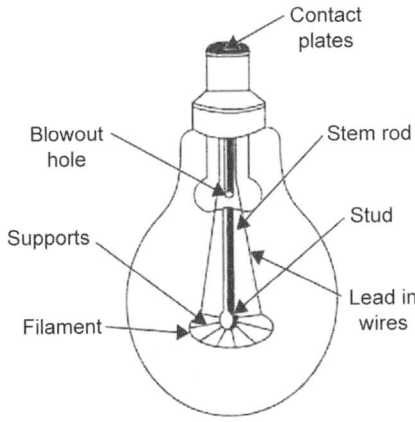

Fig. 3.14: Gas-filament lamp.

which is possible by filling the bulb with an inert gas—argon with a small percentage of nitrogen. Nitrogen is added to reduce the possibility of arcing. Krypton is the best gas for this purpose but it is so expensive that it is used only in special purpose lamps, such as miners lamps. The tungsten filament can safely be burn at temperature of 2,400 to 2,750°C according to the size of the lamp. However, due to presence of gas there is heat loss due to convection currents. This loss depends upon the surface area of filament. As such coiled coil filaments, which take much less space compared with coiled filaments are used with such lamps. A coiled coil filament is made by winding tungsten wire on a fine iron wire to form a spiral which is again wound on to a thick iron wire to form a coiled coil (iron is later on dissolved out by acid). Helical filaments also have a slower rate of tungsten evaporation. Figs 3.15a and b is shown a coiled filament and coiled coil filament respectively. This evaporation eventually causes bulb blackening because the tungsten vapour condenses as a black film on the inner surface of the bulb. In a gas-filled lamp, the hot gas carries the tungsten vapour upward. Therefore a black spot forms at the top of the bulb instead of spreading over the entire inner surface, as in a high-vacuum bulb. Chemicals called 'getters' are often placed inside the bulb to capture tungsten vapour and thereby reduce the rate of blackening. A piece of wire mesh called a collector grid may also be attached to each lead-in-wire to attract the particles of tungsten vapour. Efficiency of coiled coil lamp is about 12 lumens/watt.

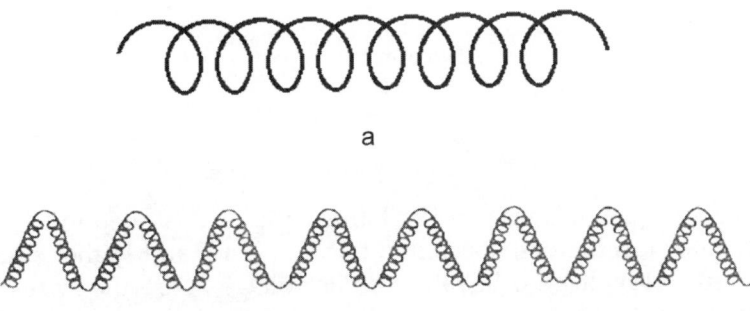

Figs 3.15a and b: (a) Coiled filament, (b) Coiled coil filament.

For low wattage lamps, however, the heat loss due to introduction of gas is more than in medium wattage lamps, so for low wattage (up to 40 watts) vacuum type lamps are used.

3.7.2.2 Halogen Lamp

The halogen lamp is the latest member in the family of incandescent lamps. It possesses numerous advantages over the ordinary incandescent lamp. As already stated, the life and efficiency of an incandescent lamp fall off with use-partly due to slow evaporation of the filament and partly due to black deposit formed on the inside of the bulb. The addition of a small amount of halogen vapour to the filling gas restores part of the evaporated tungsten vapour back to the filament by means of a chemical reaction, i.e. there is a sort of 'regenerative cycle'.

Halogen lamps possess the following advantages:
  i. No blacking of lamp, hence no depreciation of lumens output.
 ii. High operating temperature with increased luminous efficiency varying from 22 lumens/watt-33 lumens/watt.
iii. Reduced dimension of lamps-miniature size.
 iv. Long life-2000 hours.
  v. Better colour radiation.

Halogen lamps, which are being manufactured in sizes up to 5 kW, are suitable for outdoor illumination of building, playing fields, large gardens, fountains, car parks, airport runways, etc. and for lighting of public halls, factories, sports halls, photo film, and TV studio, etc.

3.7.3 Gaseous Discharge Lamps

Incandescent lamp has two disadvantages—low efficiency and coloured light. Gaseous discharge lamps have been developed to overcome these drawbacks.

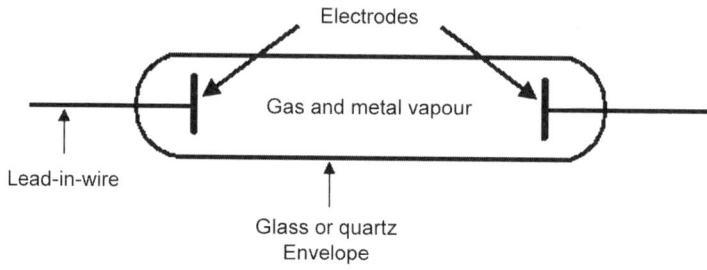

Fig. 3.16: Gaseous discharge lamp.

The basic principle of a gaseous discharge lamp is illustrated in Fig. 3.16. Gases are normally poor conductors, specially at atmospheric and higher pressures, but application of suitable voltage, called the ignition voltage, across the two electrodes can result in a discharge through the gas, which is accompanied magnetic radiation. The wavelength of this radiation depends upon the gas, its pressure and the metal vapour used in lamp. Argon gas and sodium and mercury vapours are commonly employed in the manufacture of gaseous discharge lamps.

Once the ionisation has commenced in the gas, it has a tendency to increase continuously accompanied by a fall in the circuit resistance, i.e., gaseous discharge lamp possesses a negative resistance characteristic. In order to limit the current to a safe value use of a choke or a ballast is made. The choke performs the dual functions of providing the ignition voltage initially, and limiting the current subsequently. Since due to use of choke the power factor becomes poor (0.3–0.4), therefore, in order to improve the power factor of the gaseous discharge lamp use of a condenser is made. The light spectrum obtained is, however, discontinuous (i.e., it consists of one or more coloured lines). The colour of the light obtained depends upon the nature of the gas or vapour used. Discharge lamps are of two types:

i. Those which give the light of the same colour as produced by the discharge through the gas or vapour such as sodium vapour, mercury vapour and neon gas lamps.

ii. Those which use the phenomenon of fluorescence and are known as fluorescent lamps. In these lamps, the discharge through the vapour produces ultraviolet waves which cause fluorescence in certain materials called phosphors. The inside of a fluorescent lamp is coated with a phosphor which absorbs invisible ultraviolet rays and radiates visible rays. Example is fluorescent mercury-vapour tube.

The gaseous discharge lamps are, in general, considered superior to metal filament lamps. However, they suffer from the following drawbacks:

i. High initial cost and poor power factor.

ii. Starting is somewhat complicated requiring starters in some cases and transformer in others.

iii. These take time to attain full brilliancy.

iv. Ballasts are necessary for stabilising the current since, such lamps have negative resistance characteristic.

v. Light output fluctuates at twice the supply frequency. The flicker causes stroboscopic effect.

vi. These lamps can be used only in particular position.

3.7.3.1 Sodium-Vapour Discharge Lamp

The sodium vapour discharge lamp consists of a bulb containing a small amount of metallic sodium, neon gas, and two sets of electrodes connected to a pin type base. Circuit connection is shown in Fig. 3.17a. The neon gas initiates the discharge and also helps to develop enough heat to vaporize the sodium. Since long discharge paths are necessary, therefore, the discharge envelope is usually bent into U shape. The lamp operates at a temperature-like 300°C and in order to conserve the heat generated and assure the lamp operating at normal air temperatures the discharge envelope is enclosed in a special vacuum envelope designed for this purpose. The lamp must be operated horizontally, or nearly so, to keep the it idiom well spread out along the tube, although some small lamps may be omitted vertically, lamp cap up. Care should be taken in handling these lamps, particularly when replacing inner U-tube, for if it is broken and sodium comes in contact with moisture fire will result. The sodium-vapour lamp is only suitable for alternating current, and therefore, require choke control. This requirement is met by operating the lamp from a stray field, step-up, tapped autotransformer with an open circuit secondary voltage of 470 to 480 volts. The

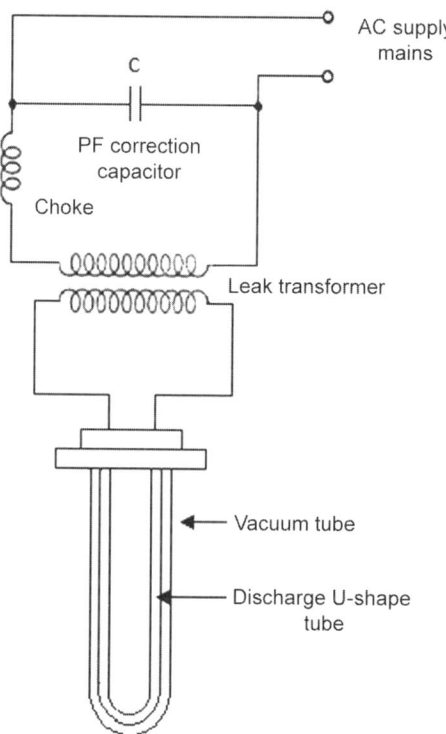

Fig. 3.17a: Circuit connection of a sodium vapour lamp.

uncorrected power factor very low, about 0.3, and capacitor must be used to improve the power factor to about 0.8.

When the lamp is not in operation, the sodium is usually in the form of solid deposited on the side walls of the tube, therefore, at first when it is connected across the supply mains the discharge takes place in the neon gas and gives red-orange glow. The metallic sodium gradually vapourizes and then ionizes, thereby producing the characteristic mono-chromatic yellow light, which makes objects appear as grey. The lamp will come up to its rated light output in approximately 15 minutes. It will restart immediately should the power supply be momentarily interrupted since the presence of vapour is quite low and the voltage is sufficient to restrike the arc.

The wiring diagram of sodium vapour lamps is given in Fig. 3.17b. The efficiency of a sodium vapour lamp under practical conditions is about 40–50 lumens/watt. The major application of this type of lamp is for high way and general outdoor lighting where colour discrimination is not required, such as street lighting, park, rail yard, storage yard, etc. Such lamps are manufactured in 45, 60, 85 and 140 watt ratings. The average life is about 3000 hours and is not affected by voltage variations. At the end of this period the light output will be reduced by 15% due to aging.

The lamp fails to operate when

   i. the filament breaks or burns out,

  ii. the cathode stops to emit electrons,

 iii. the sodium particle, may concentrate on one side of the tube,

 iv. the lamp tube is blackened owing to sodium vapour action on the glass in which case the output will be reduced.

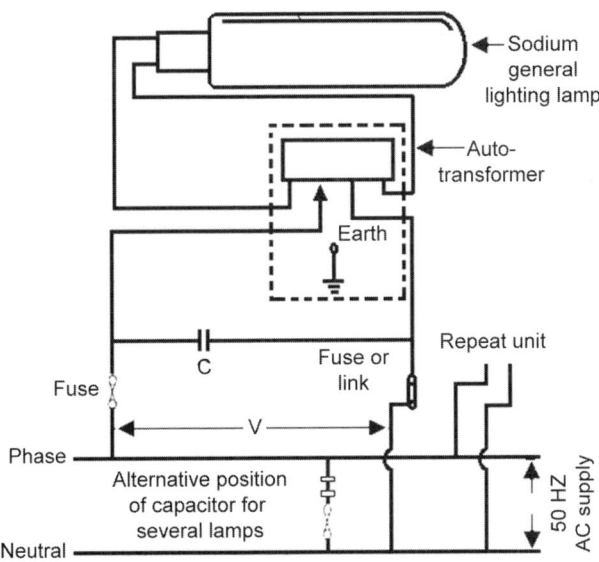

Fig. 3.17b: Wiring diagram of sodium vapour lamp.

3.7.3.2 *High-Pressure Mercury-Vapour Discharge Lamp*

The extensive use of the mercury vapour discharge lamp depends entirely upon the versatility of the mercury vapour because with changes in pressure, temperature, voltage and other characteristics, spectral quality and efficiency of the lamp changes.

The mercury-vapour discharge lamp is similar in construction to the sodium-vapour discharge lamp. It consists of a discharge envelope enclosed in an outer bulb of ordinary glass. The discharge envelope may be of hard glass or quartz. The space between the bulb is partially or completely evacuated to prevent heat loss by convection from the inner bulb. The outer bulb absorbs harmful ultraviolet rays. The inner bulb contains argon and a certain quantity of mercury. In addition to two main electrodes a starting (auxiliary) electrode is connected through a high resistance (as shown in Fig. 3.18a) is also provided. Circuit connection of mercury vapour discharge lamp is shown in 3.18b. The main electrodes are made of tungsten wire in the shape of helix. In this case no separate heater is required for the cathode which is heated by the constant bombardment of the heavy mercury ions.

The lamp has to have auxiliary equipment for use with standard mains voltage, and the necessary connections are shown in Fig. 3.18c. The choke is provided to limit the current to a safe value. This choke lowers the power factor, so a capacitor is connected across the circuit to improve the power factor. These lamps must be operated vertically, since if they are used horizontally convection will cause the discharge to touch the glass bulb, which will fail. Lamps which are intended to operate horizontally are fitted with a magnetic device which will hold the luminous column central.

When the supply is switched on, full mains voltage is applied between the auxiliary electrode and neighbouring main electrode: this breaks down the gap and a discharge through the argon gas takes place. This enables the main discharge to commence. As

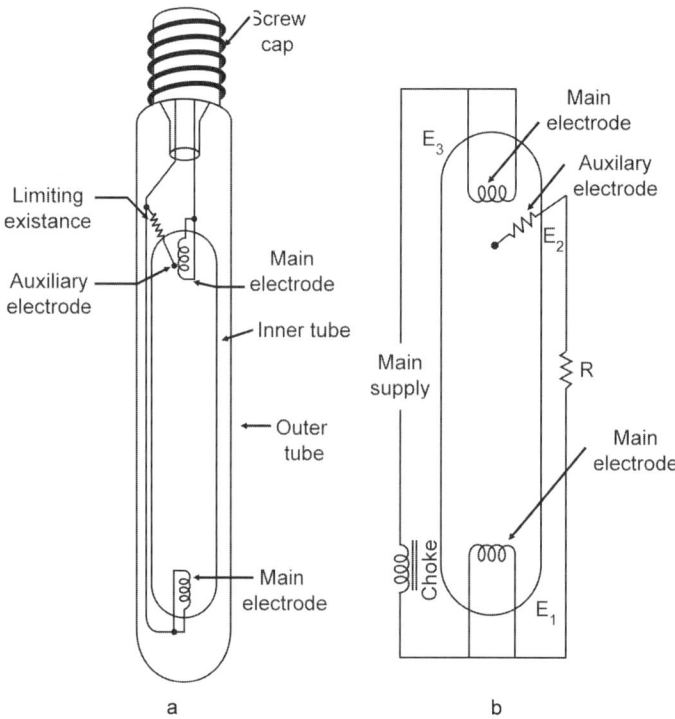

Figs 3.18a and b: (a) Mercury vapour discharge lamp, (b) Circuit connection of mercury vapour discharge lamp.

the lamp warms up mercury is vaporised, increasing the vapour pressure and the luminous column becomes brighter and narrower. The lamp requires 4 or 5 minutes to attain full brilliancy. If the supply is interrupted, the lamp must cool down and the vapour pressure be reduced before it will start. This takes 3 or 4 minutes. The temperature of operation inside the inner bulb is about 600°C. It gives greenish blue colour light, which causes colour distortion. The efficiency is about 30–40 lumens/watt. These lamps are manufactured in 250 and 400 watt ratings for use on 200–250 volts AC supply mains. The pressure of vapour in these lamps is 2–3 atmosphere. Lamps of this type are used for general industrial lighting, railway yards, ports, work area, shopping centres, etc., where greenish-blue colour light is not objectionable. The lamp described above is MA type. Another type, which is manufactured in 300 and 500 watt rating for use on AC as well as DC supply mains, is MAT type. This is similar to MA type except that choke is not used as a ballast. Space between two tubes instead of being evacuated consists of a tungsten filament in series with a discharge tube which acts as a ballast. When the supply is switched on, it operates as a filament lamp, its full output being given by the outer tube. At the same time the discharge or inner tube begins warming up and at a particular temperature a thermal switch operates cutting a part of the filament and thereby increasing the voltage across the discharge tube. The filament contributes a considerable portion of red rays. The combination of the rays from the filament and the blue radiations from the discharge tube produce a useful colour. As the filament acts as a resistance, the overall power factor of the lamp is about 0.95 and therefore, capacitor is not required.

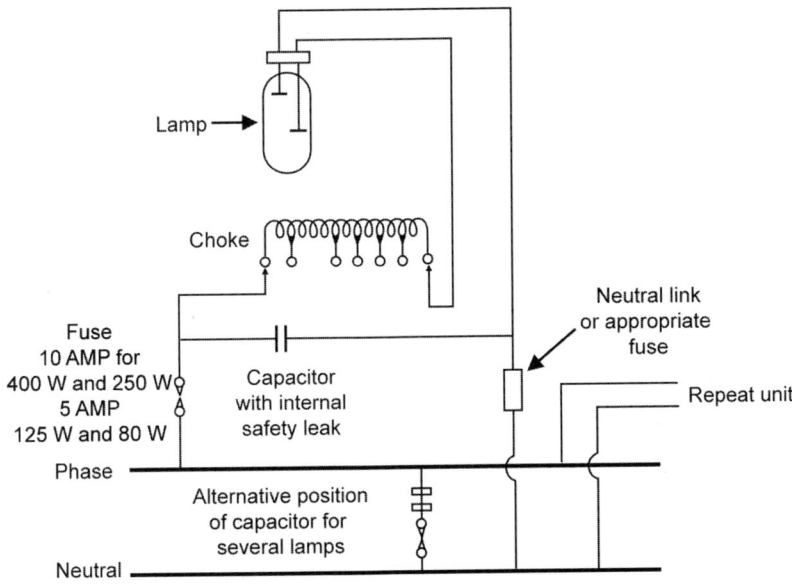

Fig. 3.18c: Wiring diagram of mercury vapour discharge lamp.

Lower wattage lamps, such as 80 and 125 watts, are manufactured in a different design and using high vapour pressure of about 5–10 atmosphere. These are known as MB types. These operate in is manner similar to MA type except that resistance in series with starting electrode is large and outer bulb is of quartz, (not of ordinary glass) in order to withstand high temperature so that these lamps can be used in any position.

3.7.3.3 Mercury Iodide Lamps

These lamps are similar in construction to high pressure mercury vapour lamps but in addition to mercury, a number of iodides are added which fill the gaps in the light spectrum, and thus, improve the colour characteristic of the light. Their efficiency is also higher (75—90 lumen/watt). A separate ignition device, in addition to the choke, is required for the mercury-iodide lamp. Such lamps are suitable for application in the fields of flood-lighting, industrial lighting and public lighting.

3.7.3.4 Neon Lamp

It is a cold cathode lamp and consists of a glass bulb filled with neon gas with a small percentage of helium, these lamps give orange pink coloured light. Electrodes are of pure iron and are spaced only few mm apart so that lamps can be made for voltages as low as 110 volts AC or 150 DC. For use on AC the electrodes are of equal size. On DC the gas glows near the negative electrode, therefore, negative electrode is made larger in size. The efficiency of neon lamp lies between 15–40 lumens/watt. Owing to discharge of the gas between the electrodes in the form of an arc, it may cause the current drawn by the lamp to increase indefinitely. This is prevented by connecting a high resistance of few thousand ohms in series and mounting it in the cap. The lamp of this type is of the size of an ordinary incandacent lamp. The power consumption is of the order of 5 watts. Fig. 3.19 shows a circuit of neon lamp.

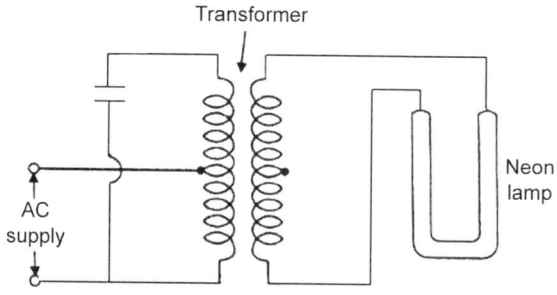

Fig. 3.19: Neon lamp.

Neon lamps are used as indicator lamps, night lamps, for determination of polarity of DC mains and in larger sizes as neon tubes for the purpose of advertising.

☞ Neon Tubes

The popularity of high-voltage neon lighting arose almost entirely from its use in advertising, for signs, or in the decorative treatment of buildings, but later the lighting field became important. The neon tube, which is used in varying lengths up to about 9 metres, may be bent into almost any desired shape during manufacture. It consists of a length of glass tubing containing two electrodes, normally cylindrical in shape, of iron, steel, or copper.

The true neon tube contains neon, but the term is now used also for tubes with fillings of other rare gases. By varying the composition of glass and adding different substances to neon gas different colours such as orange, red, yellow, green, etc. are obtained. The diameter, of the tube vary, and common sizes of 10, 15, 20 and 30 mm carry currents of 25, 35, 60, and 150 mA respectively. Voltage required may vary from 300 to 1,000 V per metre of tube length and for starting the discharge a striking voltage, about 1½ times this value is required. Such voltage is obtained by making use of a step-up transformer having a high leakage reactance so that it gives a drooping characteristic. The usual operating voltage it 6,000 volts.

The tubes are mounted either on a wooden frame or a metal base. These are matched with step-up transformers by connecting suitable tappings for the rated current. Connections between letters are made by nickel wires, the glass tubings being slipped over them. The power factor of neon tubes is quite low and is improved by using capacitors. The capacitors can, however, be placed only on the low voltage side of the transformer.

☞ Faults

Flickering may be due to:
  i. Transformer secondary voltage too low; this can be rectified by adjustment of the transformer tappings.
  ii. Reduction a gas pressure in tube due to absorption of the gas into the electrodes, the tube may be removed and refilled by the manufacturers.

Neon tubes installed in the open requires frequent cleaning, say 4 times per year. Care must be taken that the key of the opened lock switch is removed and held by the operator while work is being done.

3.7.4 Fluorescent Tubes

Fluorescent lighting has a great advantage over other light sources in many applications. The tubes can be obtained in a variety of lengths, with illumination in a variety of colours. It is possible to achieve quite high lighting intensities without excessive temperature rise and, owing to the nature of light sources, the danger of glare is minimised. It must not, however, be thought that fluorescent lamps can be used indiscriminately without giving careful consideration to the type of reflector to be employed or to the correct positioning of the light source. An exposed tube in the line of vision is usually as bad as a tungsten lamp would be. The efficiency of fluorescent lamp is about 40 lumens per watt, about three times the efficiency of an equivalent tungsten filament lamp. The fluorescent tube consists of a glass tube 25 mm in diameter and 0.38–1.52 m in length. The inside surface of the tube is coated with a thin layer of fluorescent material in the form of a powder.

The coating materials used depend upon the colour effect desired and may consist of zinc silicate, cadmium, silicate or calcium tungstate. These organic chemicals are known as phosphors which transforms short-wave invisible radiations into visible light. By mixing the various powders light of any desired colour including day light can be obtained. The tube contains small quantity of argon gas at a pressure of 2.5 mm of mercury and one or two drops of mercury. It is provided with two electrodes coated with electron emissive material. A starting switch is provided in the circuit, which puts the electrodes directly across the supply mains at the time of starting, so that electrodes may get heated and emit sufficient electrons. A stabilising choke is connected in series with it, which acts as a ballast in running condition and provides a voltage impulse for starting. A capacitor is connected across the circuit to improve the power factor.

Fluorescent tubes are available in the following sizes:

Table 3.1	Specification of Fluorescent tubes		
Length	*Wattage*	*Length*	*Wattage*
38 cm	14 W	122 cm	40 W
46 cm	15 W	152 cm	65 W
61 cm	20W	152 cm	80 W
100 cm	25 W		

The fluorescent lamp in common with most electrical devices may cause a certain amount of radio-interference. This interference may be caused due to one of the following factors:

1. Direct radiation from the bulb to the antenna. This effect diminishes rapidly as the radio is separated from the lamp. Thus, for example, at a radius of 2.8 metres interference from this cause is negligible.
2. Line radiation from the electric supply line to the antenna.
3. Line feedback from the lamp through the line to the radio.

The radio-interference effect may be reduced to a minimum by connecting a small capacitor (0.05 µF) across starter terminals, as shown in Figs 3.20a and b.

The starting switches are of two types, namely the thermal type and the glow type.

The connections of a fluorescent tube incorporating a thermal type starter are shown in Fig. 3.20a. The thermal starter is a current operated device and consists of two metallic strips and a heater coil. The bimetallic strips are in contact with each other when the lamp is not in operation. When the supply is switched on, the two electrodes get connected in series through the thermal switch, the relatively large current rising them to incandescence. The current also flows through the heater element as a result of which bimetallic strips break contact. This causes interruption in the current flowing through the circuit, which further results in a high voltage surge across the electrodes of the tube, which is enough to strike the arc between the electrodes. This arc is then maintained by the normal lamp voltage. Bombardment of the electrode's surface by the positive mercury ions maintains their temperature so that they continue to emit electrons. Thus, the tube is put in operation. The thermal switch is now generally obsolete because of its more complicated construction, greater cost and greater power loss.

The connections of a fluorescent tube incorporating a glow type starter are shown in Fig. 3.20b. The glow type starter is a voltage operated device and consists of two bimetallic electrodes enclosed in a glass bulb filled with a mixture of helium and hydrogen. Normally the contacts are open. When the supply is switched on, the potential across bimetallic electrode causes a small glow discharge at a small current not enough to heat up the tube electrodes (filament). This discharge is enough, however, to heat the bimetallic strips of the switch causing them to bend and make contact. The result is a large current through the electrodes, their temperature being raised to incandescence and the gas in the immediate neighbourhood is ionised. After one or two seconds the bimetallic strips cool down and the contacts open. This opening of

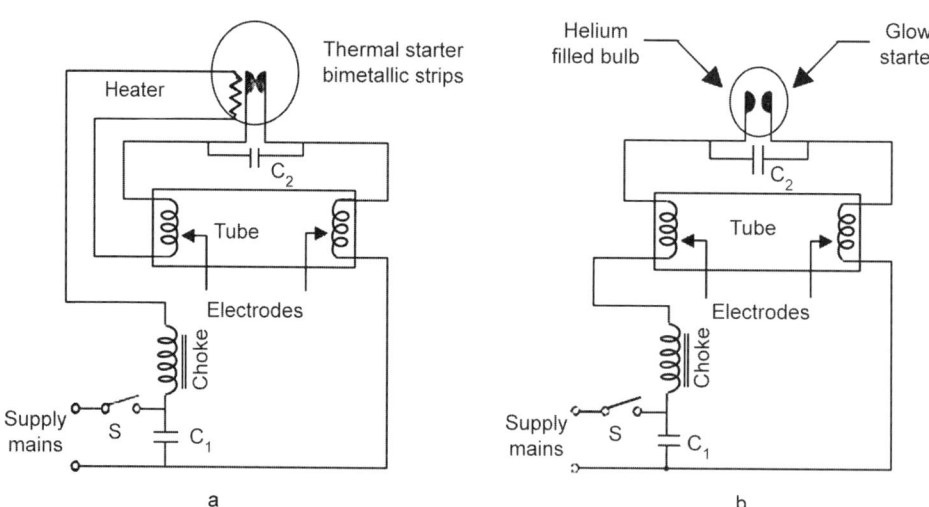

Figs 3.20a and b: Connection diagram for fluorescent tube: (a) Incorporating a thermal type starter, (b) Incorporating a glow type starter.

contacts in series with the choke causes a momentary high voltage, which is sufficient to start the discharge in the main tube. The starter ceases to glow as the voltage is now too low. A small capacitor is placed inside the starter to suppress arcing and radio-interference.

☞ Startless Fluorescent Lamp Circuit

Nowadays starterless circuits, as instant start, rapid start, etc. for such tubes are also employed (Fig. 3.21). The primary winding of an auto-transformer is connected in parallel with the tube and receives practically full mains voltage. The filaments are provided with current from secondary tappings. When the lamp has started the transformer receives the normal tube voltage and the filament currents are correspondingly reduced.

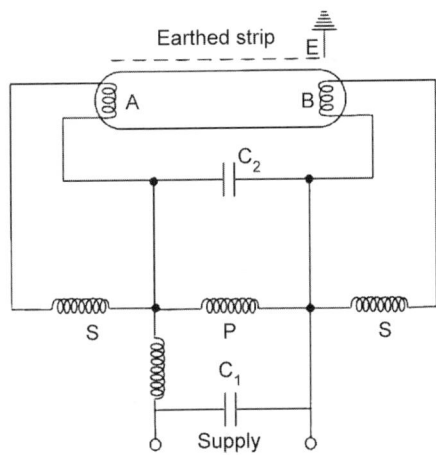

Fig. 3.21: Starterless circuit.

☞ Stroboscopic Effect

At the usual alternating current supply frequency of 50 Hz, a discharge lamp will be extinguished 100 times per second. Although this effect seldom noticeable in normal conditions. It is possible that this may, in some instances, give rise to a stroboscopic effect, that is moving objects such as rotating parts of machinery, illuminated by this light, may appear to be moving in disunity or jerks, or rotating more slowly than their actual speed.

In actual applications, however, where this effect might cause annoyance, it can be practically eliminated in a three lamp unit by connecting each lamp on a separate phase of a 3-phase system and it can be greatly reduced in a two lamp unit by the use of a two lamp control unit, which employs a condenser in the ballast of one of the lamps. (Fig. 3.22a). The current through the lamps is almost 90° out of phase and under these conditions light output of one of the lamps is at a maximum. This method has an additional advantage of giving an overall power factor of nearly unity for the unity of two lamps. In this arrangement one circuit remains at a low power factor at about 0.5 lagging, while the other circuit incorporating a series capacitor C remains at a power factor of about 0.5 leading.

☞ Fluorescent Lamps for DC Supply

In the above discussions, it has been assumed that the supply to the fluorescent lamps is AC But if, the available supply is DC some special accessories and circuit modification will have to be made.

    i. The choke coil holds a low impedance under DC and therefore, a ballast resistance is connected in series with the choke in order to limit the current.

    ii. On systems below 220 V, starting becomes less certain on DC Only thermal type starters should be used.

    iii. The positive end becomes relatively dark on account of the tendency of the mercury vapour to migrate towards the negative end of the tube. In order to

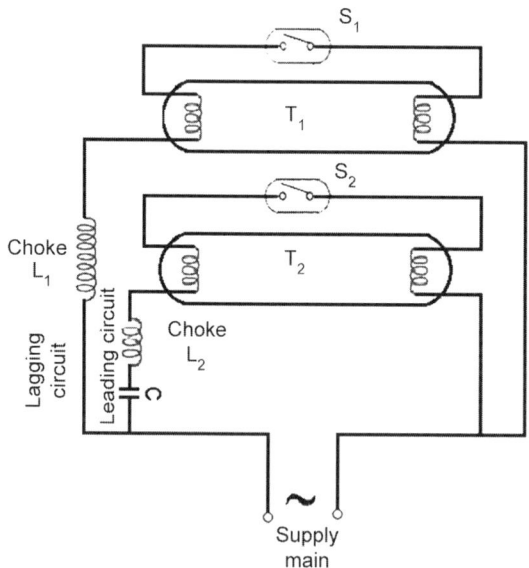

Fig. 3.22a: Two lamp control unit.

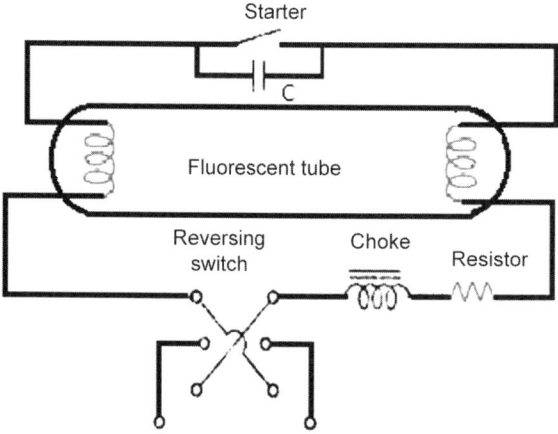

Fig. 3.22b: Operation of fluorescent tube on DC.

overcome this defect a reversing switch is included in the circuit between the supply and the fitting, as shown in Fig. 3.22b.

In DC operation of fluorescent tube there is no problem of power factor correction and no stroboscopic effect. Its disadvantages are—low efficiency due to power loss ballast series resistance, increased cost of the ballast resistance and reversing switch and less life of the tube (about 80% of that with AC operation).

☞ *Colour of Fluorescent Light*

As already mentioned the colour of fluorescent light depends upon the fluorescent powder used and the vapour pressure in the lamp. A few important variations of white colour with their applications are given below:

i. **Daylight:** This is suitable for industrial applications where checking of material is carried out at high illumination levels (say 1,000 lux or more) and for normal lighting service.

ii. **Warm-white:** This is mostly used for street lighting where colour rendition, due to low level of lighting (say 10 lux or less), is of little importance.

iii. **Warm-white deluxe:** It is used for illuminating rooms of large gathering such as shops restaurants, and dwelling houses. It combines well with the light of incandescent lamps, as in show-windows, etc.

iv. **White:** It is useful for use in offices, drawing offices, schools and factories. It harmonizes exceptionally well with the daylight, and therefore, light is a good supplement to daylight.

v. **White-deluxe:** It is very suitable for use in offices, schools and shops where colour rendition is important.

☞ *Useful Lamp Life*

In general, fluorescent lamps lose their usefulness because of reduction in light output before they fail to operate. Darkening of the bulb occurs because of the effect of mercury on the fluorescent coating and because of the material given off by electrodes. The latter specially causes darkening at the ends of the bulb late in life. The rate of depreciation in light output diminishes throughout the life; the first hundred hours produce approximately as much darkening as the following 1,000 hours. Rated output is based on conditions at 100 hours.

Frequent starting of lamps may take more life out of the electrodes than long hours of burning because momentarily there is a higher than normal voltage drop at the electrodes which causes the active material to sputter or evaporate off. If a lamp be started once a minute, for example, the hours of burning will be shorter than normal, but if we switched on and burned continuously, its life will be longer than normal. When the active material on the electrodes is nearly exhausted, the voltage required for starting will rise and may equal or exceed the available supply. This may occur after the lamp has been started thousands of times or burned beyond its rated life. Sometimes the end of life is indicated by the lamp flashing momentarily and then going out.

The normal life of a fluorescent lamp is 7,500 hours. The average life is for three burning hours per switching operation. The actual life may vary from 5,000 to 10,000 hours depending upon the operating conditions. Light output is reduced by 15–20% after 4,000 hours of operation and it is, therefore, a good practice to replace the fluorescent lamp after 4,000–5,000 burning on economical grounds.

☞ *Performance Curves*

Performance curves are shown in Fig. 3.23. The effect of voltage variation in the case of fluorescent lamps is less marked as compared to the incandescent lamps. However, their life and performance are adversely affected both by low and high voltage. With increased voltage there is a greater heating of electrodes and they lose emissive material by evaporation. With reduced voltage, the current reduces causing sputtering at the electrodes shortening their life.

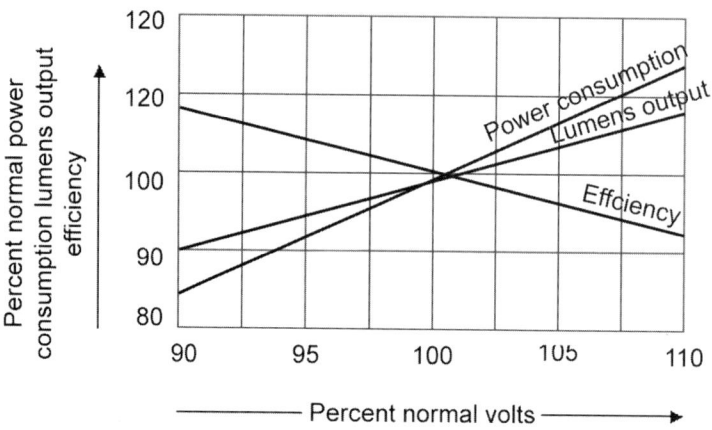

Fig. 3.23: Performance curves of fluorescent lamp.

The best performance of fluorescent lamps is obtained at 20–25°C operating temperature. It decreases rapidly when a lamp is operated at a lower temperature or is exposed to cold wind drafts. The fluorescent lamps may be enclosed in outer envelops for their protection from cold weather in exposed locations. For operation at high temperature, the fittings with provision for air circulation should be employed.

☞ *Operating Instructions*

    i. A fluorescent lamp should start up with little blinking. Blinking indicates a defective starter, low voltage or a defective tube. Both the choke and the starter will get damaged due to overheating, if this condition is allowed to persist. The tube or the starter must be replaced immediately.

   ii. Correct voltage must be ensured.

  iii. The starter should be replaced every time a tube is changed. A bad starter shorten the life of the tube.

  iv. Frequent switching operations should be avoided.

☞ *Merits and Demerits*

As already stated the fluorescent lamps have got merits of high luminous efficiency, long-life, low running cost, low glare level and less heat output. The demerits are stroboscopic effect, small wattage requiring large number of fittings and magnetic hum associated with the choke causing disturbance. The problem of noise can be solved to some extent by mounting chokes on resilient pads.

☞ *Faults*

We will now discuss various faults which inflict fluorescent lamps and indicates the possible cause with remedies therefore of, as follows:

Table 3.2	Troubleshooting & maintenance for fluorescent lamps		
Sl. no.	Symptoms	Possible causes	Remedies
1.	Blinking on and off	1. Low voltage	If required change tapping ballast.
		2. Loose contact	Check all connections.
		3. Defective starter	Replace starter.
		4. Defective tube	Replace tube.
		5. Wrong connections	Check all connections.
2.	End of a tube remains lighted	1. Defective starter	Change starter.
3.	Failure of lamp to light up	1. Defective tube	Change tube.
		2. Defective starter	Change starter.
		3. Defective choke	Change choke.
		4. Defective and loose holders	Turn tube through 90°.
4.	End blackening (early life)	1. Too low or too high voltage	Adjust ballast tapping.
	(normal life)	1. Mercury deposits at the ends	It will evaporate as the lamp operates.
5.	Burn out electrodes	1. Control unit not in circuit or choke short-circuited	Test with new a new choke.
6.	Dark streaks along lamp	1. Mercury globules	Rotate tube through 180°.
7.	Snaking spiraling (with glow type starter)	1. Insufficient heating due to either starter opening quickly or wrong tappings made on the ballast	Renew starter or ballast. In case of new lamp, it cures after few days.

Table 3.3	Comparison between fluorescent tubes and filament lamps		
Sl. no.	Aspect	Fluorescent tubes	Incandescent lamps
1.	Luminous efficiency	40 lm/Watt	10 lm/Watt
2.	Cost	Initial cost is more but more economical	Initial cost is less.
3.	Starting	Starting troubles may be there	No starting troubles
4.	Heating effect	No heat is evolved	Lot of heat is evolved
5.	Life	10,000 hours	1,000 hours
6.	Illumination	Diffused light	Shadows obtained
7.	Variety colours	Large variety of colours obtained	Can produce only a few colours
8.	Brightness	Less	More
9.	Stroboscopic effect	Yes (objectionable)	No
10.	Maintenance cost (overall)	Low	High

3.7.5 Compact Fluorescent Lamp

A compact fluorescent lamp (CFL), also called compact fluorescent (Fig. 3.24) light, energy-saving light, and compact fluorescent tube, is a fluorescent lamp designed to replace an incandescent lamp; some types fit into light fixtures formerly used for

Fig. 3.24: Compact fluorescent lamp.

incandescent lamps. The lamps use a tube which is curved or folded to fit into the space of an incandescent bulb, and a compact electronic ballast in the base of the lamp.

Compared to general-service incandescent lamps giving the same amount of visible light, CFLs use one-fifth to one-third the electric power compared to others, and last eight to fifteen times longer. A CFL is a bit costly than an incandescent lamp, but can compensate over five times its purchase price in electricity costs over the lamp's lifetime. Like all fluorescent lamps, CFLs contain toxic mercury which complicates their disposal. In many countries, governments have established recycling schemes for CFLs and glass generally.

The working principle of a CFL bulb remains the same as in other fluorescent lighting: electrons that are bound to mercury atoms are excited to states where they will radiate ultraviolet light as they return to a lower energy level; this emitted ultraviolet light is converted into visible light as it strikes the fluorescent coating on the bulb (as well as into heat when absorbed by other materials such as glass).

CFLs radiate a spectral power distribution that is different from that of incandescent lamps. Improved phosphor formulations have improved the perceived color of the light emitted by CFLs, such that some sources rate the best "soft white" CFLs as subjectively similar in color to standard incandescent lamps.

3.7.6 LED

A light-emitting diode (LED) is a two-lead semiconductor light source. It is a p–n junction diode, which emits light when triggered. When a specific voltage is applied to the leads, electrons are able to recombine with electron holes within the device, releasing energy in the form of photons. This effect is called electroluminescence, and the color of the light (corresponding to the energy of the photon) is determined by the energy band gap of the semiconductor. An LED (Fig. 3.25) is often small in area and integrated optical components may be used to shape its radiation pattern.

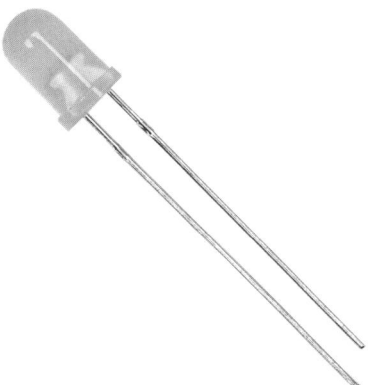

Fig. 3.25: LED

A P–N junction can convert the absorbed light energy into its proportional electric current. The same process is reversed here (i.e. the P-N junction emits light when electrical energy is applied on it). This phenomenon is generally called electroluminescence, which can be defined as the emission of light from a semi-conductor under the influence of an electric field. The charge carriers recombine in a forward P-N junction as the electrons cross from the N-region and recombine with the holes existing in the P-region. Free electrons are in the conduction band of energy levels, while holes are in the valence energy band. Thus, the energy level of the holes will be lesser than the energy levels of the electrons. Some part of the energy must be dissipated in order to recombine the electrons and the holes. This energy is emitted in the form of heat and light.

White LED lamps now compete with CFLs for high-efficiency house lighting.

3.8 REQUIREMENTS OF GOOD LIGHTING

Good lighting makes seeing easy and is comfortable the eye, following factors affect good lighting:

a. **Illumination level:** Degree of illumination is important as it gives the necessary brightness to the object to be viewed. The Table 3.4. gives the recommended illumination level.

b. **Contrast:** Background brightness against which the eye locates the object is the contrast. In case the contrast is too much, eye gets irritated. Proper contrast is then important.

c. **Shadow:** Long and sharp shadows cause rapid fatigue of eye. This should be avoided. Avoidance of long and sharp shadows is possible by

  i. using high mounting heights of lamps

  ii. employing large number of lamps

  iii. employing wide surface source of light (e.g. fluorescent lamps).

d. **Glare:** If the light enters directly to the eye in large quantities, it produces glare. The glare may be due to the source itself or may be due to presence of polished surface causing reflection of light and being incident directly on the eye. Glare

Sl. no.	For	Recommended illumination level	Degree of illumination
Table 3.4	**Recommended illumination level**		
1.	Entrance way, toilets, suburban roads, yards, thorough fares, ordinary city roads, factory, etc.	Up to 50 lux	Very low illumination level.
2.	Staircase, corridors, bedrooms, parking areas, warehouses, electrical equipment rooms, bathrooms, etc.	Up to 100 lux	Low illumination level.
3.	Living room, reception room, lecture hall, reading hall, restaurant, lounge, museum, hospital indoor, assembly hall, etc.	Up to 150 lux	Medium illumination level.
4.	Kitchen, bakeries, libraries, work benches, machine room, etc.	Up to 200 lux	High illumination level.
5.	Office, sustained reading, typing, cash counter, drawing office laboratories, indoor game place, etc.	Up to 300 lux	Very high illumination level
6.	Seeing, testing laboratory, instrument repair and manufacturing. Drawing board, electronic work, etc.	Up to 700 lux	Extremely high illumination level.

can be avoided by increasing the height of the lamp and by using proper reflections.

e. **Colour rendering:** Light produced by sun or incandescent source consists of all wavelength in visible spectrum. Radiations of different wavelengths produce sensation of different colours. Thus, when a white light hits an object, the object absorbs all the radiations except one which will, after reflection, enter he eye producing the sensation of the colour of the object. As for example, if a white light falls on a body of blue colour, it will absorb all components of wavelengths of the white light except the wavelength corresponding to blue colour. Colour rendering roperty of incandescent light and fluorescent light is better than that of sodium or mercury vapour lamps.

3.9   DIFFUSING AND REFLECTING SURFACES: GLOBES AND REFLECTORS

When light falls on polished metallic surfaces or silvered surfaces, then most of it is reflected back according to the laws of reflection (Fig. 3.26a) i.e. the angle of incidence is equal to the angle of reflection. Only a small portion of the incident light is absorbed and there is always the image of the source. Such reflection is known as specular reflection.

However, as shown in Fig. 3.26b, if light is incident on coarse surfaces like paper, frosted glass, painted ceiling, etc., then it is scattered or diffused in all directions, hence no image of the source is formed. Such reflection of light is called diffuse reflection. A perfect diffuser is one that scatters light uniformly in all directions and hence appears equally bright from whatever direction it is viewed. A white blotting paper is the nearest approach to a diffuser.

By reflecting factor of a surface is meant the ratio $= \dfrac{\text{Reflected light}}{\text{Incident light}}$

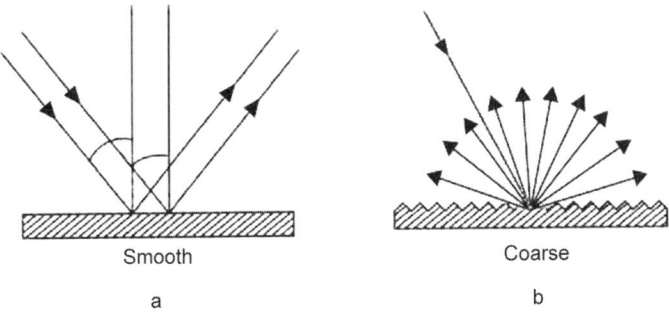

a   b

Figs 3.26a and b: (a) Specular reflection (b) Diffuse reflection.

It is also known as reflection ratio or coefficient of reflection of a surface.

If the light is incident on a transparent surface, then some of it is absorbed and greater percentage of it passes through and emerges on the other side.

To avoid direct glare from electric arcs and incandescent filament lamps, they are surrounded more or less completely by diffusing shades or globes. In addition, a reflector may also be embodied to prevent the escape of light in directions where it serves no useful purpose. In that case, so far as the surroundings are concerned, the diffusing globe is the source of light. Its average brilliancy is lower the more its diffusing area. Depending on the optical density, these globes absorb 15 to 40% of light emitted by the encircled bulb. The bulbs may also be frosted externally by etching or sandblasting but internal frosting is better because there is no sharp scratching or cracks to weaken the glass.

In domestic fittings, a variety of shades are used whose main purpose is to avoid glare. Properly designed and installed prismatic glass shades and holophane type reflectors have high efficiency and are capable of giving accurate predetermined distribution of light.

Regular metallic reflection is used in search-light mirrors and for general lighting purposes. But where it is used for general lighting, the silvered reflectors are usually fluted to make the illumination as uniform as possible.

Regular cleaning of all shades, globes, and reflectors is very important otherwise the loss of light by absorption by dust, etc., collected on them becomes very serious.

Various types of reflectors are illustrated in Figs 3.27 to 3.31. Fig. 3.27 shows a holophane stiletto reflector used where extensive, intensive or focussing light distribution is required.

The optical combination of a lamp, reflector and a lens plate, as shown in Fig. 3.28, provides a high degree of light control. Multiple panels can be conveniently incorporated in fittings suited to different architectural schemes.

The dispersive reflector of Fig. 3.29 is suitable for practically all classes of industrial installations. The reflector is a combination of concave and cylindrical reflecting surfaces in the form of a deep bowl of wide dispersive power. It gives maximum intensity between 0° and 45° from the vertical.

The concentrating reflector of parabolic form shown in Fig. 3.30 is suitable for situations requiring lofty installations and strongly-concentrated illumination as in public halls, foundries and power stations, etc. They give maximum intensity in zones from 0° to 25° from the vertical.

The elliptical angle reflector shown in Fig. 3.31 is suitable for the side lighting of switchboards, show windows etc., because they give a forward projection of light in the vertical plane and spread the light in the horizontal plane.

Fig. 3.27: Holophane stiletto reflector.

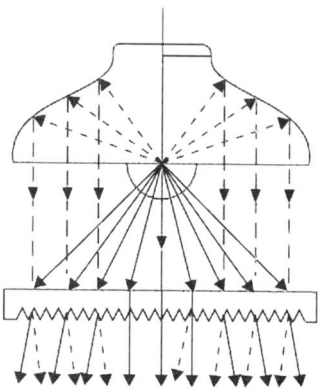

Fig. 3.28: Optical combination of a lamp, reflector and a lens plate.

Fig. 3.29: Dispersive reflector.

Fig. 3.30: Concentrating reflector of parabolic form.

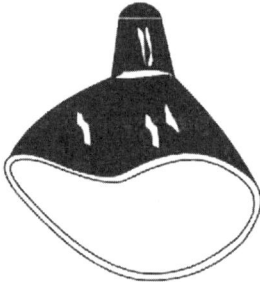

Fig. 3.31: Elliptical angle reflector.

3.10 TYPES OF LIGHT FITTINGS OR LUMINARIES

Light fittings or luminaries are classified according to the way light reaches the object such as:

1. **Direct fittings:** Here 90 to 100% light flux of lamp and fitting is in lower hemisphere and 0–10% of light flux in upper hemisphere Fig. 3.32a. These fittings are efficient, economical and have tunneling effect, i.e. ceiling remains dark.

2. **Semi direct fittings:** Here 60 to 90% of light flux of lamp is in lower hemisphere and 10 to 40% of light in upper hemisphere Fig. 3.32b. These fittings reduce the tunneling effect.

3. **General lighting give fittings:** Here 40 to 60% of light flux is in both the hemisphere, Fig. 3.32c. These fittings soft light with little shadows. Since quite large amount of light reaches objects after reflection from walls and ceiling, room decoration should be in light color and kept in good condition. Also mounting height should be much above eye level to avoid glare.

4. **Semi indirect fittings:** Here 10 to 40% of light is in lower hemisphere and 60 to 90% of light flux is in upper hemisphere, Fig. 3.32d. These fittings give soft shadows with low efficiency.

5. **Indirect fittings:** Here 0–10% of light is in lower hernia sphere and 90 to 100% of light flux is in upper hemisphere, Fig. 3.32e. These fittings give practically no shadows and no glare even with high intensity sources.

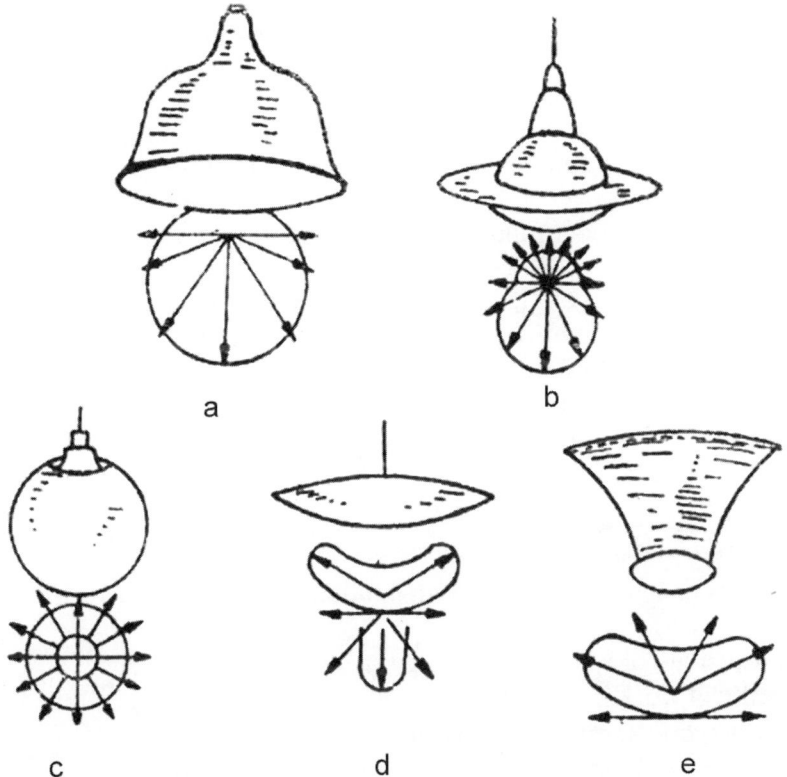

Figs 3.32a to e: (a) Direct fittings, (b) Semi direct fittings, (c) General diffusing lighting, (d) Semi indirect fittings, (e) Indirect fittings.

This system of indirect lighting is most expensive as large proportion of light is lost in ceiling which is required to have matt surface of high reflection factor.

3.11 METHODS OF LIGHTING CALCULATION

Out of several methods employed for lighting calculations, some of them are:
1. Watt per square metre method.
2. Lumen or light flux method.
3. Point to point or inverse square law method.
1. **Watt per square metre method:** This method is very handy for rough calculation or checking. It consists in making an allowance of watt per square metre of area to be illuminated according to the illumination desired on the assumption of an average figure of overall efficiency of the system.
2. **Lumen or light flux method:** The lumen or flux method of calculation for interior lighting is most widely used. The calculation is based on the average illumination required on the working plane. This method is useful where the symmetry in the layout of lighting fittings ensures that the illumination at any point does not differ much from the average value. The average illumination is given as:

$$\text{Average } E = \frac{\text{Flux received on the working plane}}{\text{Area of the working plane}} = \frac{\phi \times N \times UF \times MF}{??}$$

where ϕ is flux output of each lamp, N the number of lamp, UF the utilisation factor, MF is the maintenance factor, A the area of working plane.
3. **Point to point or inverse square law method:** This method is applicable where the illumination at a point due to one or more sources of light is required, the candle power of the sources in the particular direction under consideration being known.

When a polar curve of lamp and its reflector giving candle powers of the lamp in different directions is known, the illumination at any point within the range of the lamp can be calculated from the inverse square law. If two or more than two lamps are illuminating the same working plane, the illumination due to each can be calculated or added.

This method is not much used (because of its complicated and cumbersome applications); it is employed only in some special problems, such as flood lighting, yard lighting, etc.

3.12 FACTORY LIGHTING

In an industrial establishment an adequate amount of light produces the following good effects:
1. The productivity of labour is increased.
2. The quality of work is improved.
3. Number of work stoppages are reduced.
4. Accidents are reduced.

A factory-lighting installation, in common with indoor equipment, should provide the following:
  i. Adequate illumination on the working plane;
 ii. Good distribution of light;
iii. Simple and easily cleaned fittings;

iv. Avoid glare (from the lamp itself as well as from any polished surface, which may be within the line of vision).

General lighting: In factories and workshops the usual scheme is to mount a number of lamps at a sufficient height so that uniform distribution of light over the working the plane is obtained. In large machine shops the height is governed by the necessity of keeping the lamps above the travelling crane.

Local lighting: On some points fairly intense illumination is required. For this purpose local lighting can be provided by means of adjustable fittings attached to the machine or bench in question or mounted on portable floor standards. Such lamps should be mounted in deep reflectors to avoid the glare.

Emergency lighting: It is very desirable to provide auxiliary lighting from the sources other than the main electric supply preferably from batteries or from small petrol driven generator set. If however, emergency light circuits are operated from main electric supply, these should be completely separated from main lighting circuit.

3.13 STREET LIGHTING

The street lighting entails the following main objectives:

i. To make the traffic and obstructions on the road clearly visible in order to promote safety and convenience.

ii. To enhance the community value of the street.

iii. To make the street more attractive.

The following two general principles are employed in the design of street lighting installations:

1. Diffusion principle.

2. Specular reflection principle.

1. **Diffusion principle:** In this case the lamps fitted with suitable reflectors are employed. The design reflectors is such that they may direct the light downwards and spread as uniformly as possible over the surface of the road. In order to avoid glare the reflectors are made to, have a cut-off between 30 and 45° so that the filament is not visible except underneath.

   The diffusing nature of the road surface causes the reflection of a certain proportion of the incident light in the direction of the observer, and therefore the road surface appears bright to the observer.

   For calculating the illumination at any point on the road surface, point-to-point or inverse-square law method is employed. Over certain proportions of the road the surface is illuminated from two lamps and the resultant illumination is the sum of the illuminations due to each lamp.

2. **Specular reflection principle:** Here, the reflectors are curved upwards so that the light is thrown on the road at a very large angle of incidence. In this method, the requirement of a pedestrian, who requires to see objects in his immediate neighbourhood, is also fulfilled.

   This method is more economical, in comparison to diffusion method of lighting. However, it has the demerit that it produces glare for the motorists.

☞ *Illumination Level for Street Lighting and Mounting Height of Lamps*

In class A installations (e.g., important shopping centres and road junctions) the illumination level of 30 lm/m² is required, whereas for poorly lighted suburban streets, illumination level of 4 lm/m² is sufficient. For an average well-lighted street an illumination level of 8 to 15 lm/m² is required. Normal spacing for standard lamps is 50 metres with a mounting height of 8 metres.

For street lighting purposes, mercury vapour and sodium discharge lamps have been found to have certain particular advantages; the most important of these is the low power consumption for a given amount of light, which inspite of the higher cost of the lamps, makes the overall cost of an installation with discharge lamps less than that employing filament lamps.

The colour and monochromatic nature of light produced by discharge lamps does not matter much in street light installations. Lamp posts should be fixed at the junction of roads.

3.14 FLOOD LIGHTING

It means 'flooding' of large surfaces with the help of light from powerful projectors. Flooding is employed for the following purposes:

1. For aesthetic purposes as for enhancing the beauty of a building by night, i.e. flood lighting of ancient monuments, religious buildings on important festive occasions, etc.
2. For advertising purposes, i.e. flood lighting, huge hoardings and commercial buildings.
3. For industrial and commercial purposes as in the case of railway yards, sports stadiums and quarries, etc.

For small buildings, rather uniform flood lighting is used. Flood lights can be placed on other buildings nearby or on suitable posts at distances of not more than about 60 metres. Light should fall nearly perpendicular to the building.

Large or tall buildings are illuminated non-uniformly. Flood lights should be so located that contours and features of the building are well-defined. If any shadows are cast, they should enhance the beauty of the building or movement.

As far as possible the projectors should not be visible to the passers by. In some cases the projectors may be housed in ornamental stands.

According to the beam spread, the projectors are classified as follows:

1. Narrow beam projectors, with beam spread between 12 and 25°, are used for distance above 70 metres.
2. Medium angle projectors, with beam spread between 25 and 40°, are used for shorter range between 30 and 70 metres.
3. Wide angle projectors, with beam spread between 40 and 90°, are used for ranges between 3 and 30 metres.

Due to more accurate control of light required in narrow beam projector, its lamp has more compact filament whereas medium and wide angle projectors can make use of ordinary filament lamps. From view point of economy, use of wide angle projector with high wattage lamp is encouraged over narrow beam projector with low wattage lamp.

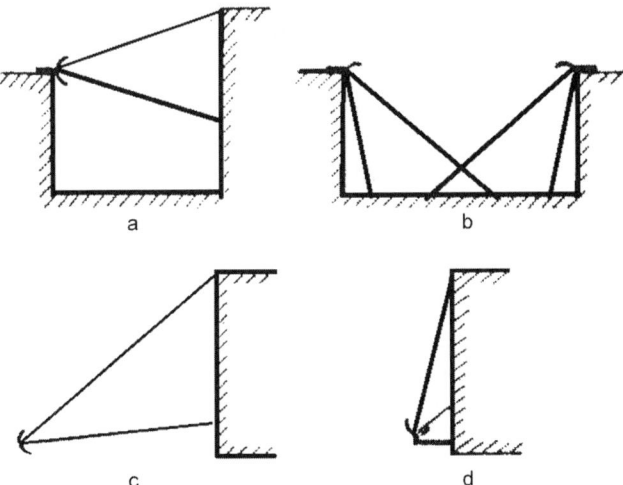

Figs 3.33a to d: (a) Mounting the projector top of an adjacent building, (b) Beam striking the surface at an angle, (c) Mounting the projector forecourt of the building, (d) Mounting the projector foot of the building.

☞ Design of Flood Lighting Installation

One of the most important factors which affects the choice of projector is the location of the projector lamps. Following are four possible locations of projector which can be encountered in practical cases:

    i. Mounting the projector on the top of an adjacent building, beam striking the surface normally Fig. 3.33a.

    ii. Same as (i) but beam striking the surface at an angle other than 90° Fig. 3.33b.

    iii. Mounting the projector in the forecourt of the building. Fig. 3.33c.

    iv. Mounting the projector at the foot of the building Fig. 3.33d.

In first three methods symmetrical projectors are used whereas in the last case we have to select asymmetrical projector. In the first case since the beam strikes the surface normally, we get uniform illumination. We have however, to pay the rent for locating the projectors and access has to be made to them for switching and maintenance purposes. Overhead projectors as in 3.33b are desirable for reducing glare and getting soft shadows. This is mostly suitable for yard lighting, quarrying, excavation or Construction work. Location of projector as in Fig. 3.33c is possible where there is enough front court yard space. Projectors may be mounted on poles or kept on the ground. Where there is no enough front court yard, projectors are mounted on the brackets fixed to the wall itself and we have to use asymmetric projectors. This becomes a necessity in view of high value of incident angle. This results in diminishing of illumination at the top edge of ellipse unless C P in that direction is increased. Hence the need of asymmetric projector. These projectors have to be spread out along the surface to be illuminated. As against this in first there methods, these can be kept at one place and therefore, this reduces wiring work. Beam angle of the projector is decided keeping in view the distance of projector from the surface. After having arrived at the location of the projector and beam spread allowing for sufficient overlap of circles of illumination, we can decide upon the number of projectors needed. Next important factor affecting the design is the illumination level of the surface. This in turn depends upon the reflecting property of the surface. It must be made clear here that it is not the illumination level of the surface to which eyes respond but it is the brightness of the surface with which we are concerned. Illumination level when multiplied with the reflection factor of the surface gives us brightness of the surface. Other factors affecting the level of illumination are the cleanliness of the surface and brightness of surroundings. If the surface is less clean, it will require more of illumination. If the brightness in the surroundings of the floodlit building is high. Either due to high level of street lighting or nearness of other floodlit buildings, it will require more illumination.

After having decided about the number of projectors required and level of illumination, the size of the lamp can be obtained by applying following relation:

$$\phi = \frac{A \times E \times \text{Waste light factor}}{\text{Beam factor} \times N \times \text{Depreciation factor}}$$

where ϕ = Lumen output of lamp

N = No. of projectors

A = Area of surface to be illuminated, m², and

E = Illumination level required in lumen/m².

3.15 PHOTOMETRY

We shall now discuss the comparison and measurement of candle powers.

The candle power of a source in any given direction is measured by comparison with a standard or substandard source employing photometer bench and some form of photometer. The experiment is performed in a dark room with dead black walls and ceiling in order to eliminate errors due to reflected light.

The photometer bench consists essentially of two steel rods which carry the stands or saddles for holding the two sources, the carriage for the photometer head and for any other apparatus employed in making measurements. One of the bar carries a bar strip, graduated in centimetres and millimetres. The carriages which slides upon the bench have except that carrying the photometer head, a circular table which can be rotated in a horizontal plane and clamped in any position. The circular table is provided with a scale graduated in degrees round its edge so that the angle of rotation of lamp from the direction of the axis of bench can be measured. The bench should be rigid so that the source being compared may be free from vibrations and the carriage holding the photometer head should be capable of moving smoothly and with very little effort. The photometer head acts as a screen for comparison of illumination of the standard source and the source under test. There are different types of photometers, which can be used for the purpose. Some of them will be described here.

The principle of most of the methods of measurement is based upon the inverse square law. The standard source, whose candle power is known (say S) and the source under test whose candle power is to be determined, are set on the bench at a distance apart with some type of screen in line with, and between, them, as shown in Fig. 3.34. The photometer head or screen is moved in between the two fixed sources until the illumination on both sides of the screen is same. If the distances of the standard source S and source under test T from photometer head are I_1 and I_2 respectively then according to inverse square law.

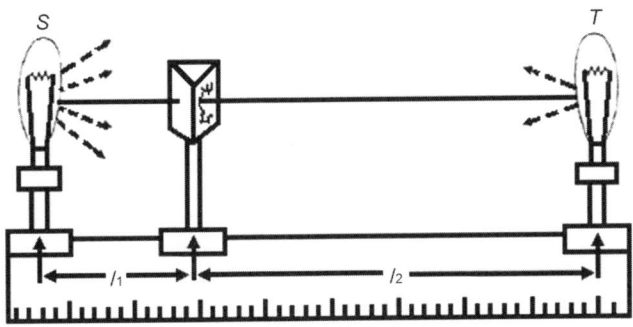

Fig. 3.34: Measurement of candle power.

$$\frac{\text{Candle power of source under test}}{\text{Candle power of standard source}} = \frac{l_2^2}{l_1^2}$$

or $\qquad$ Candle power of source under test $= S \times \dfrac{l_2^2}{l_1^2}$

Since the square of distances are involved in measurement, therefore, distances should be measured accurately. In order to obtain distance exactly two points are determined at which there is a perceptible difference in illumination from the two sides and the point half-way between them is then taken as the position of equal illumination.

Most of the photometer heads consist of some device by means of which the illumination of two surfaces, side by side — one illuminated by standard source and other by the source whose candle power is to be determined may be compared under exactly similar conditions without movement of the eye.

The photometers which are most common in use are the Bunsen and the Lummer Brodhun photometers. These photometers are best for use if the sources to be compared give light of same or approximately similar colours. In case the light from the two sources to be compared differ in colour, a Flicker photometer is best suited.

1. **Bunsen grease spot photometer:** Bunsen grease spot photometer head is shown in Fig. 3.35.

    It consists of a piece of tissue paper with grease spot in the centre, held vertically in a carrier between the two light sources to be compared. The position of the carrier is then adjusted until the semi-transparent spot and the opaque parts of the paper are equally bright, i.e., the grease spot is invisible. The distances of the tissue paper from both light sources are measured. The candle power of the source under test is then determined from the relation.

$$\frac{\text{Candle power of source under test}}{\text{Candle power of standard source}} = \left(\frac{\text{distance of source under test from photometer}}{\text{distance of standard source from photometer}}\right)^2$$

The Bunsen Grease spot photometer can be used in other way also. In this method of use, the photometer is fixed at one position and one side of the paper is illuminated first by test lamp and then by a standard lamp and distance of

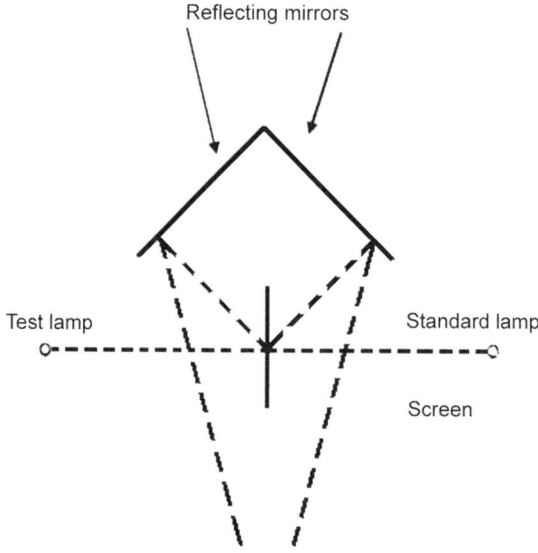

Fig. 3.35: Bunsen grease spot photometer head.

the lamp from the photometer head is measured in each case on disappearing of grease spot.

$$= \text{Candle power of standard lamp} \times \frac{l_2^2}{l_1^2}$$

The simplest and best method of use of Bunsen Grease spot photometer for comparison of illumination on either side of the screen is to use two mirrors placed behind the paper and view the illumination on both the sides through the mirrors simultaneously by standing in a plane passing through the photometer head and perpendicular to the lines joining lamps. The position of the photometer head for equal contrast in illumination between the opaque and transparent portions of the paper on the two sides is determined. Now the candle power of the lamp under test can be calculated as before.

2. **Lummer Brodhun heads:** There are two types of Lummer Brodhun heads: (i) one is the equality of brightness type and (ii) contrast type. Contrast type is more accurate and, therefore, more used in photometric measurements.

   i. *Equality of brightness type photometer head:* The photometer head consists of a plaster of paris screen, two mirrors M_1 and M_2, a compound prism P and a telescope. The compound prism is made-up of two right angled glass prisms, one of which has principal surface as spherical one with a small flat portion at the centre. This small flat portion at the centre makes optical contact with the flat surface of the other prism, as shown in Fig. 3.36.

   Each side of the screen S is illuminated by light entering from one of the sources to be compared. The light rays coming from the test lamp and standard lamp are reflected by the screen on to the mirrors M_1 and M_2 respectively which in turn reflects the same to the compound prism P. Only that portion of light reflected by mirror M_1 which falls on the flat surface of

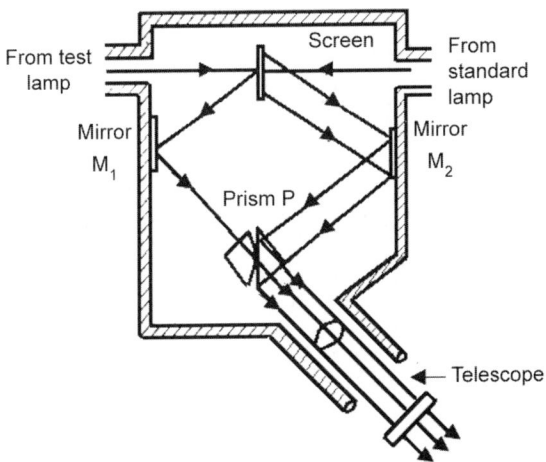

Fig. 3.36: Lummer Brodhun photometer head (equality of brightness type).

the spherical prism, passes straight into the telescope while the rest of it is reflected back. The portion of light reflected by mirror M_2, which falls on the surface of contact between the two prism passes straight through the compound prism while the rest of it is reflected through the telescope. Thus, the observer sees the centre portion of the circular area illuminated by the test lamp and outer ring illuminated by the standard lamp.

While carrying out measurements, the photometer head is moved until the dividing line between the centre portion and surrounding disappears. The disappearance of this dividing line indicates the equality of brightness. Now the distance of the photometer head from the two lamps is noted and the candle power of lamp under test can be calculated by using inverse square law principle.

ii. *Contrast type of Lummer Brodhun photometer head:* The compound prism, in the contrast type photometer head, is made-up of right angled prisms with their principal surfaces face to face. One of the prism (the left hand one in Fig. 3.37a) has its hypotenuses surface etched away at X, Y and Z so that it forms a pattern similar to that shown in the figure.

The light falling on the compound prism from both sides of the screen will pass through the un-etched portions of the prism at the junction but will be reflected by the etched surfaces. The etched portion will have difference in illumination as compared with the un-etched portion in case the illumination of the surfaces of reflections are different, as shown in Fig. 3.37b. G_1 and G_2 are two glass plates which give a little reflected light in order to maintain some difference between illumination of the etched and un-etched areas for all the positions of the photometer head. The difference is about 8% in balance position of photometer head. The photometer head will be in balance position when the difference in illumination of etched and the un-etched portions will be same on each half of the circular area. In case the balance position is disturbed, the difference in illumination of area Z and its surrounding area will decrease while the difference in illumination

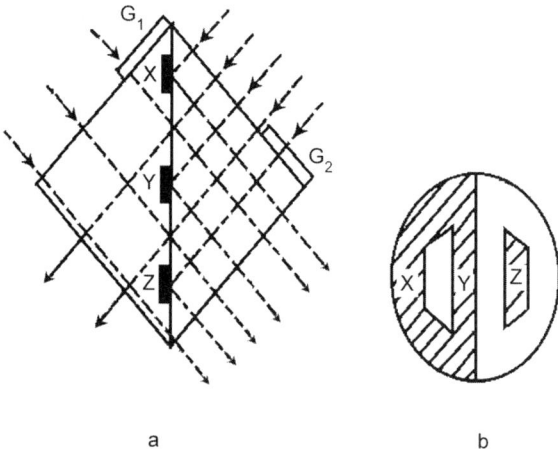

Figs 3.37a and b: (a) The compound prism, (b) Lummer Brodhun photometer head.

of area XY and the inner trapezium will increase. The photometer head is moved until the equal contrast is obtained. This type of head gives accuracy within 1% when lights of similar colours are compared.

2. **Flicker photometers:** These photometers are employed when the two sources giving light of different colours are to be compared. The working of these photometers depend upon the fact that if two illuminated surfaces are presented to the eye with rapid alternations, the flicker disappears when the surfaces are of equal brightness. These photometers are not affected by the difference in colour of two lights to be compared as much as photometers of the steady comparison type are affected since the colour difference between two alternating fields of light disappears at a lower speed of alternation than the speed at which difference in brightness disappears. The speed of alternation should be kept as low as possible at which the disappearance of the flicker can be obtained for the small variation in brightness.

3.16 INTEGRATING SPHERE

The MSCP is usually measured by means of an integrating photometer, the most accurate form of which consists of a hollow sphere (as originally proposed by Ulbricht) whose diameter is large (at least 6 times) as compared to that of the lamp under test. The interior surface of the hollow sphere is whitened by means of a special matt white paint. When the lamp is placed inside the sphere (not necessarily at its centre) then due to successive reflections, its light is so diffused as to produce a uniform illumination over the whole surface. At some point, a small matt opal-glass window, shaded from the direct rays of the source, is made in the hollow sphere. The brightness of the matt opal glass is proportional to that of the interior surface of the sphere. By using a suitable illumination photometer, the illumination of the window can be measured which can be used to find out the total flux emitted by the source.

Total flux = illumination (1m/m^2) × surface area of the sphere (m^2)

$$\text{MSCP} = \frac{\text{total flux}}{4\pi} \text{ candela}$$

☞ *Theory*

In Fig. 3.38 is shown a light source L of luminous intensity I candela and having a total flux output of FL placed at the centre of an integrating sphere of radius r and reflection factor ρ. Let E_A and E_B represent the illuminations at two points A and B, each of infinitely small area da and db respectively and distance m apart. We will now consider total illumination (both direct and reflected) at point A.

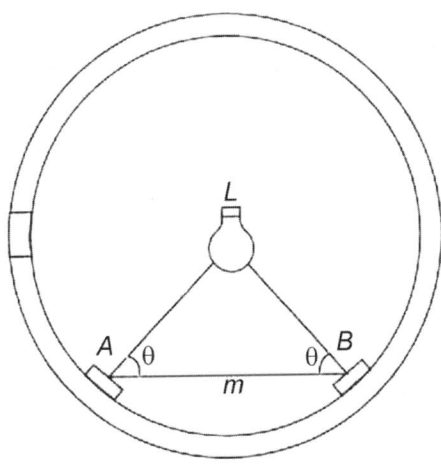

Fig. 3.38: Integrating sphere.

Obviously, E_A directly due to $L = \dfrac{1}{r^2}$

EB directly due to $L = \dfrac{1}{r^2}$

Luminous intensity of B in the direction of A is, $I_B = \dfrac{\rho E_B A_B}{\pi}$ Candela

where A_B = projected area of B at right angles to the line $BA = db \cos\theta$

$$\therefore \qquad\qquad I_B = \frac{\rho I db \, \cos\theta}{\pi r^2} \text{ Candela}$$

Hence, illumination of A due to B is $= \dfrac{I_B \cos\theta}{m^2} = \dfrac{\rho I d \cos^2\theta}{\pi r^2 m^2}$

Now, as seen from Fig. 3.36, $m = 2r \cos\theta$

∴ illumination of A due to B becomes

$$= \frac{\rho I db \, \cos^2\theta}{\pi r^2 \times 4r^2 \cos^2\theta} = \frac{\rho}{4\pi r^2} \times \frac{I}{r^2} \times db = \frac{\rho}{S} \cdot E_B \cdot db = \frac{\rho F_B}{D}$$

where F_B = flux incident on B and Λ = surface area of the sphere

Hence, total illumination due to first reflection $= \sum \dfrac{\rho F_B}{S} = \dfrac{\rho F_B}{S}$

Now, consider any other point C. Illumination on B due to point $C = \rho\, F_L/S$. The illumination on A due to C as reflected from B.

$$= \left[\rho \cdot \left(\frac{\rho F_L}{S} \right) \times \frac{db\,\cos\theta}{\pi} \right] \times \frac{\cos\theta}{m^2} = \frac{\rho F_L}{S} \times \frac{\rho \cdot db\,\cos\theta}{\pi} \times \frac{\cos\theta}{4\pi^2 \cos^2\theta}$$

$$= \frac{\rho F_L}{S} \times \frac{\rho \cdot db}{S}$$

Total illumination due to two reflections $= \sum \frac{\rho F_L}{S} \times \frac{\rho \cdot db}{S} = \frac{\rho^2 F_L}{S}$ $\left(\because \sum ab = S \right)$

Continuing this way, it can be proved that total illumination at point A from all reflections from all points

$$= \frac{\rho F_L}{S} \left(1 + \rho^2 + \rho^3 + \ldots + \rho^{n-1} \right) = \frac{\rho F_L}{S} \left(\frac{1}{1-\rho} \right)$$

Hence, total illumination at A from direct and reflected lights is

$$= E_A + \frac{\rho F_L}{S} \left(\frac{1}{1-\rho} \right)$$

If A is shielded from lamp L, then its illumination is proportional to F_L because $\frac{\rho}{S} \left(\frac{1}{1-\rho} \right)$ is a constant factor. Obviously, if either brightness or illumination at one point in the sphere is measured, it would be proportional to the light output of the source. This fact is made use of while using this sphere as a globe photometer.

WORKED EXAMPLES

EXAMPLE 3.1: A lamp with a reflector is mounted 12 in above the centre of a circular area of 24 metres diameter. If the combination of the lamp and reflector gives a, uniform CP of 960 over the circular area, determine the maximum and minimum illumination produced on the area. [Pb. Univ. Elec. Drives and Utilisation of Elec. Fiiergy Nov; 1983]

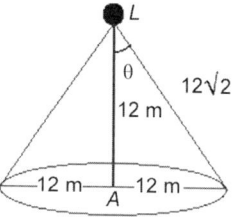

Fig. 3.39: A lamp with a reflector.

Solution: Candle Power of the lamp, $CP = 960$

The maximum illumination will occur directly below the lamp, i.e. at point A (Fig. 3.39) and

$$\frac{-CP}{h^2} = \frac{960}{12^2} = 6.67 \text{ lux}$$

Minimum illumination will occur at the periphery of the circular area, i.e., at point B and

$$\frac{-CP}{h^2}\cos^2\theta = \frac{960}{12^2} \times \left(\frac{12}{\sqrt{12^2+12^2}}\right)^3$$

$$\frac{-960}{12^2} \times \left(\frac{1}{\sqrt{2}}\right)^3 = 2.36 \text{ lux}$$

● **EXAMPLE 3.2**: To lamp posts are 16 m apart and are fitted with a 100 CP lamp each at a height of 6 m in above ground. Calculate the illumination on the ground (a) under each lamp (b) midway between the lamps.

Solution:
Candle power of each lamp = 100 cp
Height of lamp from the ground, h = 6 metres
Distance between the two lamp posts = 16 metres
Illumination midway between the lamps,
E_C = Illumination due to lamp L_1 + Illumination due to lamp L_2

$$= \frac{100}{6^2} \times (0.6)^3 + \frac{100}{6^2} \times (0.6)^3$$

$$= 1.2 \text{ lux}$$

Illumination under either of the lamps, say under lamp L_2,
E_B = Illumination due to lamp L_1 + Illumination due to lamp L_2

$$= \frac{100}{6^2} \times \left(\frac{6}{17.09}\right)^3 + \frac{100}{6^2} = 2.9 \text{ lux}$$

● **EXAMPLE 3.3**: Two similar lamps having uniform intensity of 500 candle power in all directions having below tthe horizontal are mounted at height of 4 metres. What must be the maximum spacing between the lamps so that the illumination the ground midway between the lamps shall be at least one-half the illumination directly under the lamps.

Solution:: Candle power of lamp = 500 CP
Height of lamps from the ground, h = 4 metres
Let the maximum spacing between the lamps be of d metres
Illumination midway between the lamps,
E_C = 2 × Illumination due to either lamp

$$= 2 \times \frac{500}{4^2} \times \frac{4^3}{\left(4^2+\dfrac{d^2}{4}\right)^{3/2}} = \frac{4000}{\left(4^2+\dfrac{d^2}{4}\right)^{3/2}}$$

Illumination under either of the lamp, say under lamp L_2,

E_B = Illumination due to lamp L_1 + Illumination due to lamp L_2

$$= \frac{500}{4^2} \times \frac{4^3}{\left(4^2 + d^2\right)^{3/2}} + \frac{500}{4^2} = \frac{2000}{\left(4^2 + d^2\right)^{3/2}} + 31.25$$

Since $E_c = \frac{1}{2} E_B$

☯ **EXAMPLE 3.4**: Estimate the number and wattage of lamps which would be required to illuminate a workshop space 60×15 m by means of lamps mounted 5 metres above the working plane. The average illumination required is 100 lux. Coefficient of utilisation = 0.42; Maintenance factor = 0.8; Luminous efficiency = 16 lmIW; space-height ratio = unity.

Solution: Given : A = $60 \times 15 = 900$ m² ; E = 100 lux ; UF = 0.42 ; MF = 0.78 ; Luminous efficiency = 16 lm/W

$$\text{Gross lumens required} = \frac{E \times A}{UF \times MF} = \frac{100 \times 900}{0.42 \times 0.78} = 274725$$

$$\text{Total wattage required} = \frac{274725}{16} = 17170 \text{ W}$$

For a space-height ratio of unity, only three lamps can be mounted along the width of the room. Similarly, 12 lamps can be arranged along the length of the room.

Total number of lamps = $12 \times 3 = 36$

$$\text{Wattage of each lamp} = \frac{17170}{36} = 477 \text{ W}$$

We will take the nearest standard lamp of 500 W. The arrangement of the lamps will be as shown in the Fig. 3.40.

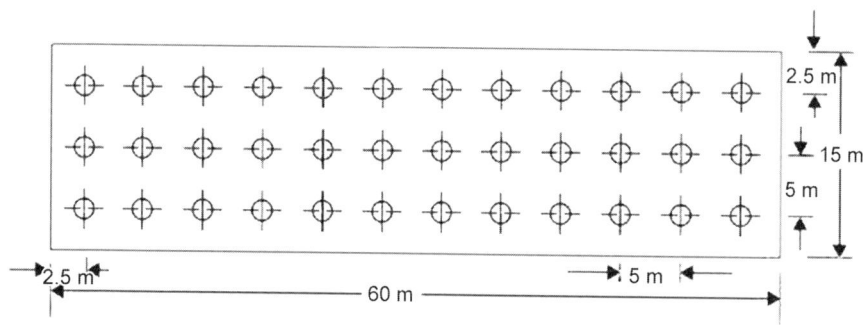

Fig. 3.40: Arrangement of the 36 lamps.

☯ **EXAMPLE 3.5**: A hall 30 m long and 12 m wide is to be illuminated and illumination required is 50 metre-candles. Five types of lamps having lumen outputs as given below are available:

Watts	100	200	300	500	1000
Lumens	1615	3650	4700	9950	21500

Taking a depreciation factor of 1.3 and utilisation coefficient of 0.5, calculate the number of lamps needed in each case to produce required illumination. Out of the above five types of lamps select most suitable type and design a suitable scheme and make a sketch showing location of lamps. Assume a suitable mounting height and calculate space-height ratio of lamps.

Solution: Given : $A = 30 \times 12 = 360$ m²; $E = 50$ metre-candles; $UF = 0.5$;

$$MF = \frac{1}{DF} = \frac{1}{1.3}$$

$$\text{Gross lumens required} = \frac{A}{UF} \cdot \frac{E}{MF} = \frac{360 \times 60}{0.5 \times (1/3)} = 46800 \text{ lumens}$$

$$\text{No. of 100 W lamps required} = \frac{46800}{1615} = 29$$

$$\text{No. of 200 W lamps required} = \frac{46800}{3650} = 13$$

$$\text{No. of 300 W lamps required} = \frac{46800}{4700} = 10$$

$$\text{No. of 500 W lamps required} = \frac{46800}{9950} = 5$$

$$\text{No. of 1000 W lamps required} = \frac{46800}{21500} = 2$$

Let the mounting height be 5 metres.

Most suitable type of lamps will be 300 W lamps. 10 lamps will be arranged in two rows, each row having 5 lamps giving spacing of 6 metres along length as well width of the hall with space height ratio $\frac{6}{5}$, i.e. 1.2

The arrangement of lamps is shown in Fig. 3.41.

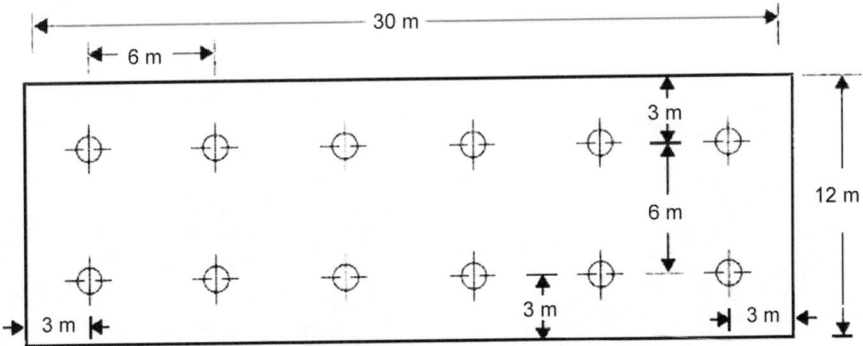

Fig. 3.41: Arrangement of the 10 lamps.

QUESTIONS

1. What is the range of wavelengths to which eye is sensitive? On what physical basis does the colour of light perceived by eye depend?
2. Discuss (i) Inverse square law and (ii) Lambert's cosine law.
3. Explain the various types of lighting schemes with relevant diagrams.
4. What has the texture of surface got the importance in illumination problems?
5. What do you understand by polar curves as applicable to light source? Explain.
6. What are the main faults of a lighting system and how these are overcome?
7. Why tungsten is selected as the filament material and on what factors its life depends?
8. What advantages and disadvantages has sodium and high pressure mercury lighting got over that of filament lamp?
9. Compare fluorescent and filament lamps on basis of quality of light, capital and running costs.
10. What are the advantages of fluorescent lighting over plain mercury discharge lighting?
11. What are the advantages of coiled coil filament gas filled lamp?
12. Explain with sketch the principle of working of a sodium vapour tamp and enumerate its advantages and disadvantages as source of light.
13. What is the effect of increasing the space height ratio of amp over recommended values of the illumination?
14. Describe various light fittings used for indoor lighting giving special applications of each one of them what are the qualities sought in the fittings meant for outdoor installation?
15. What are drawbacks of direct lighting system and how these are overcome?
16. Why it is economical to use few 'large sources of light mounted high for industrial use than many sources of low output?
17. Explain the Rosseau's construction for calculating MSCP of a lamp.
18. When the use of discharge lamps in factories is to be avoided?
19. What are advantages and disadvantages of direct indirect and general diffusing lighting systems?
20. What are relative merits of projector and dispersive reflector scheme of outdoor lighting?
21. Write short notes on:
    i.   Bunsen photometer head
    ii. Lummer-Brodhun photometer head
    iii. Flicker photometer head.
22. Explain how you will measure the candle power of a source of light.
23. What are the objects to be achieved for good street lighting and what are main faults to be avoided?
24. What different types of outdoor flood lighting installations are known to you?

ELECTRIC HEATING

4.1 INTRODUCTION

The application of electrically produced heat is on a rise because of its economical aspect and availability of electrical energy. Electrical heating is widely used both for domestic and industrial applications. Domestic applications include (i) room heaters (ii) immersion heaters for water heating (iii) hot plates for cooking (iv) electric kettles (v) electric irons (vi) popcorn plants (vii) electric ovens for bakeries and (viii) electric toasters, etc. Industrial applications of electric heating include (i) melting of metals (ii) heat treatment of metals like annealing, tempering, soldering and brazing, etc. (iii) moulding of glass (iv) baking of insulators (v) enamelling of copper wires, etc.

4.2 ADVANTAGES OF ELECTRICAL HEATING

As compared to other methods of heating using gas, coal and fire, etc., electric heating is far superior for the following reasons:

1. **Cleanliness:** In the absence of dust and the ash of the fuel, charge never gets contaminated.
2. **Absence of flue gases:** In the absence of flue gases and soot, atmosphere and charged are not polluted. We thus get clean and hygienic operation.
3. **Ease of control:** We can easily control the temperature of charges either manually or made fully automatic. We can also adopt any heating and cooling cycle.
4. **Economical:** Electric heating is economical because electric furnaces are cheaper in their initial cost as well as maintenance cost since they do not require big space for installation or for storage of coal and wood. Moreover, there is no need to construct any chimney or to provide extra heat installation.
5. **Better working condition:** As the radiation losses are low, working with electric furnaces is convenient and cool. Moreover these furnaces are noiseless in operation also.
6. **Ease of adaption:** Electric heating can be adapted easily to our requirements of heating such as heating local spots, or heating material uniformly.
7. **Very high efficiency of utilization:** In electrical heating source of heat can be brought directly to the point where heat is required thereby reducing losses and

increasing the efficiency. Further there are no products of combustion which involves heat losses in their removal from combustion chamber.

8. **Uniform heating:** In all the methods of heating by burning fuels, heat is to be conducted from the outer surface of material to inside of the material. Thus inside core of the material comparatively remain cold. By electric heating, it is possible to generate heat inside the core of the material, thereby uniformity in heating is closed achieved.

9. **Heating of nonconducting material:** It is possible, only with electric heating, to get nonconducting material heated uniformly throughout the section. This is possible because heat is generated inside the material itself.

10. **Safety:** Electric heating is quite safe because it responds quickly to the controlled signals.

4.3   HEATING METHODS

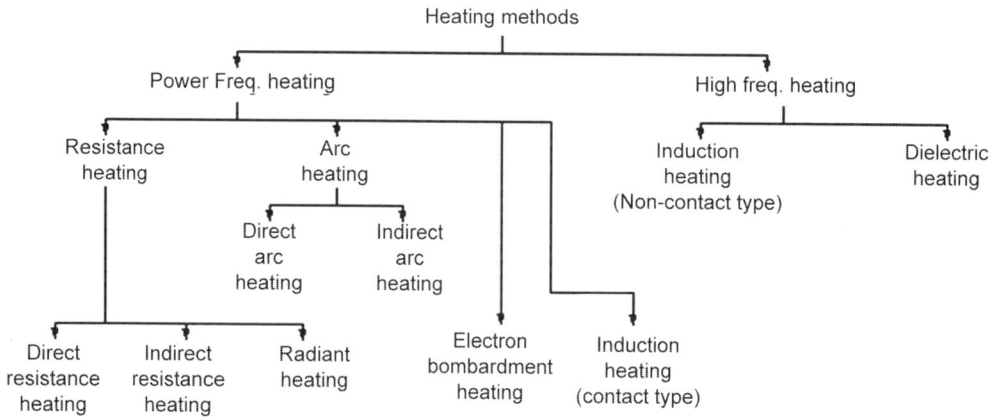

4.4   RESISTANCE HEATING

Where electrical current passes through a resistance, power loss takes place therein, which appears in the form of heat.

Power loss $P = I^2R$ Watts.

$$= VI \text{ Watts.}$$

$$= \frac{v^2}{R} \text{ Watts.} \tag{4.1}$$

where R is the effective resistance of the element.

Unlike induction and dielectric heating, resistance heating works equally well with low voltage and low frequency supply even with DC. In resistance heating, all the energy given to the heating element is converted to heat and there is loss of energy only in delivering the converted energy from heating element to the heating load. In high frequency heating, there is loss of energy at two stages first while converting power frequency supply and second while transferring high frequency power to the heating load. Thus over-all efficiency of resistance heating is high. This makes resistances heating competitive with other methods of heating employing fuels. In addition to above, resistance heating methods is favoured on account of uniform power demand giving a good load factor and almost unity power factor.

4.5 DIRECT RESISTANCE HEATING

In direct resistance heating the material (or charge) to be heated is treated as a resistance and current is passed through it. The charge may be in the form of powder, small solid pieces or even can be liquid. The two electrodes are inserted in the charge and connected to either AC or DC supply (Fig. 4.1). Obviously, two electrodes will be required in the case of DC or single-phase AC supply but there would be three electrodes in the case of 3-phase supply. When the charge is in the form of small pieces, a powder of high resistivity material is sprinkled over the surface of the charge to avoid direct short circuit. Heat is produced when current passes through it. This method of heating has high efficiency because the heat is produced in the charge itself.

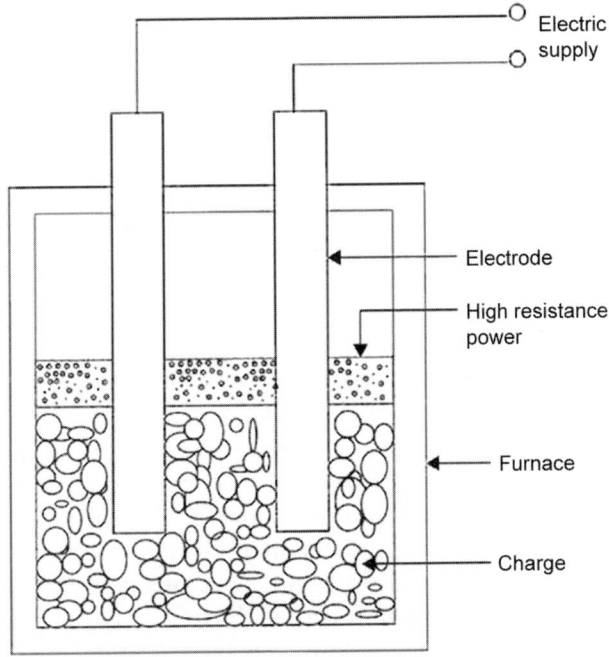

Fig. 4.1: Direct resistance heating.

4.6 INDIRECT RESISTANCE HEATING

Here electric current is passed through an element of high resistances. Passage of an electric current through resistance produces a power loss, specifically I^2R loss, which manifests itself in the form of heat. Heat is then transferred from the heating element to the charge mainly by radiation and convection and rarely by conduction. Sometimes, resistance is placed in a cylinder which is surrounded by the charge placed in the jacket as shown in the Fig. 4.2. This arrangement provides uniform temperature. Moreover, automatic temperature control can also be provided. Example of indirect resistance heating will be found in room heater, hair drier, soldering iron, flat iron, immersion heater, hot plate, frying pan electric kettle, electric toaster, electric water heater, electric oven, etc.

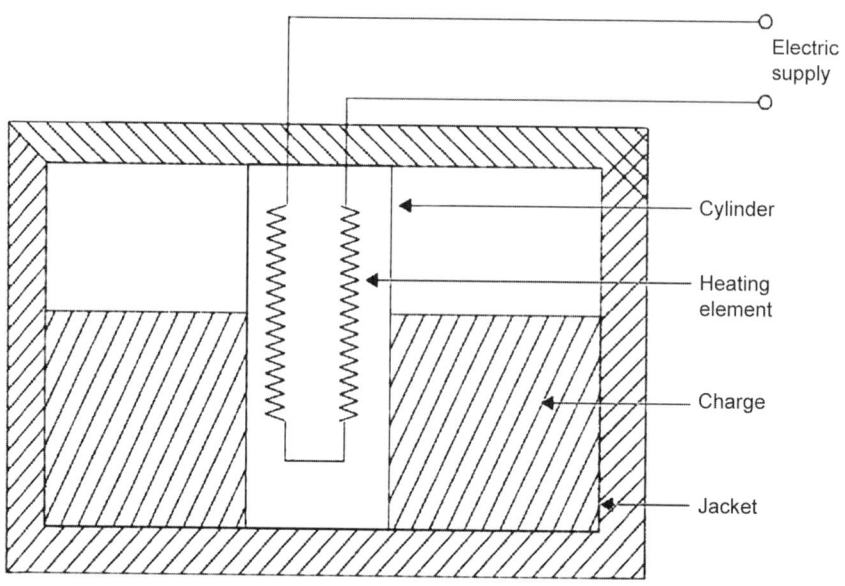

Fig. 4.2: Indirect resistance heating.

4.6.1 Types of Furnaces

One way of classifying resistance furnaces is based on the working temperature. Low temperature furnaces up to 300°C often called ovens, are mostly used for drying varnish coatings, vulcanizing and hardening of synthetic material. These are also used for tempering hardened steel pieces. Medium temperature furnaces having working temperature between 300 and 1050°C are used for annealing, normalising of steel and nonferrous metals, melting of nonferrous metals and stove enameling. High temperature furnaces having working temperature between 1050 and 1350°C are used for hardening purposes. Typical furnace employing heating elements and having provision of carrying out in particular atmosphere is shown in Fig. 4.3.

Resistance heating using metallic resistors is widely used for heat treatment, but the temperature associated with metallic resistors are not high enough for melting most of nonferrous metals and cast iron. Graphite can be however be used at much higher temperature. If satisfactory means are provided to reduce the rate of oxidation to minimum to compensate for increased resistance as wear takes place and to replace the graphite resistor easily and quickly when its useful life is complete, full benefit can be taken of the advantages of resistances heating over higher temperature working ranges. Rocking resistor furnace as manufactured by m/s. Birlefco of England using graphite resistors is shown in Fig. 4.4.

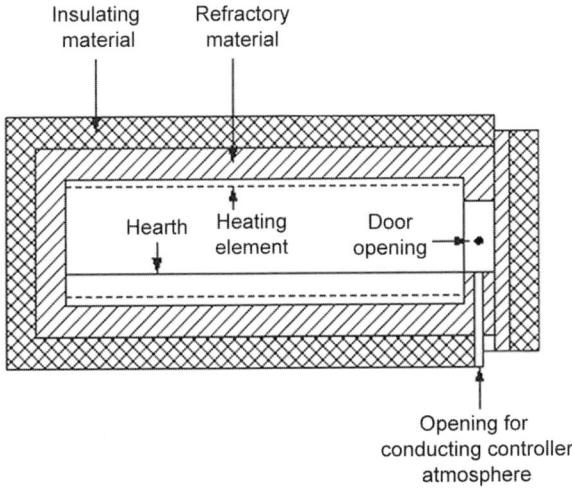

Fig. 4.3: Resistance furnace.

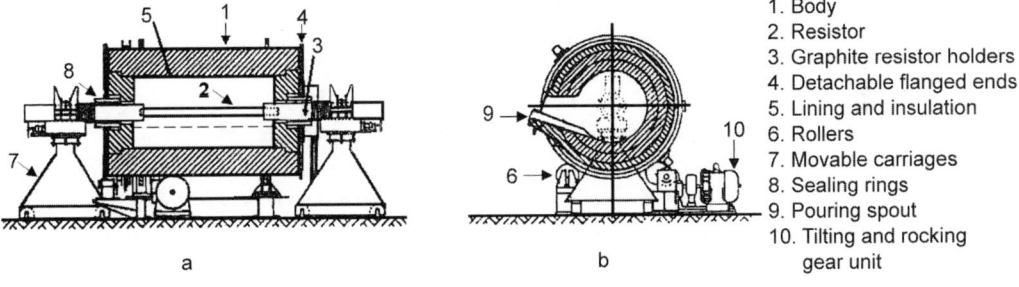

1. Body
2. Resistor
3. Graphite resistor holders
4. Detachable flanged ends
5. Lining and insulation
6. Rollers
7. Movable carriages
8. Sealing rings
9. Pouring spout
10. Tilting and rocking gear unit

Figs 4.4a and b: (a) Rocking resistor furnace using graphite resistors, (b) Cross section.

It consists of a cylindrical shell (1) with detachable flanged ends (4) for the ease of lining. The furnace is designed to rotate on roller (6) to enable the body to be rocked backwards and forwards. The refractory lining (5) is built into the shell. To reline the furnace, the shell is lifted off the rollers and turned vertically on end. The top end cover is removed and the lining material rammed round suitable formers. Graphite resistor (2) is arranged along the horizontal axis and is held in place by two large diameter holders (3) also made of graphite. One of the holder which is detachable has the resistor fitted in a plug socket while a conical connection is made with the other. The graphite holders pass through close fitting refractory tube and properly designed sealing rings (8) prevent the ingress of the air. The holders are mounted in free moving spring loaded carriages (7) so that good electrical contact is always maintained. The carriages are separated and remain stationary as the furnace rocks. Robust connection are made to the holders and large flexible cables are used to carry the heavy electrical current which is supplied from a special transformer with tapping on the primary winding so that the voltage can be adjusted on the secondary to compensate for the increased resistance as the resistor wears. It is thus possible to maintain almost constant power input. One end carriage is specially arranged so that it can quickly be run out away from the furnace shell, which enables the resistor to be changed in a few minutes. The carriage may also be run out when the furnace is charged so as to avoid possible damage to the resistor. To equalize lining wear, the body may be reversed for charging. To equalize lining wear, the body may be reversed in its rollers.

On the passage of current, graphite resistor is heated throughout its entire length and thus heat is evenly distributed in the furnace and radiated on the cold charge and furnace lining. As soon as charge melts, the furnace is rocked backwards and forwards to absorb the heat from the refractory lining and to mix the metal. Provision is usually made for automatic control of the rocking motion, the amplitude of which is adjustable. The white hot covering is necessary to prevent oxidation. These reducing condition permit very high yield when alloy such as cadmium, copper containing a readily oxdisable element, have to be made.

Unlike Ajax-Wyatt furnace, rocking resistor furnace can be completely emptied. This furnace is considered superior to rocking arc furnace on following accounts.

  i. Life of the furnace lining is more as intense local temperature of the electric are is avoided.

 ii. Due to even temperature distribution, heavy loss of volatile elements is avoided such as zinc in brasses.

iii. It has better power factor (PF) resulting in lower electrical bill.

iv. There is no noise due to arc.

4.6.2 Requirement of a Good Heating Material

Indirect resistance furnaces use many different types of heating elements for producing heat. A good heating element should have the following properties:

1. **High specific resistance:** When specific resistance of the material of the wire is high, only short length of it will be required for a particular resistance (and hence heat) or for the same length of the wire and the currrent, heat produced will be more.

2. **High melting temperature:** If the melting temperature of the heating element is high, it would be possible to obtain higher operating temperatures.

3. **Low temperature coefficient of resistance:** In case the material has low temperature coefficient of resistance, there would be only small variations in its resistance over its normal range of temperature. Hence, the current drawn by the heating element when cold (i.e. at start) would be practically the same when it is hot.

4. **High oxidising temperature:** Oxidisation temperature of the heating element should be high in order to ensure longer life.

5. **Positive temperature coefficient of resistance:** If the temperature coefficient of the resistance of heating element is negative, its resistance will decrease with rise in temperature and it will draw more current which will produce more wattage and hence heat. With more heat, the resistance will decrease further resulting in instability of operation.

6. **Ductile:** Since the material of the heating elements has to have convenient shapes and sizes, it should have high ductility and flexibility.

7. **Mechanical strength:** The material of the heating element should posses high mechanical strength of its own. Usually, different types of alloys are used to get different operating temperatures. For example, maximum working temperature

of constant an (45% Ni, 55% Cu) is 400°C, that of nichrome (50%, Ni 20% Cr) is 1150°C, that of Kantha (70% Fe, 25% Cr, 5% Al) is 1200°C and that of silicon carbide is 1450°C. With the passage of time, every heating element breaks open and becomes unserviceable.

4.6.3 Materials of Heating Elements

Depending upon the service conditions such as maximum operating temperature, there are different types of alloys for heating elements. These alloys mainly fall into four classes as follows in Table 4.1.

Table 4.1	Properties of some commercial heating elements			
Particulars	*Nichel-chromium Iron*	*Nichel-chromium*	*Iron chromium aluminium*	*Nichel-copper*
Composition	60% Ni, 16% Cr, 24% Fe.	80% Ni, 20% Cr.	70% Fe, 25% Cr, 5% Al.	45% Ni, 55% Cu.
Commercial name		Nichrome	Kanthal	Eureka or Constantan
Maxm. Working temperature	950°C	1150°C	1200°C	100°C
Specific resistance	110 $\mu\Omega/cm^3$	109 $\mu\Omega/cm^3$	110 $\mu\Omega/cm^3$	19 $\mu\Omega/cm^3$
Specific gravity	8.28	8.36	7.2	8.88

Nichel-chromium Iron alloy is the cheapest and most economical and stronger for temperatures up to 950°C. **Nichel-chromium** alloy has not only good resistance to oxidation up to but at the same time it has sufficient strength. On the other hand **Iron chromium aluminium** alloy has very good resistance to oxidation at high temperature but lacks strength. **Nickel copper** alloy is frequently used for heating elements operating at low temperatures. It has virtually zero resistance temperature coefficients which is most important property.

Various forms of silicon carbide are used for temperature up to 1400°C. For still higher temperatures platinum, molybdenum or carbon can be used as resistor.

4.6.4 Causes of Failure of Heating Elements

Some of the factors responsible for failure of heating elements are:
1. Formation of hot spots.
2. General oxidation of the element and intermittency of operation.
3. Embrittlement caused by the rain growth.
4. Contamination of element.
1. **Formation of hot spots:** Hot spots are the points in heating element which are at a higher temperature than the main body of the element. Hot spots may be due to any of the following causes:
    a. High rate of local oxidation may reduce the cross-section of the element wire, thereby increasing the resistance at that spot. This will produce more heat locally, giving rise to breakdown of the element.
    b. Shielding of element by supports, etc. Will reduce the local heat loss by radiation and cause a rise of temperature of shielding portion of the element.

Therefore, minimum number of supports without producing distortion of the element should be used.

c. Due to too high an element temperature, insufficient support for the element or the selection of wrong material, sagging and warping of the element may result. This sagging and warping of the element may result. This sagging and warping may cause uneven spacing of section, thereby producing hot spots where sections are closest together or even there may be actual shorting of adjacent section of an element.

2. **Oxidation and intermittency of operation:** At high temperature, oxide scale is formed on the surface of the heating element which is continuous and tenacious. This oxide layer is so strong that it prevents further oxidation of the inner metal of the element. However, if the element is used quite often, the scale is subjected to thermal stresses produced by frequent cooling and heating. Thus, the oxide scale, therefore, cracks and flakes off, exposing further fresh metal to oxidation. This will produce increased local oxidation producing hot spots.

3. **Embrittlement due to grain, growth:** All heating alloys, containing, tend to form large brittle grains at high temperatures. When cold the elements are very brittle and liable to rupture easily on slightest handling and jerks.

4. **Contamination and corrosion:** Element may be subjected to dry corrosion produced by their contamination with the gases of the controlled atmosphere prevailing in annealing furnaces or fumes from flux used in brazing furnaces or oil fumes produced by the heat treatment of components contaminated with lubricant.

Of all the condition mentioned above which determine the life of the element most critical are the temperature of the hottest point and ratio of intermittent to continuous working.

4.6.5 Temperature Control of Resistance Furnaces

Since voltage of supply and resistance of elements are independent parameters and current through element are dependent parameter, there are basically types of temperature control firstly by varying the applied voltage of elements secondly by varying the resistance of the elements thirdly by varying the ratio on and off time of supply.

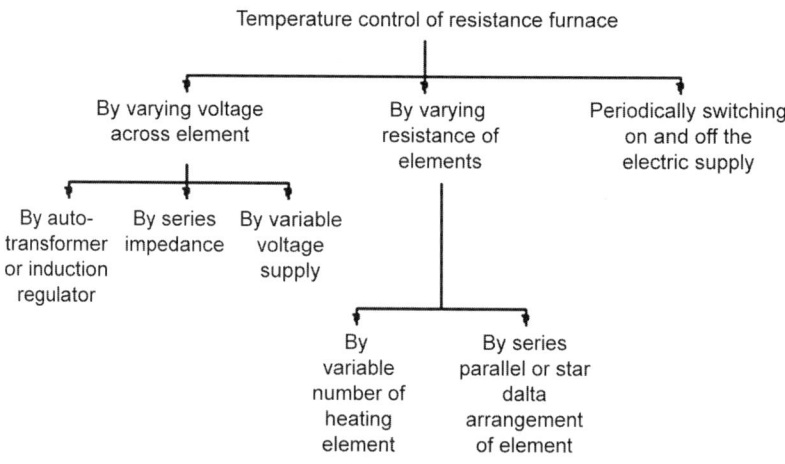

1. **Autotransformer or induction regulator:** By providing different taps on the autotransformer (Fig. 4.5a) or by varying the position of the rotor of an induction generator (Fig. 4.5b), we get variable voltage supply for the furnace.

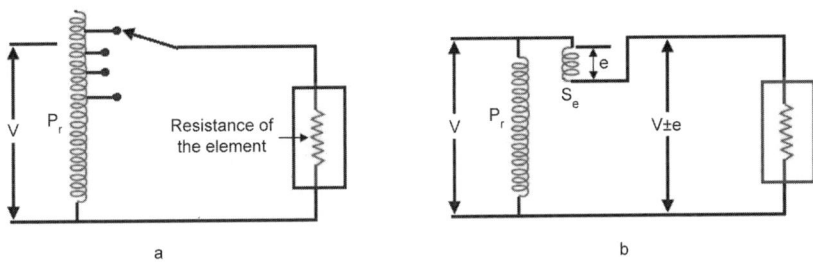

a   b

Figs 4.5a and b: (a) Different taps on the auto-transformer, (b) Varying the position of the rotor of an induction generator.

2. **Series impedance:** By connecting impedance in series with the element, we apply reduced voltage to the element and, therefore reduced current (Fig. 4.6). This method is wasteful and therefore used only for small furnaces.

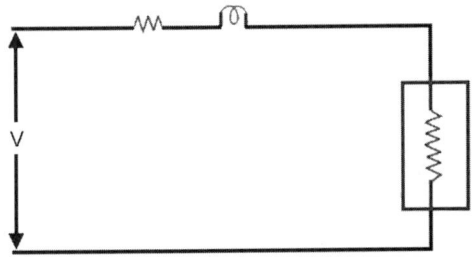

Fig. 4.6: Connecting impedance in series with the element.

3. **Variable voltage supply:** If the furnace is very big, there may be independent generating set for the furnace and we can give variable voltage supply to it.

4. **Variable number of elements:** If R be the resistance of one element and n be the number of elements working in parallel, total resistance of elements is $\dfrac{R}{n}$.

$$\text{Heat produced in furnace} = \frac{V^2}{\dfrac{R}{n}} = \frac{V^2 n}{R} \qquad (4.2)$$

As shown in Fig. 4.7 by putting more and more parallel resistance in the circuit, we can change the temperature. With this method we get uneven heating unless elements not in use are well-distributed. This mean complicated wiring.

5. **Series parallel or star delta arrangement of elements:** In case of single phase supply, the heating elements can be put in series (for low temperature) or in parallel (for high temperature) by means of series parallel switch (Fig. 4.8).
   In case of 3 phase supply, elements can be put in star (for low heat) or in delta (for high heat) by means of delta switch (Fig. 4.9).

6. **Periodically switching on and off the electric supply:** This method of temperature control is used in case of small ovens. Here supply to the oven is

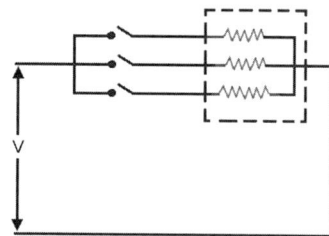

Fig. 4.7: Parallel resistance in the circuit.

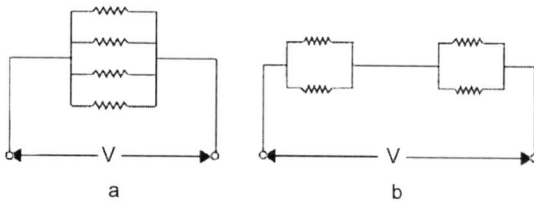

Figs 4.8a and b: (a) Heating elements in series, (b) Heating elements in series parallel.

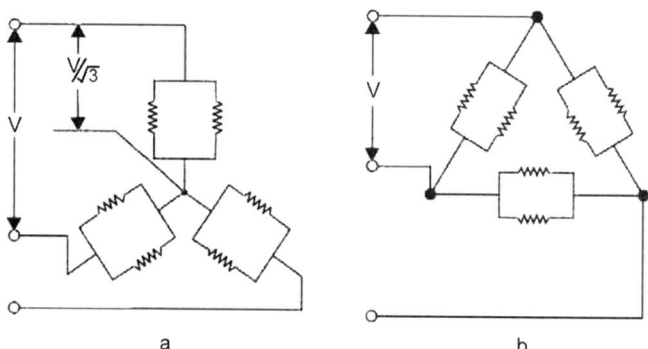

Figs 4.9a and b: (a) Heating elements in star, (b) Heating elements in delta.

given to through the thermostat switch which switches on and switch off the supply at particular temperature. Final temperature attained is proportional to the ratio of $\left(\dfrac{\text{time interval switch remains on}}{\text{total time interval of the on off cycle}}\right)$. Advantage of this method is that it is more efficient than series impedance method.

4.6.6 Deign of Heating Element

Normally, wires of circular cross-section or rectangular conducting ribbons are used as heating elements. Under steady-state conditions, a heating element dissipates as much heat from its surface as it receives the power from the electric supply. If P is the power input and H is the heat dissipated by radiation, then $P = H$ under steady-state conditions.

As per Stefan's law of radiation, heat radiated by a hot body is given by

$$H = 5.72\ eK\left[\left(\frac{T_1}{100}\right)^4 - \left(\frac{T_2}{100}\right)^4\right] W/m^2$$

where T_1 is the temperature of hot body in °K and T_2 that of the cold body (or cold surroundings) in °K.

e = emissivity. Its value is 1 for black body and 0.9 for heating elements. 0.1, 0.15, 0.25, 0.65 for polished aluminium, copper, cast iron and black iron plate respectively.

K = radiating efficiency. Its value depends upon the disposition of heating elements. Its value is 1 for single element and may go down as 0.5 for many elements.

Both e and K are dimensionless.

Now, $P = \dfrac{V^2}{R}$ and $R = \rho\dfrac{1}{A} = \rho\dfrac{1}{\pi 2/4} = \dfrac{4\rho l}{\theta d^2}$

$$\therefore\ P = \frac{V^2}{4\rho l/\pi d^2} = \frac{\pi d^2 V^2}{4\rho l}\ \text{or}\ \frac{1}{d^2} = \frac{\pi V^2}{4\rho P} \tag{4.3}$$

Total surface area of the wire of the element = $(\pi d) \times 1$

If H is the heat dissipated by radiation per second per unit surface area of the wire, then heat radiated per second = $(\pi d) \times 1 \times H$ (4.4)

Equating (4.3) and (4.4), we have

$$P = (\pi d) \times 1 \times H\ \text{or}\ \frac{\pi d^2 V^2}{4\rho l} = (\pi d) \times 1 \times H$$

$$\text{or}\quad \frac{d}{l_2} = \frac{4\rho H}{V^2} \tag{4.5}$$

4.7 RADIATING HEATING

From equation 4.3, it is clear that greater the temperature of the element, more will be the rate of heat transfer. However, the maximum operating temperature of nichrome wire is limited to about 1150°C on account of oxidation and melting down of element wire. Tungsten has high melting point of 3500°C and in vacuum, there is no danger of oxidation. In radiation of heating, heating element consists of tungsten filament lamps together with reflectors to direct all the heat on the charge. These lamps are operated at 2300°C instead of 3000°C giving greater portion of infrared heat rays. These lamps are operated from low voltage supply of 115 V. This is due to the fact that for same heating, current taken will be more if voltage is less. Thus we will need thick and robust filament. Further, operating the filament at 2300°C increases the life of the filament. Advantage of radiate heating is that we not only get high heat transfer rate and consequence reduction in heating time but heat absorption remains more or less constant whatever the charge temperature. As against this, in low temperature oven, heat absorption falls off very considerably as the temperature of charge increases.

Radiant heating is mainly used for drying enamel or painted surfaces. High concentration of radiant energy enables heat to penetrate the coating of paint or enamel to a depth sufficient to dry it out without wasting energy in the body of the work piece.

When voltage between two electrodes, separated by an air gap, is increased, a stage is achieved when voltage gradient in the air gap is such that the air in the gap becomes good conductor of electricity. Arc is said to exist when electric current passes through air gap. It should be noted that a very high voltage is required to establish an arc across an air gap but to maintain an arc, small voltage may be sufficient. An alternate way of producing an arc is to short circuit two electrodes momentarily and on withdrawing them back, we get an arc. With this method of striking an arc, we do not require high voltage. An electric is provides a large quantity of power in a small volume. This concentration of heat developed a high temperature of about 3500°C from carbon arc, forms a good heating source. This principle is made use of in an electric arc furnace.

Other advantages of electric arc furnace for steel making over conventional method are as follows.

i. It produces steels demanding high purity and exacting analysis. This is because in arc furnace wide variety of condition can be precisely controlled.

ii. Arc furnace can operate on 100% steel scarp which is cheaper than pig iron whereas the cupola/converter practice requires a proportion of pig iron in cupola charge.

iii. Capital cost of electric arc furnace shop is approximately two-third of the capital cost of an open hearth shop for the same output of bulk steel.

4.8.1 Direct Arc Furnace

In direct arc furnace, arc is established between electrodes and charge and electric current flow through the body of the charge, developing heat due to electric resistance of the charge, developing heat due to electric resistance of the charge (through relatively small in amount) in addition to the heat radiated from the arc. There are mainly two types of direct arc furnace, namely, those with non-conducting bottom Fig. 4.10 and those with conducting bottom Fig. 4.11. The former type has been much in use.

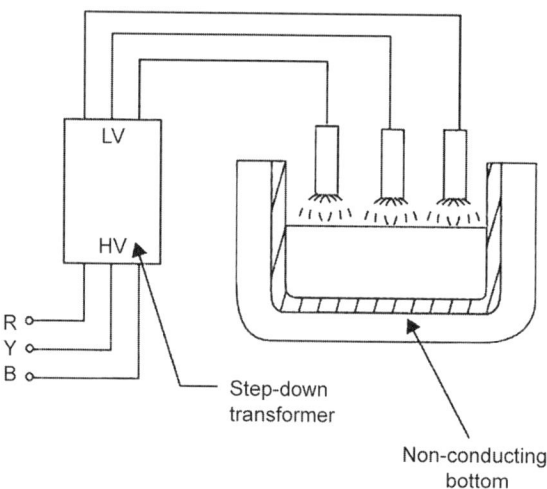

Fig. 4.10: Direct arc furnace with non-conducting bottom.

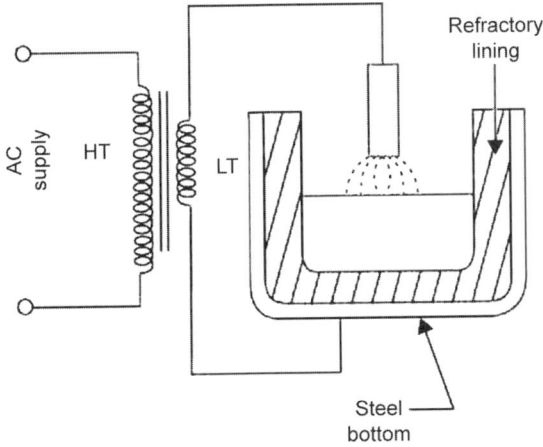

Fig. 4.11: Direct arc furnace with conducting bottom.

4.8.1.1 Construction

It consists of steel plate shell in the outer region usually cylindrical in shape. Recently conical shaped steels, with walls flaring outward as they rise, have been used. It is claimed that this design gives better protection to the refractories. Bottom, side walls and roofs of the furnace are lined with fire bricks, magnetite bricks and silica blocks. The bottom is lined with magnetite mix or ground ganister mix depending upon whether we want basic or acidic lining. Detailed account of the lining of furnace is beyond the scope of this book. All the furnace are provided with a door on one side through which heat can be worked and the tap hole and spout on the other side to take out charge and slag. The furnace rests on platform which can be tilted for pouring. Tilting mechanism consists of a motor which drives a pinion which meshes with a semicircular toothed rack at the bottom of platform. The furnace may be door charged or charged from top. Door charging is adopted for small furnace. There are three designs of top charging-tilt top, swing top and gantry lift type. In tilt top design, roof along with electrodes is lifted to clear off the shell and is tilted in vertical plane to make room for charging from top. In swing top design, roof after clearing off from the shell is swung in horizontal plane. In gantry type design, roof with electrodes is mounted on gantry crane which travels on the charging platform. When furnace is to be charged, the electrodes are to be raised to clear off the shell, then the roof is lifted and by means of gantry cranes, moved to one side of the furnace. In all top charged furnaces, there is not only saving in charging time of large furnaces but this also simplifies the repair of parts of the surface, side walls such as tap hole and slag door arches. Direct arc furnaces are of three phase type as shown in Fig. 4.12.

The electrodes pass through circular opening in the roof spaced at the vertices of an equilateral triangle. Water cooled rings are placed around the electrode to act as seal. The electrodes are filled with water cooled wedge clamps with the arms that extend over the furnace from vertical mast. The clamps form the electrical connection between the power supply and electrodes. These furnaces are made having inside diameter of shell 7 to 22 ft., capacity 3 to 100 tons using transformers 1.5 to 25 MVA.

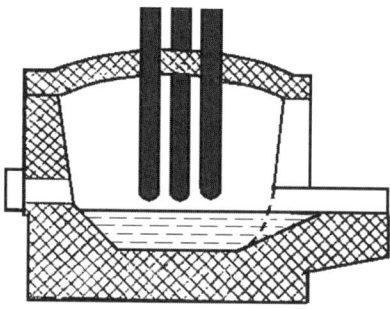

Fig. 4.12: Direct arc furnaces are of three phase type.

Modern trend in the furnace design is to sectionalize three parts of the furnace, *viz.* the hearth, the wall and the roof with the view to replacing and renewing rapidly any section which becomes worn out. Not only the roof is made removable but new design enables the cylindrical wall section to be lifted off the hearth for maintenance. This reduces the time that a furnace is out of commission for repairs.

After the metal is melted, refining is required. Subjecting the melt to the section of the slag for considerable time. However if the melt is subjected to stirring action the refining period could be shortened. Stirring of metal bath is carried out electromagnetically by passing low frequency current through a coil located under the furnace.

4.8.1.2 Electrodes

There are mainly three types of electrodes, namely carbon electrode, graphite electrode and soderberg electrodes. Material of the electrode (carbon and graphite) is selected on account of their infusibility, electrical conductivity, insolubility, chemical inertness, mechanical strength and resistance to thermal shock. Though carbon and graphite are essentially same they differ very much in the physical and electrical properties. Carbon electrodes are amorphous while graphite electrodes are obtained by heating the carbon electrodes to a very high temperature. This also brings about thorough volatilization and removes most of the impurities. Specific resistance of graphite electrode is one-fourth that of carbon electrode and for same size has more current carrying capacity that carbon electrode. Carbon electrode are very cheap and cost less than one-half as much for same weight as graphite electrodes. Due to low current carrying capacity, carbon electrodes are seldom used for furnaces having transformers above 3 MVA.

Carbon electrodes are used with small furnaces for manufacture of Ferro alloys, aluminum, calcium carbide, phosphorous, etc. by submerged arc process. Graphite electrodes are used in large furnaces.

Soderberg electrode are continuous type of electrodes made of special paste which is baked by the heat of furnaces and the heat of conducting current. The paste is contained in thin steel cylinder. As electrode is consumed in the furnace, electrode along with casting is lowered into the furnace and a new section of casting is wielded at top of the last section. These electrodes are used where rapid movement of electrodes is not required such as in reduction furnaces.

4.8.1.3 Reactors

An electric circuit is said to be unstable when small change in the operation conditions (cause) of the circuit, such condition are produced as to reinforce the original cause, thereby increasing the magnitude of original small change successively, till dangerous conditions to the circuit itself, are produced.

An electric circuit with constant voltage supply and an arc load is basically unstable. This is due to the fact that any increase in the arc current will increase its diameter which will reduce its resistance. This will further be increasing the arc current. In this way, cause helps the effect which in turn helps the cause. From above, the cause of instability is analyzed to be the negative volt ampere characteristic of the arc, i.e. voltage drop in the arc decreasing with the increase in the arc current. In Fig. 4.13, curve A gives the voltage ampere characteristic of an arc, B is the voltage ampere characteristic of a reactor and C is characteristic of a circuit having both reactor and a load. On the right side of dotted line the resultant volt ampere characteristic is rising and, therefore, stable operation. On the left side of this line, it indicates an unstable operation.

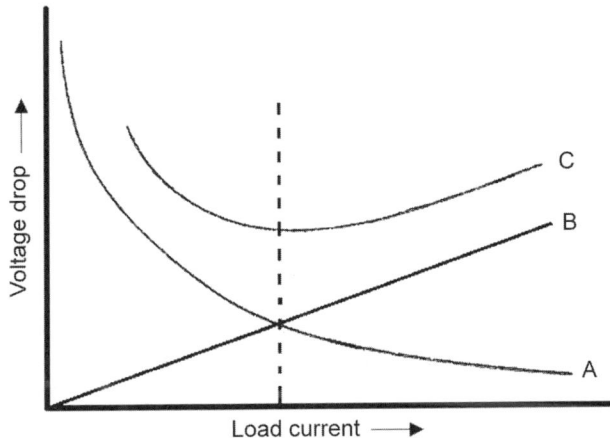

Fig. 4.13: Voltage ampere characteristic.

Besides reactor working towards stabilizing an arc, it serves as a safety device. At the start of the melt, the resistance of the secondary circuit is very low. Therefore, quite often short circuit conditions are produced on the secondary circuit. The reactor limits the magnitude of this short circuit current surges.

4.8.1.4 Power Supply to the Arc Furnaces

Power supply for electric arc furnaces is of low voltage high current type. This is due to following reasons.

1. Heating effect is proportional to square of the current, therefore, heavy currents are needed.
2. High voltage between electrodes and charge produces high voltage gradient between them. Nitrogen of furnace atmosphere gets ionized and is absorbed by the charge. This intruding effect on the charge produces embrittlement.

3. High currents and low voltages keep the electrodes very near to the charge as the arc is small in length. Thus arc is away from the roof and therefore, life of the roof refractory is increased.

4.8.1.5 Conditions for Maximum Output

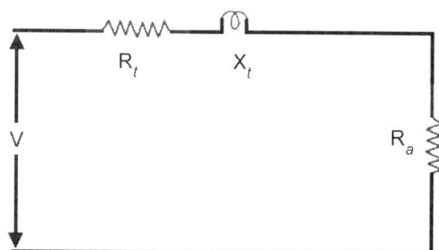

Fig. 4.14: Equivalent circuit of an electric arc furnace.

Equivalent circuit of an electric arc furnace is shown in Fig. 4.14 in which

$R_t =$ Total resistance of transformer referred to secondary side + the resistance of lead and electrode.

$X_t =$ Total reactance of transformer referred to secondary side + reactance of stabilizing reactor and reactance of leads.

$R_a =$ Resistance of arc.

Arc current, $I_a = \dfrac{V}{\left(R_a + R_t\right)^2 + X_t^2}$

Power loss in the arc $P = \dfrac{V^2}{\sqrt{\left(R_a + R_t\right)^2 + X_t^2}} \times R_a$

$$= \dfrac{V^2 R_a}{R_a^2 + 2R_a R_t + R_t^2 + X_t^2}$$

$$= \dfrac{V^2}{R_a + 2R_t + \left(\dfrac{R_t^2 + X_t^2}{R_a}\right)}$$

Power loss P is maximum when denominator is minimum.

$$\dfrac{d}{dR_a}\left[R_a + 2R_t + \left(\dfrac{R_t^2 + X_t^2}{R_a}\right)\right] = 0$$

$\therefore \qquad\qquad 1 + 0 - \left(\dfrac{R_t^2 + X_t^2}{R_a}\right) = 0$

$\therefore \qquad\qquad R_a = \sqrt{R_t^2 + X_t^2}$

Power loss in the furnace is maximum when resistance of the arc is numerically equal to the impedance of whole electrical circuit referred to secondary side excluding resistance of arc. This is a case when applied voltage is considered constant.

Power factor at maximum power loss

$$\text{Cos}\theta = \frac{R_a + ?_t}{\sqrt{(R_a + R_t)^2 + X_t^2}}$$

$$= \frac{R_a + R_t}{\sqrt{R_a^2 + 2R_a R_t + R_t^2 + X_t^2}}$$

Putting $R_t^2 + X_t^2 = R_a^2$ we get

$$\text{Cos}\phi = \frac{R_a + R_t}{\sqrt{2R_a^2 + 2R_a R_t}} = \frac{R_a + R_t}{\sqrt{2R_a(R_a + R_t)}} = \sqrt{\frac{R_a + R_t}{2R_a}}$$

$$\text{Cos}\phi = \frac{1}{\sqrt{2}}\sqrt{1 + \frac{R_t}{R_a}}$$

Neglecting R_t in comparison with R_a

$$\text{Cos}\phi = \frac{1}{\sqrt{2}} = 0.707$$

Therefore, if R_t is neglected, maximum power takes place when the power factor at secondary side is 0.707. Since current and potential transformer that operate the power factor meter are installed on the primary circuit of power transformer, therefore, if R_t is not neglected and due to phase angle errors of CT and PT maximum arc power input to furnace occurs at PF reading on the power factor meter from 0.75 to 0.80. Fig. 4.15 gives performance characteristics of a typical arc furnace.

For any given voltage, if we increase the arc current, KVA will go on increasing in straight line. KW input to whole circuit increase till PF of 0.707. Beyond this current, KW of whole circuit decreases KW loss in arc will be less than the circuit KW by the amount of loss in R_t. From the above discussion we should not be misled into belief of getting increased heat by increasing the current setting beyond those that give optimum KW loss as such practice results not only in high KVA demand and lower PF but also in actual loss of heat. As a thumb rule it is not economical to operate an arc furnace with primary side PF under 0.8.

4.8.1.6 Power Transformer

Rate of absorption of heat by charge is the function of the temperature gradient within the mass of charge. This rate of absorption decreases continuously as the average temperature of charge increases. Thus at the start of the melt, rate of heat absorption is more and it gradually decreases as melt proceeds further. Later the rate of heat absorption by the charge is largely dependent upon the area exposed to radiation. Hence with the charge consisting of heavy pieces of scrap, the rate of heat absorption will be less than that with light scrap with large exposed area.

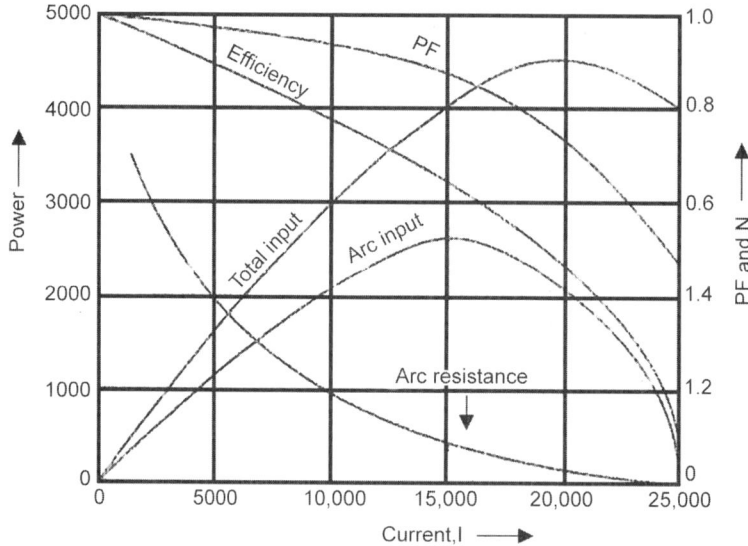

Fig. 4.15: Performance characteristics of a typical arc furnace.

We know that the rate of heat production in the furnace is proportional to square of voltage. If the rate of heat production is less than the maximum rate of heat absorption, heat will take more time. On the other hand if rate production is more than rate absorption, it will reduce the life of side walls and roof of furnace. Therefore, the importance of voltage control can very well be realized.

Further at start, in the case of top charge furnace, since the furnace is charged full with scrap, it requires operation at reduced voltage for a few minutes till electrodes are buried in the charge. Thereafter furnace is to be worked at top voltage until bulk of the scrap has been melted down. After the charge is melted down, operating voltage of the furnace is reduced. On load tap change gear is provided to select suitable operating voltage within comparatively fine limits without disconnecting the transformer from the supply.

It is not easy mater for the operator to determine when precisely he should reduce his voltage. If the reduction is too soon, then he is unable to attain full output of the furnace. If however he is too late in making the reduction, excessive wear of the refractory will occur. There can however, be a scheme where temperature of the refractory can be made to control the furnace voltage and therefore power from this.

As arc furnace transformer has to be special design having mechanical rigidity to enable the windings an electrical insulation to withstand the heavy mechanical stresses set-up by the high current surges. Furnaces transformer virtually consists of two transformers. The first transformer would be a regulating transformer perhaps an autotransformer. All tap changing is done on this resulting transformer. The second transformer is a fixed ratio transformer. For PF correction, capacitors are connected across the output terminals of the regulating transformer so that as the voltage on the secondary of transformer varies so does the voltage across the capacitors to give PF correction proportional to the load.

Average rating of small foundry furnaces in the range of 3 to 5 tons would have a transformer rating of 600 to 700 KVA per ton. The melting achieved with these ratings is from 1 to 1½ hours. The well-known characteristics of foundry furnace, namely fairly heavy surging during the melting down period, call for a transformer design which must be extremely robust in order to take care of electromechanical stresses and thermal stresses. The latter are particularly important in the case of foundry furnace transformer since with this type of unit there is very little refining time and the transformer therefore has no chance of cooling down immediately after the melting down period has been completed. On the other hand, large bulk producing furnaces of rating of about 100 to 150 tons, the electrical rating of transformer per ton of furnace capacity is comparatively low between 200 to 300 KVA per ton; consequently the melting rate of the furnace is reduced to 2¼ to 2½ hours. During the refining period, furnace operates at a very much reduced electrical input approximately to about one-third full-load. This gives the transformer an opportunity of cooling down before the next melting period commences. The thermal overload capacity of the transformer therefore need not be as great as in the case of foundry furnace transformer.

Foundry furnace installation may require between 400 and 600 switching operation per week. For such installation where fault level of the system does not exceed 250 MVA. It is usually quite satisfactory to install an oil circuit breaker. For large arc furnace installation requires for bulk steel production, where fault level is probably not less than 500 MVA and switching operations between 50 and 100 per week, both oil type and air blast type circuit breakers can be used. Air blast breaker is gaining in popularity mainly because it is more easily maintained and will handle considerably more switching operations without attention to the contacts than will the oil circuit breaker.

4.8.1.7 Heat Control

It has been proved above that for a given voltage, there is a particular value of arc resistance, thereby electrode position, which gives optimum KW loss in the arc. Thus, the importance of the electrode control can be well understood.

One of the methods of electrode control employs electrohydraulic system. There are three current transformers, one on each phase. They feed small motors which control hydraulic valves which introduce water under pressure into the main cylinder whose piston operates the electrode arms. The system for each electrode is quite independent. The water under pressure for this cylinder is obtained from an air vessel, half of which is filled with water and rest with compressed air. Whenever pressure in the vessel falls below 80 lbs./sq. inch, the automatic contactors start the pump motor which pumps water into the air vessel. When pressure of air reaches 110 lbs/sq. inch, contactors open and motor stops. The automatic control is worked by diaphragm switch. The water from the air vessel is taken to the electrode lifting cylinder through the valves controlled by the motor connected to the current transformers as desired. Electrohydraulic system of control is losing ground to rotary control system because an all-electric control can be looked after by one skilled electrician. On the other hand electrohydraulic control requires both electrical and mechanical maintenance and therefore two skilled classes of technicians.

Second method employs variable voltage control of electrode winch motor by means of rotating controls commonly known by various trade names such as Rotator, Regale or Manga Volt. Such controlled type generator has as conventional type DC shunt generator, self-excited shunt field winding and one or two separately excited control field windings. Resistance of the shunt field windings is equal to critical field resistance, i.e. resistance is equal to the slope of air gap line in the magnetization characteristic. In other words, field is said to be tuned. Electrode control scheme is shown in the Fig. 4.16. Rotating control machine has two control windings a and b. Winding 'a ' is a current control field of rotating control machine and is connected to the CT through plate rectifier , winding 'b' is potential control field and is energized from the voltage drop across the arc, i.e. voltage between electrode and bottom shell of the furnace. When supply is given, voltage is applied to the potential control field 'b' which causes the generator voltage to build up in the direction so that electrode winch motor lowers down the electrode. When electrode touches the charge, control field winding 'b' is de-energized. When second electrode strikes the metal, heavy current flows and control winding is energized and the generator voltage is built-up in such a direction as to cause electric motor to lift electrode. As the arc is lengthened its voltage increases and current decreases, until a balance is reached between potential and current control fields. In Fig. 4.16 control of one phase is only shown. All variable voltage generators of three phases are mounted on the same shaft and driven by one induction motor.

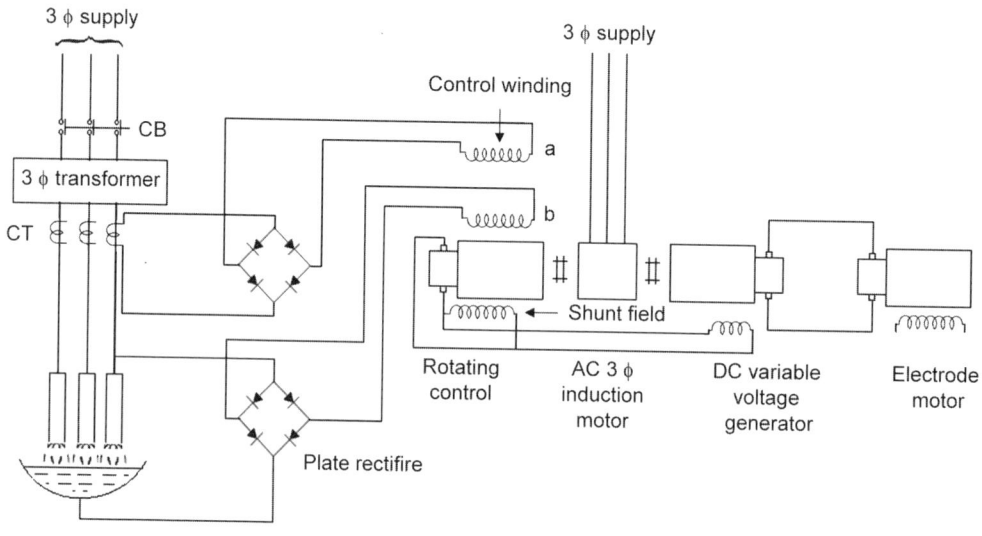

Fig. 4.16: Electrode control scheme.

With this method of electrode control, electrode motor torque is directly proportional to the average unbalance between electrode current and voltage across the arc and gives smooth acceleration and retardation and increase speed of response. Due to increased speed of response there are smaller load fluctuation, decrease of electrode breakage and decrease of KWH. Consumption per ton of ingots.

4.8.1.8 Advantage of Direct Arc Furnace

1. Heat is not received by the charge by radiation from the arc, but some heat is also generated by the flow of current through the resistance of charge. Therefore, we obtain very high temperature of heat.

2. Since the arc is in contact with the charge, due to electromagnetic induction, there is stirring action, in the charge. This gives through mixing. Therefore, the melt, we get, is more uniform in composition.

4.9 INDIRECT ARC FURNACE

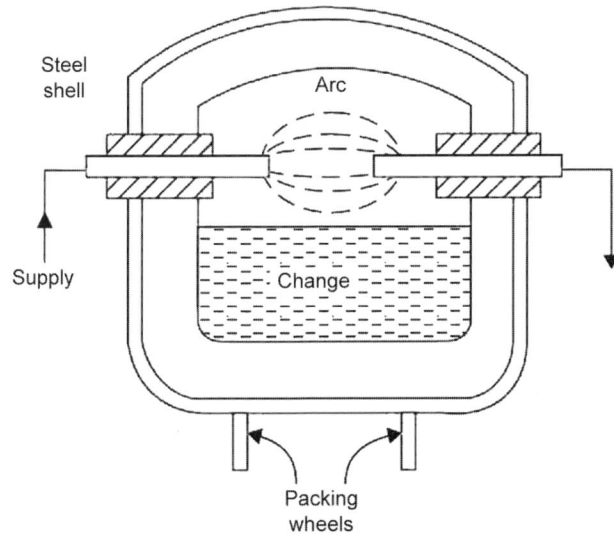

Fig. 4.17: Indirect arc furnace.

Here arc exists between two electrodes and heat is radiated from the arc to the charge Fig. 4.17. Maximum temperature attained by the charge is low as compared to that attained in direct arc furnace. As no current flows through the charge, there is no stirring action. Therefore, it becomes necessary to rock these furnace. These indirect furnaces are therefore made of cylindrical or spherical shell properly lined with refractory bricks, etc. from inside. Shell is supported horizontally on rollers. These rollers are connected by heavy reduction gears with reversible motor controlled by series of time relay switches by mean of which the furnace is rocked backward and forwards, angle progressively increasing from 15 to 20° with the start to 160° as melting proceeds. The objects of rocking the furnace is not only to bring the charge as quickly as possible into contact with the heating lining and to uncover parts of the charge below the surface which have not received heat radiation directly from the arc but this incidentally ensures long-life of the refractory lining. To make the rocking possible, there are two electrodes one from each side of cylindrical or spherical shell. In other words indirect arc furnaces are single phase type. To avoid heavy unbalanced load on 3 phase supply line, these indirect arc furnaces are made of low capacity say less than one ton. However special balancing circuit permits the load to be distributed equally over three phases.

This furnace is also sometimes called rocking arc furnace. The charge in this furnace is heated not only by radiation from the arc between electrode tips but it is also heated by conduction from the heat refractory during the rocking action. It should be noted that since arc does not strike direct to the charge, the risk of carbon pick-up is virtually eliminated.

The furnace is started by bringing the electrodes into solid contact to strike the arc and then withdrawing them to give the correct power input. Power input is regulated by adjusting the arc length by moving the electrodes which in turn are controlled by lead screws, either towards or away from each other. For manual control lead screws are moved by hand wheels. For automatic control lead screws are actuated by small motor controlled by relays in the furnace power circuit.

Another useful operational feature of this furnace is its shell changing. When original lining is in need of repairs shell can be lifted off its rollers by overhead crane after electrodes are withdrawn clear of the ports and new shell placed there. This therefore avoids outrage of furnace due to lining repair. Shell changing may also be required to workout heats of different metals requiring different types of furnace lining.

☞ *Advantages*

Rocking arc furnaces have got following advantages:

i. **Flexibility:** Single furnace can handle large or small heat of widely different analysis. Interchangeable furnace shells can be reserved for different alloys and can be substituted in a matter of minutes.

ii. **High melting speed:** Large areas of heated refractory come into contact with the charge by rocking action. This enables high melting rate to be attained without endangering the refractory. A large output of metal from a rapid succession of heat promotes the maximum utilization of foundry floor space and labour.

iii. **Economy:** Heart losses are small and power consumption is low. This is because melting is rapid and takes place in a completely closed chamber. High output rate gives low labour costs so that total production costs per ton of molten metal arc highly competitive.

iv. **Low metal losses:** Metal losses due to oxidation and volatilization are exceedingly low because furnace chamber is closed and carbon arc generates a reducing atmosphere above the metal.

v. **Sound castings:** The agitation caused by the rocking action of the furnace and the absence of combustion gases ensure freedom from blow holes, inclusion and segregations. High pouring temperature may readily be attained. These two factors are responsible for sound castings in thin and intricate designs.

☞ *Uses*

Rocking arc furnaces are ideally suited for making casting of alloy iron for heat resisting, abrasion resisting and similar special purposes. It is also suitable for nonferrous castings of copper, bronze, gun metal, nickel alloys, etc. particularly for hydraulic and other pressure fittings.

4.10 ELECTRON BOMBARDMENT HEATING

In electron bombardment heating, controlled stream of electrons is produced between an emitter and the material to be heated. When the fast moving electrons hit the material they release their KE which appears as heat in the material to be heated. Power released by the electron beam

$$P = nqV \text{ Watts} \tag{4.6}$$

where *n* is number of charged particles hitting per second, *q* the charge in coulombs per particle and *V* the voltage through which charges have been accelerated from rest.

Fig. 4.18 shows the layout of various essential parts of a versatile heating device where provision exists for accurate adjustment of the beam on the specimen. Emitter which is frequently a tungsten filament is heated by low voltage supply. In order that electrons emitted should have high kinetic energy, they have to be imparted necessary acceleration.

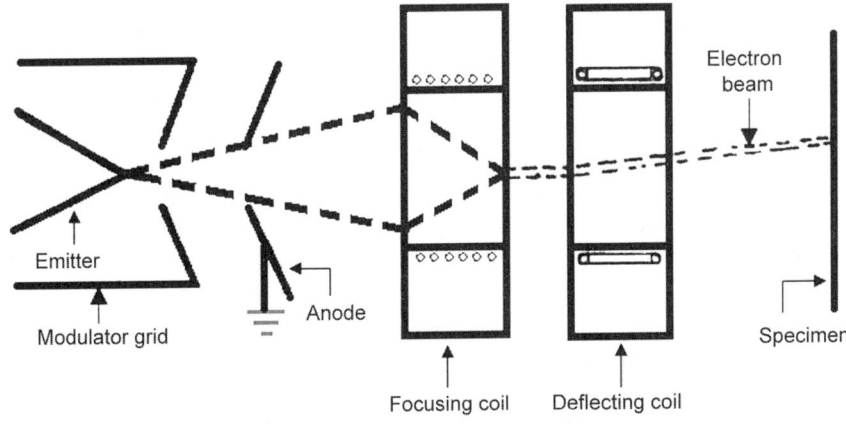

Fig. 4.18: Layout of various essential parts of a versatile heating device.

This is achieved by applying high DC voltage between emitter and anode which is usually kept at earth potential. An aperture is provided in the anode plate to allow the accelerating electrons to pass through to enable them to be applied to the specimen. In the absence of modulator grid, the stream of electrons is not well-defined with the result that large area of specimen will be heated. The area to be heated is limited by providing metal screen called modulator surrounding the cathode. This screen is kept at slightly negative potential with respect to emitter and a well-defined stream of electron passes through the aperture in the modulator. By adjusting the current through focusing lens, the area heated by the beam can be varied in ratio of more than 1:100. Direction of the beam can be adjusted by varying the current through the definition coil. This is how accurate choice of region to be heated is made. Power of the beam can be changed by changing the KE of the colliding electron of the beam. This is achieved by either varying the emission of cathode or by varying the accelerating voltage between cathode and anode. There is however limit to this voltage as when it exceeds 20 to 30 kV, it increases the intensity of X-rays produced form which shielding has to be provided. Since the mean free path of electrons is too low to be useful at high pressure. It becomes necessary to carry out heating by electron bombardment only in vacuum at pressures much below 10^{-3} mm of mercury. Electron beam need not be circular. We can have line focus beams of any other shaped input. Power density possible with this method of heating is very high and power density of 5×10^5 kw/sq. cm is said to have been achieved. The main difficulty in electron beam melting furnaces is the pressure rise on account of volatilization of the impurities and the metal itself. This can cause electrical discharge breakdown. As such these furnaces are provided with adequate pumping capacity and surge limiting devices are incorporated in the power supply.

Application of this type of heating will be found in floating zone melting, growing of silicon crystals, electron beam melting electron beam wielding. Floating zone melting is adopted for reactive and refractory material which if heated in crucible get contaminated. Specimen to be zone melted are made in the form of rod by arc melting or powder pressing. Very narrow region of the vertical rod is melted or powder pressing. Very narrow region of the vertical rod is melted which is held attached to the patent metal by the forces of surface tension. This way zone can be moved along the rod. Zone melting is used in laboratory for refining, homogenizing the alloys and growing single crystals. Electron bombardment heating is used in crystal pulling furnace also. One difficulty with growing single crystals of silicon is that this material is easily contaminated by the impurities originating from the material of the crucible or the heating filament. Unlike heating the material in crucible in eddy current heating, here molten pool of silicon is formed by electron bombardment. This molten pool rests on underlying parent material. The beam power is adjusted by moving the filament assembly axially with respect to modulator in such a way that we obtain stable molten pool. Single crystal seed is then dipped into the molten pool which after being wetted by the melt is pulled or withdrawn at a speed of 2 mm/min. till molten pool approaches the base of the charge. Silicon crystals produced by pulling technique in vacuum are found to have very much reduced oxygen content as determined by optical transmission measurements.

4.11 HIGH FREQUENCY HEATING

The main and conspicuous difference between conventional method of heating and high frequency method of heating is that in the former case heat is transferred either by convection, conduction, or radiation while in the latter case conversion of electromagnetic energy into heat energy takes place inside the material.

High frequency heating can be applied mainly to two classes of material firstly, conducting materials ferromagnetic of non-magnetic; secondly, insulating material. Heating process to first class of material is called induction heating and to the second class of material as dielectric heating.

4.12 INDUCTION HEATING

When alternating current flows in a conductor, it provides reversing magnetic field. emf will be induced in the secondary of a transformer, the only difference between the two process is that in transformer, electrical energy transferred from primary, by electromagnetic induction is utilized outside the secondary winding whereas in induction heating, this energy is utilized in the secondary winding, i.e. in heating the charge itself.

We are quite familiar with eddy current loss in transformer cores. This loss is dissipated in the form of heat in the transformer cores. If such a loss in the form of a heat is utilized in heating metals, it becomes useful. Consider a piece of metal forming the core of the coil in Fig. 4.19. When an alternating voltage is applied to the coil, an alternating flux is set-up and to the rate of change of flux eddy currents flow in the metal. The direction of the flux is along the axis of the coil. The induced voltage in the core on account of the rate of change of the flux is in a plane at right angle to the direction of the flux. The eddy currents due to the induced voltage are also in this plane. The metal core therefore provides closed paths at right angles to the flux. These closed paths are like the short circuited secondaries having resistance and inductive reactance. The power loss of the eddy currents in the core produces heat.

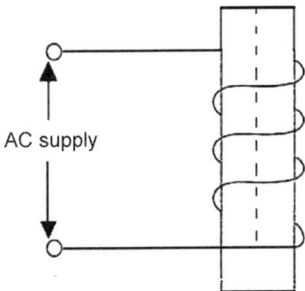

Fig. 4.19: Induction heating.

The heat in the disc can be increased by:

    i. High coil current;
    ii. Larger number of coil turns;
    iii. High frequency supply;
    iv. Close spacing between the coil and work;
    v. The disc may be magnetic material (higher permeability);
    vi. Higher electrical resistivity of the disc (magnetic material).

In case the charge to be heated is non-magnetic the heat generated is due to eddy current losses, whereas, if it is a magnetic material, there will be hysteresis losses in addition. The hysteresis loss is proportional to frequency whereas eddy current loss is proportional to square of frequency when operating at low frequency. At high frequency the heating due to hysteresis becomes very small as compared to eddy currents. This is due to higher temperature attained by the charge at higher frequencies when the material ceases to posses magnetic properties. It is well known that above curie temperature the magnetic materials lose their magnetic properties. Similarly eddy current losses also do not follow f^2 law as frequency is increased higher and higher. In fact at 10 kHz the total heating may vary directly as the frequency and even drop to one-half power at frequencies of the order of 500 kHz.

The depth of penetration of induced current into the disc.

$$d = \frac{1}{2\pi} \sqrt{\frac{\rho \times 10^9}{\mu f}} \text{ cm.}$$

Where ρ = specific resistance of molten charge in ohm-cm.

    f = frequency in Hz; μ = permeability of the charge.

This shows higher the frequency of supply the lower the depth of penetration.

The depth is $\propto \dfrac{1}{f}$.

Types of Induction Furnaces

There are mainly two types of induction furnaces: (i) Core type or low frequency induction furnace; (ii) Coreless type or high frequency induction furnace.

Core type furnaces are of three different construction, i.e. into (i) Direct core-type furnace (ii) Vertical core-type or Ajax Wyatt furnace and (iii) Indirect core-type furnace.

4.12.1 Direct Core-type Furnace

It is shown in Fig. 4.20 and is essentially a transformer in which the charge to be heated forms a single-turn short-circuited secondary and is magnetically coupled to the primary by an iron core. The furnace consists of a circular hearth which contains the charge to be melted in the form of an annular ring. When there is no molten metal in the ring, the secondary becomes open-circuited thereby cutting off the secondary current. Hence, to start the furnace, molted metal has to be poured in the annular hearth. Since, magnetic coupling between the primary and secondary is very poor, it results in high leakage and low power factor. In order to nullify the effect of increased leakage reactance, the furnace is operated at low frequencies of the order of 10 Hz or so. If the transformer secondary current density exceeds 500 A/cm^2 then, due to the interaction of secondary current with the alternating magnetic field, the molten metal is squeezed to the extent that secondary circuit is interrupted. This effect is known as **"pinch effect"**.

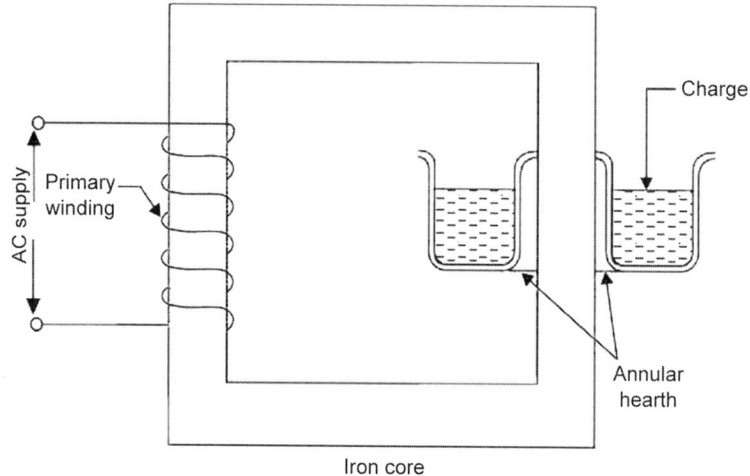

Fig. 4.20: Direct core-type furnace.

☞ *Drawbacks*

1. It has to be run on low-frequency supply which entails extra expenditure on motor-generator set or frequency convertor.
2. It suffers from pinching effect.
3. The crucible for charge is of odd shape and is very inconvenient for tapping the molten charge.
4. It does not function if there is no molten metal in the hearth, i.e. when the secondary is open. Every time molten metal has to be poured to start the furnace.
5. It is not suitable for intermittent service.

However, in this furnace, melting is rapid and clean and temperature can be controlled easily. Moreover, inherent stirring action of the charge by electromagnetic forces ensures greater uniformity of the end product.

4.12.2 Vertical Core-type Furnace

It is also known as Ajax-Wyatt furnace and represents an improvement over the core-type furnace discussed above. As shown in Fig. 4.21, it has vertical channel (instead of a horizontal one) for the charge, so that the crucible used is also vertical which is convenient from metallurgical point of view. In this furnace, magnetic coupling is comparatively better and power factor is high. Hence, it can be operated from normal frequency supply. The circulation of the molten metal is kept up round the Vee portion by convection currents as shown in Fig. 4.21. As Vee channel is narrow, even a small quantity of charge is sufficient to keep the secondary circuit closed. However, Vee channel must be kept full of charge in order to maintain continuity of secondary circuit. This fact makes this furnace suitable for continuous operation. The tendency of the secondary circuit to rupture due to pinch-effect is counteracted by the weight of the charge in the crucible. The choice of material for inner lining of the furnace depends on the type of charge used. Clay lining is used for yellow brass. For red brass and bronze, an alloy of magnetia and alumina or corundum is used. The top of the furnace is covered with an insulated cover which can be removed for charging. The furnace can be tilted by the suitable hydraulic arrangement for taking out the molten metal. This furnace is widely used for melting and refining of brass and other non-ferrous metals. As said earlier, it is suitable for continuous operation. It has a PF of 0.8–0.85. With normal supply frequency, its efficiency is about 75% and its standard size varies from 60–300 kW, all single-phase.

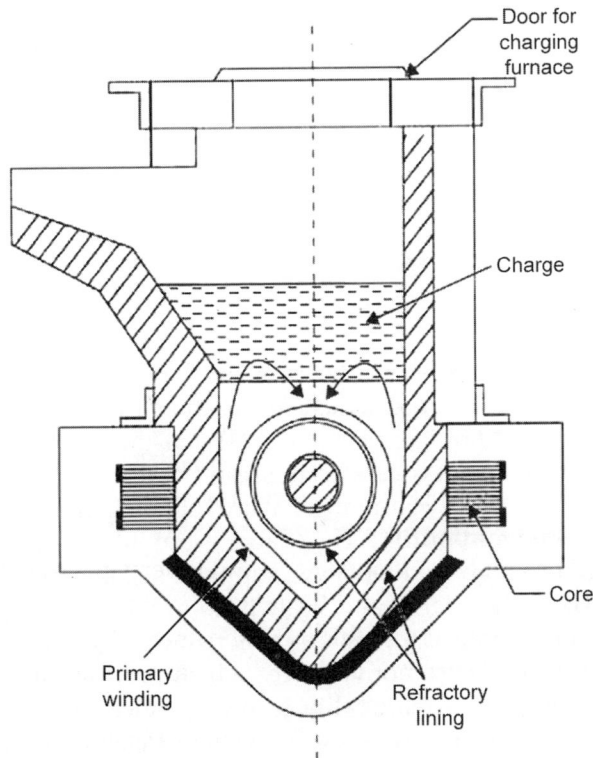

Fig. 4.21: Ajax-Wyatt furnace.

4.12.3 Indirect Core-type Furnace

In this furnace, a suitable element is heated by induction which, in turn, transfers the heat to the charge by radiation. So far as the charge is concerned, the conditions are similar to those in a resistance oven. As shown in Fig. 4.22, the secondary consists of a metal container which forms the walls of the furnace proper. The primary winding is magnetically coupled to this secondary by an iron core. When primary winding is connected to AC supply, secondary current is induced in the metal container by transformer action which heats up the container. The metal container transfers this heat to the charge. A special advantage of this furnace is that its temperature can be automatically controlled without the use of external equipment. The part AB of the magnetic circuit situated inside the oven chamber consists of a special alloy which loses its magnetic properties at a particular temperature but regains them when cooled back to the same temperature. As soon as the chamber attains the critical temperature, reluctance of the magnetic circuit increases manifold thereby cutting off the heat supply. The bar AB is detachable and can be replaced by other bars having different critical temperatures.

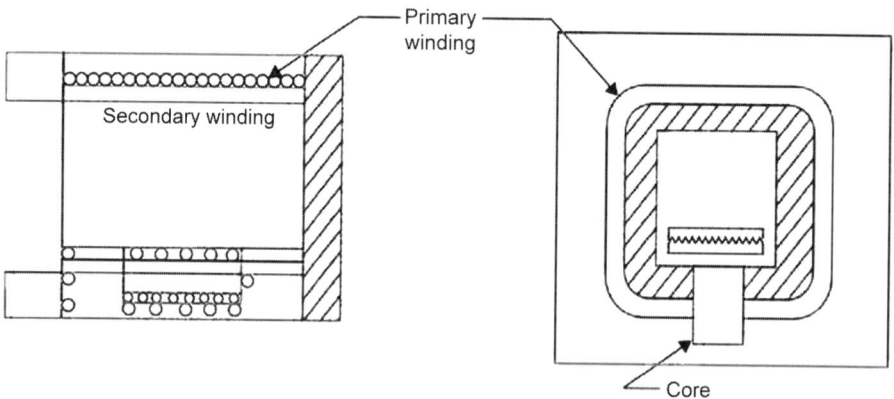

Fig. 4.22: Indirect core-type furnace.

4.12.4 Coreless Furnace

As shown in Fig. 4.23, the three main parts of the furnace are (i) primary coil (ii) a ceramic crucible containing charge which forms the secondary and (iii) the frame which includes supports and tilting mechanism. The distinctive feature of this furnace is that it contains no heavy iron core with the result that there is no continuous path for the magnetic flux. The crucible and the coil are relatively light in construction and can be conveniently tilted for pouring. The charge is put into the crucible and primary winding is connected to a high-frequency AC supply. The flux produce by the primary sets up eddy-currents in the charge and heats it up to the melting point. The charge need not be in the molten state at the start as was required by core-type furnaces. The eddy-currents also set-up electromotive forces which produce stirring action which is essential for obtaining uniforms quality of metal. Since flux density is low (due to the absence of the magnetic core) high frequency supply has to be used because eddy-current loss $W_e \propto B^2 f^2$. However, this high frequency increases the resistance of the primary winding due to skin effect, thereby increasing primary Cu losses. Hence,

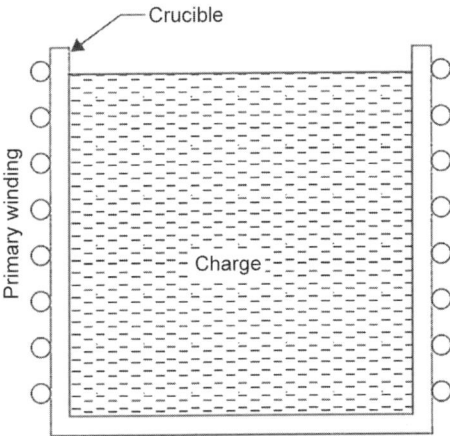

Fig. 4.23: Coreless furnace.

the primary winding is not made of Cu wire but consists of hollow Cu tubes which are cooled by water circulating through them. Since magnetic coupling between the primary and secondary windings is low, the furnace PF lies between 0.1 and 0.3. Hence, static capacitors are invariably used in parallel with the furnace to improve its PF

Such furnaces are commonly used for steel production and for melting of non-ferrous metals like brass, bronze, copper and aluminum, etc., along with various alloys of these elements. Special application of these furnaces include vacuum melting, melting in a controlled atmosphere and melting for precision casting where high frequency induction heating is used. It also finds wide use in electronic industry and in other industrial activities like soldering, brazing hardening and annealing and sterilizing surgical instruments, etc.

☞ *Advantages*

1. They are fast in operation.
2. They produce most uniform quality of product.
3. They can be operated intermittently.
4. Their operation is free from smoke, dirt, dust and noises.
5. They can be used for all industrial applications requiring heating and melting.
6. They have low erection and operating costs.
7. Their charging and pouring is simple.

4.12.5 Sources of High frequency for Induction Heating

Basically there are three types of equipment used for converting the electrical energy at low frequency to one at high frequency suitable for induction heating:

    i. The motor generator set.
    ii. The spark gap converter.
    iii. The vacuum tube oscillator.

The motor generator set consists of an induction motor coupled to a specially designed generator having both the armature and field winding on the stator. The change in reluctance produces a corresponding change in the magnetic flux flowing in the iron circuit which induces voltage in the armature winding. The voltage is proportional to the rate of change of total flux and the frequency is determined by the number of complete flux reversals per second.

The basic principle of operation of a spark gap converter is the alternate charging an discharging of a capacitor (Fig. 4.24). The discharge current is passed through the turns of the induction heating coil. The spark gap acts as valve which periodically connects the capacitor to the charging source. The valve of inductance and capacitance in the discharge circuit decide the frequency of heating. The spark gap converter produces pulses of high frequency but of decreasing magnitude.

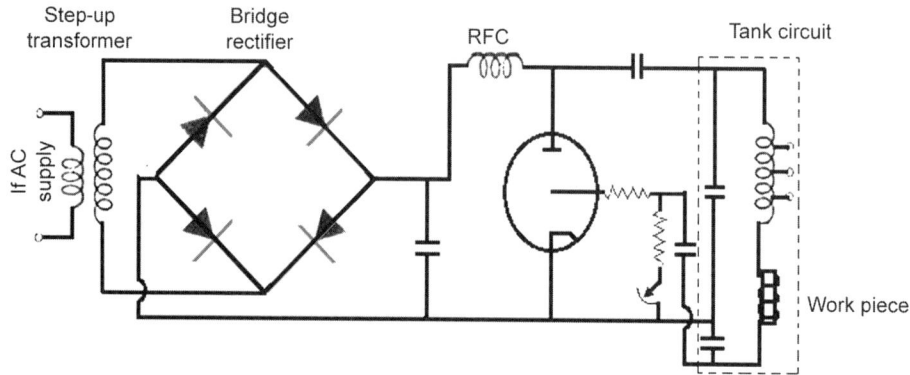

Fig. 4.24: Simplified circuit of a vacuum tube oscillator.

The vacuum tube oscillator or electronic heater as it is some times called, produces large power output at a higher frequency of the order of 600 kHz. Fig. 4.25 shows a typical oscillator. The 50 Hz supply is first converted into DC and then inverted into high frequency currents.

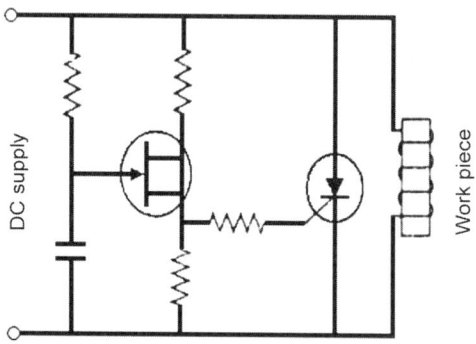

Fig. 4.25: SCR inverter circuit.

The capacitances and inductances (that of working piece also) decide the frequency of supply to the work piece. The vacuum tube oscillators are less efficient as compared to SCR as the voltage drop across the SCR is very low of the order of 1 V. The SCR operates at 90% efficiency. The SCRs are fired by pulses of gate current produced by a UJT as shown in Fig. 4.26.

The frequency of supply to the work piece depends upon the value of R and C. The smaller the product of RC, higher is the frequency across the work piece.

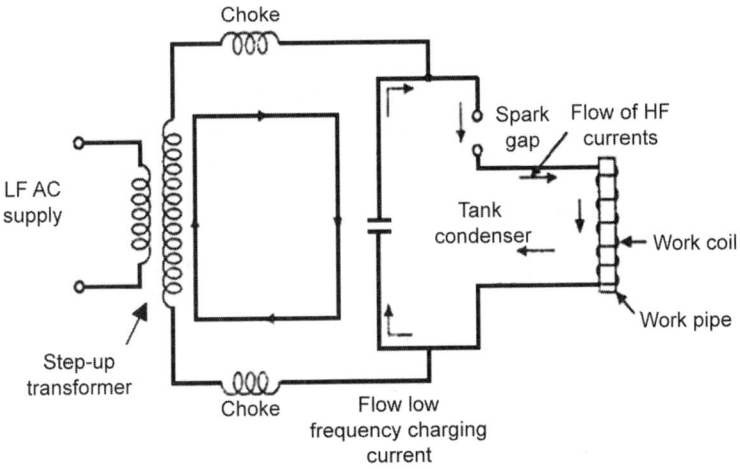

Fig. 4.26: Basic spark-gap converter circuit.

4.13 DIELECTRIC HEATING

When a ferromagnetic material is subjected to alternating magnetic field, it gets heated up due to eddy currents induced in the material and the hysteresis loss in the material. The hysteresis loss is due to reversal of magnetic field which brings about magnetic molecular friction and results in heating the material. Similarly when an insulating material is subjected to an alternating electric field, the atoms get stressed and due to the inter-atomic friction heat is produced. This loss is known as dielectric loss.

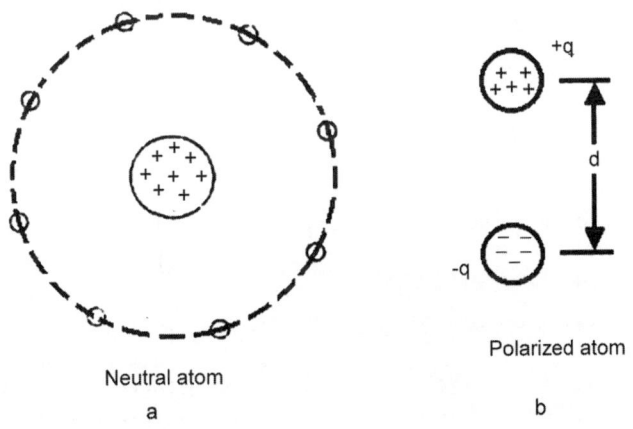

Figs 4.27a and b: Dielectric material—an atom.

We know that an atom of any material consists of a nucleus, positively charged at the centre and electrons, negatively charged surrounding the nucleus. The center of the negative charge coincides with the nucleus and hence, atom as a hole acts as a neutral particle.

Fig. 4.27a when it is subjected to an electric field, the charge distribution is distributed and the centre of negative charge does not coincide with the nucleus, the two centres are displaced by say distance d, Fig. 4.27b. The electric is said to be polarized. If the charge on the nucleus is q, the product qd is known as the dipole moment and acts in a direction from negative charge to positive charge. If the electric field is alternating, the forces within the atoms are reversed and a distortion of electron path occurs in the opposite direction. The atoms are thus in a constant state of unrest changing with each shift in polarity. It is observed that the loss increases with increase in frequency and strength of the electric field. However, if the electric field applied is very strong, it may overstress the charges and hence some electrons may be knocked off from the outermost orbit resulting in rupture of the dielectric medium. Also as far as possible no air gap should be left over between the electrode and the specimen to be heated as normally the dielectric strength of air is smaller than those of dielectric materials to be heated. Also for any voltage applied across the electrodes, the electric field intensity in the air gap is higher than in the specimen. If an air gap is left, the air will be ionized and may result in breakdown. Therefore, it is desirable in dielectric heating not to apply high voltage but to use high frequencies. Even though voltages up to 20 KV have been used but from personal safety point of view voltages between 600 to 3000 are more common.

All dielectric materials can be considered to be imperfect capacitor and can be represented by a parallel combination of a resistor R and a capacitor C, as shown in Figs 4.28a to c.

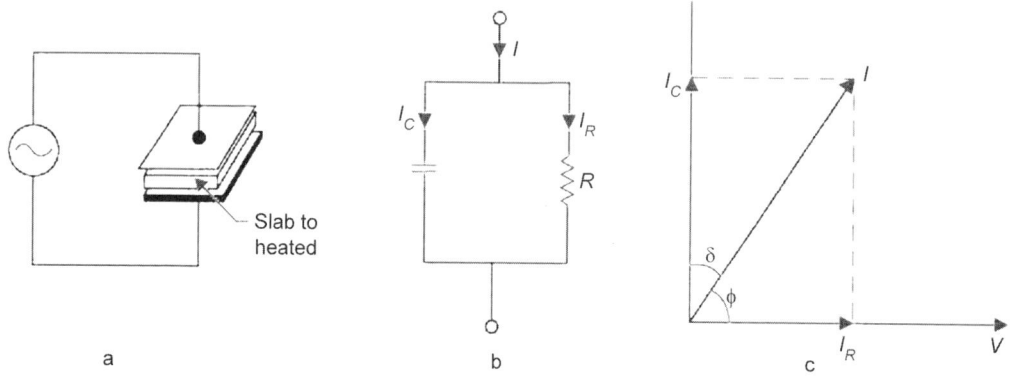

a　　　　　　　　　　　b　　　　　　　　　　　c

Figs 4.28a to c: (a) Dielectric between the parallel plates, (b) Equivalent circuit, (c) Phasor diagram.

Here current I_l through the equivalent resistance represents the leakage current, I_c represents the charging current of the dielectric.

Dielectric loss $P = \dfrac{V^2}{R}$

From the phasor diagram $\dfrac{V/R}{V\omega C} = \tan \delta$

$\therefore$

$$P = V^2 \omega C \tan \delta \qquad\qquad (4.7)$$

Normally δ is small $\tan\delta = \delta$

$\therefore$ $$P = V^2 \omega C \delta$$

where δ is in radian and is known as the loss angle. It is an indication of the state of the dielectric whether it is healthy or unhealthy. The higher the value of δ, the less healthy is the state of the dielectric material.

The capacitance of a parallel plate capacitor is given as:

$$C = \frac{\varepsilon_0 \varepsilon_r A}{d}$$

where ε_0 is the permittivity constant, ε_r is the relative permittivity constant of the medium. A is the area of the plate in sq.m, d is the thickness of the medium in metres.

$$P = V^2 2\pi f \cdot \frac{\varepsilon_0 \varepsilon_r A \delta}{d}$$

$\therefore$ (4.8)

i.e. $P \propto V^2$ and $P \propto f$

The product $\varepsilon_r \, \delta$ is known as the loss factor. The relative permittivity is an induction of the retained energy due to molecular deformation whereas δ is an indication of the amount of leakage current which will flow through the insulator and so produce a heat loss.

4.13.1 Advantages of Dielectric Heating

i. Since heat is generated within the dielectric medium itself, it results in uniform heating.
ii. Heating becomes faster with increasing frequency.
iii. It is the only method for heating bad conductors of heat.
iv. Heating is fastest in this method of heating.
v. Since no naked flame appears in the process, inflammable articles like plastics and wooden products, etc., can be heated safely.
vi. Heating can be stopped immediately as and when desired.

4.13.2 Uses of Dielectric Heating

Since cost of dielectric heating is very high, it is employed where other methods are possible or are too slow. Some of the applications of dielectric heating are as under:

i. For gluing of multilayer plywood boards.
ii. For baking of sand cores which are used in the moulding process.
iii. For preheating of plastic compounds before sending them to the moulding section.
iv. For drying of tobacco after glycerine has been mixed with it for making cigarettes.
v. For baking of biscuits and cakes, etc. in bakeries with the help of automatic machines.
vi. For electronic sewing of plastic garments like raincoats, etc. with the help of cold rollers fed with high-frequency supply.

vii. For dehydration of food which is then sealed in air-tight containers.
viii. For removal of moistures from oil emulsions.
ix. In diathermy for relieving pain in different parts of the human body.
x. For quick drying of glue used for book binding purposes.

4.14 CHOICE OF FREQUENCY

The selection of frequency for heating is important because it has a great bearing on the work to be heated and the method of its heating whether by induction heating or dielectric heating. Furnaces running on power frequency of 50 Hz can be of 1 MW capacity whereas those running on medium frequencies (500 to 1000 Hz) have a capacity of 50 kW and those running on high frequency (1 to 2 MHz) have capacities ranging from 200 to 500 kW.

1. **Induction heating:** While choosing frequency for induction heating, the following factors are considered:

   a. Thickness of the surface to be heated. Higher the frequency, thinner the surface that will get heated.

   b. The time of continuous heating. Longer the duration of heating, deeper the penetration of heat in the work due to conduction.

   c. The temperature to be obtained. Higher the temperature, higher the capacity of the generator required.

2. **Dielectric heating:** The power consumed during dielectric heating, $P = 2\pi f C V^2$ tanδ. As seen, $P \propto f V^2$. Hence, rate of heat production can be increased by increasing voltage or voltage across any specimen is limited by its thickness or because of the consideration of potential gradient, breakdown voltage and safety, etc., voltages ranging from 600 to 3000 V are used for dielectric heating, although voltages of 20 kV or so are also used sometimes.

Rate of heat production can also be increased by applying high potential but it is also limited because of the following considerations:

   a. Possibility of formation of standing waves between the surface of two electrodes having wavelength nearly equal to or more than one quarter of the wavelength of the particular frequency used.

   b. Necessity of employing special matching circuit at higher frequencies due to the fact that maximum power transfer takes place when the oscillator impedance equals the load impedance.

   c. At higher frequencies it is difficult for tuning inductance to resonate with the charge capacitance.

   d. At higher frequencies, it is almost impossible to get uniform voltage distribution.

   e. Since higher frequencies disturb nearby radio station services, special arrangement has to be made to stop radiations from the high-frequency generator used for the purpose.

Table 4.2 shows the frequencies required for various heating purposes and the type of equipment used.

SCR inverter circuits can also be used for all these applications.

Table 4.2	Types of heating, frequency and source used	
Type of heating	*Frequency used*	*Source used*
1. Induction heating		
a. Low temperature heating of metal, annealing.	50–500 Hz	Rotating generator or a diode frequency converter.
b. Melting, deep heat penetration 'through' heating.	500 to 10 kHz	-do-
c. Surface heating of metals	10 to 200 kHz	Spark gap generator.
Hardening	100 to 500 kHz	Spark gap generator, vacuum tube oscillator.
d. Heating metal pieces, wire and metal strips	400 to 1000 kHz	Vacuum tube oscillator.
2. Dielectric heating	1 to 50 MHz	Vacuum tube oscillator.

WORKED EXAMPLES

☻ EXAMPLE 4.1: A resistance oven employing nichrome wire is to be operated from 220 V single-phase supply and is to be rated at 16 kW. If the temperature of the element is to be limited to 1170°C and average temperature of the charge is 500°C find the diameter and length of the element wire. Radiating efficiency = 0.57, Emissivity = 0.9, Specific resistance of nichrome = 109 × 10⁻⁸ Ωm.

Solution: Given: $V = 220$ V; $P = 16$ kW ; $T_1 = 273 + 1170 = 1443$ K; $T_2 = 273 + 500 = 773$ K

$\eta_{rad} = 0.57$; $e = 0.9$; $\rho = 109 \times 10^{-8}$ Ωm.

l, d :

We know that, $\dfrac{l}{d^2} = \dfrac{\pi V^2}{4\rho P} = \dfrac{\pi \times 220^2}{4 \times 109 \times 10^{-8} \times (16 \times 10^3)} = 2179660$ (i)

Now, $H = 5.57\eta_{rad}e\left[\left(\dfrac{T_1}{100}\right)^4 - \left(\dfrac{T_2}{100}\right)^4\right]$ W/m²

$= 5.67 \times 0.57 \times 0.9 \left[\left(\dfrac{1443}{100}\right)^4 - \left(\dfrac{773}{100}\right)^4\right] = 115729$ W/m²

Now, total heat dissipated/sec. = Electrical power input

∴ $\pi \times l \times 115729 = 16000$ $dl = 0.044$

Or $d^2 l^2 = 0.001936$ (ii)

Multiplying (i) and (ii), we have

$0.001936 = 4219.8$

$l = 16.16$ m

$d = \dfrac{0.044}{16.16} = 2.723 \times 10^{-3}$ m $= 2.723$ *mm*

☺ **EXAMPLE 4.2**: The following data related to a 3-phase are furnace:

Quantity of steel to be melted in one hour = 4.3 tonnes

Specific heat of steel = 0.5 kJ/kg°C

Latent heat of steel = 37.2 kJ/kg

Melting point of steel = 1370°C

Initial temperature of steel = 19.1°C

Overall efficiency of steel = 50%

Input current = 5700 A

Resistance of transformer referred to secondary = 0.008 Ω

Reactance of transformer referred to secondary = 0.014 Ω

Determine the following:

   i. Average kW input to the furnace

  ii. Arc voltage

 iii. Arc resistance

 iv. Power factor of the current drawn from the supply, and

  v. Average kVA input to the furnace.

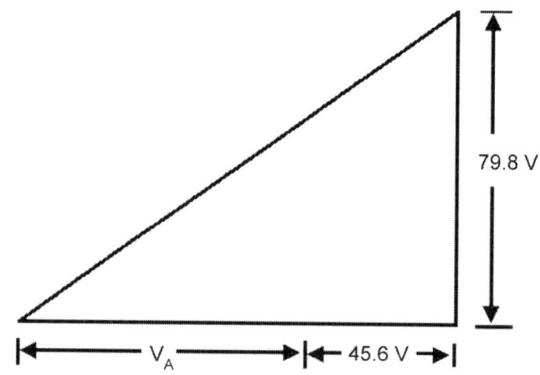

Fig. 4.29: Voltage drop due to transformer resistance and reactance.

Solution: Given: m = 4.3 tonnes = 4300 kg; Time = 1 hour; c = 0.5 kJ/kg°C; L = 37.2 kJ/kg; Melting point, t_2 = 1370°C; Initial temp. of steel, t_1 = 19.1°C; η = 50%, I = 5700 A; R = 0.008 Ω; X = 0.014 Ω

  i. Average kW input to the furnace:

Energy required to melt 4.3 tonnes of steel

$$= m \times c \times (t_1 - t_2) + mL = m \ [c(t_2 - t_1) + L]$$

$$= 4300 \ [0.5 \times (1370 - 19.1) + 37.2] = 3064395 \ kJ = \frac{3064395}{3600} = 851.2 \ kWh$$

$$\text{Average output} = \frac{\text{Total energy required in kWh}}{\text{Time of melting in hours}}$$

$$= \frac{851.2}{1} = 851.2 \ kW$$

$$\text{Average input} = \frac{\text{Average output}}{\text{Overall efficiency}} = \frac{851.2}{0.5} = 1702.4 \ kW.$$

ii. Arc voltage, V_A:

Voltage drop due to transformer resistance = 5700 × 0.008 = 45.6 V

Voltage drop due to transformer reactance = 5700 × 0.014 = 79.8 V

From Fig. 4.29:

Open circuit secondary voltage

$$= \sqrt{(V_A + 45.6)^2 + 79.8^2} \ \ volts$$

Power factor, $\cos \phi = \dfrac{(V_A + 45.6)}{\sqrt{(V_A + 45.6)^2 + 79.8^2}}$

Total power input = 3 × power drawn per phase

= 3 × current drawn per phase × secondary voltage × PF

$$1702.4 \times 10^3 = 3 \times 5700 \times \sqrt{(V_A + 45.6)^2 + 79.8^2} \times \dfrac{V_A + 45.6}{\sqrt{(V_A + 45.6)^2 + 79.8^2}}$$

$$V_A + 45.6 = \dfrac{1702.4 \times 10^3}{3 \times 5700}$$

Arc voltage, 54 V. **(Ans.)**

iii. Arc resistance, R_A:

$$R_A = \dfrac{V_A}{I} = \dfrac{54}{5700} = 0.00947 \ \Omega$$

iv. Power factor, cos:

$$\cos \phi = \dfrac{45.6}{\sqrt{(V + 45.6) + 79.8}}$$

$$= \dfrac{54 \quad 45.6}{\sqrt{(\quad + 45.6) + 79.8}} = 0.7804$$

v. Average kVA input to the furnace:

$$\text{kVA input to the furnace} = \dfrac{\text{kW input}}{\text{Power factor}}$$

$$= \dfrac{1702.4}{0.7804} = 2181.4 \ \text{kVA} = 2181.4 \ \text{kVA}.$$

☻ **EXAMPLE 4.3**: Calculate the efficiency of a high frequency induction furnace which takes 10 minutes to melt 1.8 kg of aluminium. The input to the furnace being 4.8 kW and initial temperature 15°C. Specific heat of aluminium = 0.88 kJ/kg°C; melting point of aluminium = 660°C; latent heat of fusion of aluminium = 32 kJ/kg; 1 kJ = 2.78 × 10^{-4} kWh.

Solution. Given: $m = 1.8$ kg ; Input to the furnace $= 4.8$ kW; $t_1 = 15°C$; $t_2 = 660°C$, $c = 0.88$ kJ/kg°C; $L = 32$ kJ/kg; 1 kJ $= 2.78 \times 10^{-4}$ kWh.

Efficiency of the furnace, η:

Heat required to melt 1.8 kg of aluminium $= m \times c \times (t_2 - t_1) + m \times L$

$= m [c \times (t_2 - t_1) + L] = 1.8 [0.88 (660 - 15) + 32] = 1079.28$ kJ

$= 1079.28 \times 2.78 \times 10^{-4} = 0.3$ kWh

Energy input $= 4.8 \times \dfrac{10}{60} = 0.8$ kWh

$\eta = \dfrac{\text{Output}}{\text{Input}} = \dfrac{0.3}{0.8} = $ or 37.5%. **(Ans.)**

QUESTIONS

1. Why electric heating is preferred over other form of heating?
2. Explain in brief how heating is done in the following cases?
   (i) Resistance heating, (ii) Induction heating, (iii) Dielectric heating.
3. Explain the principle of induction heating. What are the applications of induction heating?
4. With a neat sketch explain the working principle of coreless type induction furnace.
5. Explain with a neat sketch the principle of coreless type induction furnace.
6. Give relative advantages and disadvantages of direct and indirect electric arc furnances.
7. What are different methods of heat transfer explain in brief?
8. What are the advantages of radiant heating?
9. Describe various types of electric heating equipment.
10. What are the causes of failure of heating elements?
11. Explain why very high frequencies should not be used for dielectric hearing.
12. Write short notes on the Ajax-Wyatt furnace.
13. Discuss the different methods of electric heating and their relative merits.
14. State the advantages of electric heating.
15. Brifely explain the different methods of electric heating.

ELECTRIC WELDING

There is a saying, "If it's metal, weld it". Welding is the metallurgical process of combining two pieces of metallic or non-metallic substances at faces rendered plastic or liquid by applying heat or pressure or both. Filler material can be used to make the union. Depending upon the amount of heat applied; we can classify the welding process such as thermal welding, gas welding and electric welding. In this chapter, we will not deal with thermal and gas welding. Electric welding will be discussed thoroughly along with some introduction to other modern welding techniques.

There are many ways of welding. All welding processes can be summarized into two distinct categories:

1. **Fusion welding:** Here the parent material is melted. Examples are:
   i. Carbon arc welding, metal arc welding, electron beam welding, electro slag welding and electro gas welding which utilize electric energy, and
   ii. Gas welding and thermit welding which utilize chemical energy for the melting purpose.
2. **Non-fusion welding:** It does not involve melting of the parent metal. Examples are:
   i. Forge welding and gas nonfusion welding which use chemical energy.
   ii. Explosive welding, friction welding and ultrasonic welding, etc., which use mechanical energy.
   iii. Resistance welding which uses electrical energy.

Proper selection of the welding process depends on the (a) type of metals to be joined, (b) probable expenditure, (c) nature of products to be fabricated, and (d) production techniques adopted. The principal welding processes have been tabulated in Fig. 5.1.

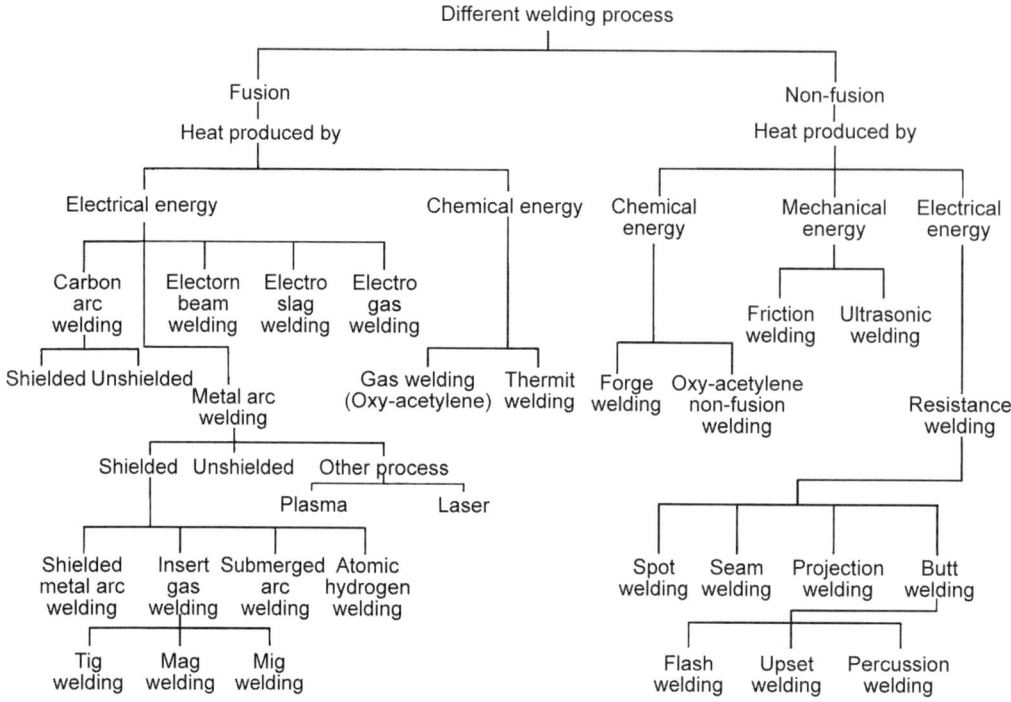

Fig. 5.1: Principal welding processes.

5.3　ELECTRIC ARC WELDING

An electric arc is the conduction of electric current through gases accompanied by heat and bright glow. An electric arc is created by short circuiting two electrodes and then drawing them apart. In the process of separating the electrodes, the area of contact of electrodes first reduces which increases the resistance, the tips of electrodes are already in red hot condition. The electrons, emerging out of the cathode and moving under accelerating force on account of voltage gradient existing between the electrodes, will on their way collide with the atoms and molecules of air, thereby ionizing the same. This is how air or gas in between electrodes create a conducting path. We will now consider in details the electrical properties of an arc.

a. **Arc stability:** Electric arc has negative resistance characteristic. As the arc current increases, voltage across it decreases, i.e. resistance of the arc decreases. Volt-amp. Characteristic of an electric arc is given in Fig. 5.2. In manual arc welding, it is impossible for a welder to hold a constant arc weld length. Decrease in arc length will result in the decrease in arc resistance and, therefore, building up of arc current. This will further reduce arc resistance. This way, effect and cause help each other till arc will be out of control. Similarly increase in the arc length will bring about increase in resistance of arc which will reduce the arc and its diameter. This will further increase the resistance of arc. Here again conditions approaching extinction of the arc will be reached. It will, therefore, to be observed that it is very difficult job for the operator to maintain a stable arc. How stability of the arc is achieved will be considered under heading "Electric supply for arc welding."

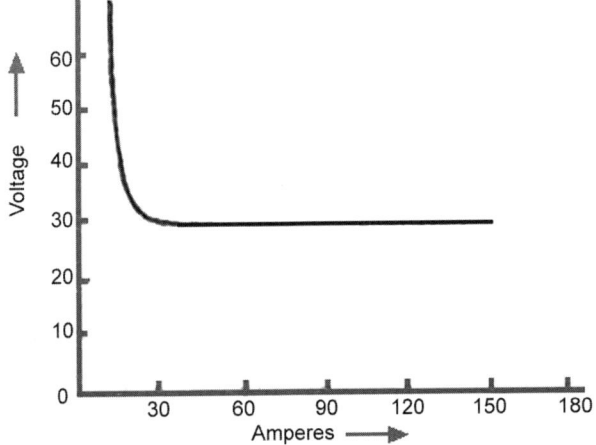

Fig. 5.2: V-I Characteristic of an electric arc.

b. We require more voltage to maintain an arc than striking it. This is because some voltage is required to ionize the air gap.

c. Electrons leave cathode and after attaining certain velocity, hit the anode where their momentum is destroyed. That is why more heat is produced in anode than cathode. As we are interested to heat the work piece to be welded, it is always connected to positive terminal of DC supply and welding electrode to the negative terminal.

d. **Arc blow:** An arc column can be considered as a flexible current-carrying conductor which can be easily deflected by the magnetic field set-up in its neighbourhood by the positive and negative leads from the DC welding set. The two leads carry currents in opposite directions and hence, set-up a repulsive magnetic force which pulls the arc away from the weld point, particularly in the welding corners, where field concentration is maximum. The deflection of the arc is called arc blow. This condition is encountered only with DC welding sets and is especially noticeable when welding with bare electrodes. It is experienced most when using currents above 200 A or below 40 A.

Due to arc blow, heat penetration in the required area is low which leads to incomplete fusion and bead porosity apart from excessive weld spatter. Arc blow can be avoided by using AC rather than DC welding machines because reversing currents in welding produces magnetic fields which cancel each other out thereby eliminating the arc blow. However, with DC welding machines, arc blow effects can be minimized by (i) welding away from the earth ground connection, (ii) changing the position of the earth connection on the work, (iii) wrapping the welding electrode cable a few turns around the work, (iv) reducing the welding current or electrode size, (v) reducing the rate of travel of the electrode and (vi) shortening the arc column length, etc.

5.4   HOW WELD METAL IS DEPOSITED?

The temperature of the core of the arc is estimated to be about 6000°C and reduces as periphery is approached. This high temperature melts the parent metal and forms a pool of molten metal. A depression due to force of the arc formed in the molten metal is called crater Fig. 5.3. Electrode wire is also melted and globules are formed.

Following wing views are held regarding transfer of these globules from electrode to the weld metal pool.

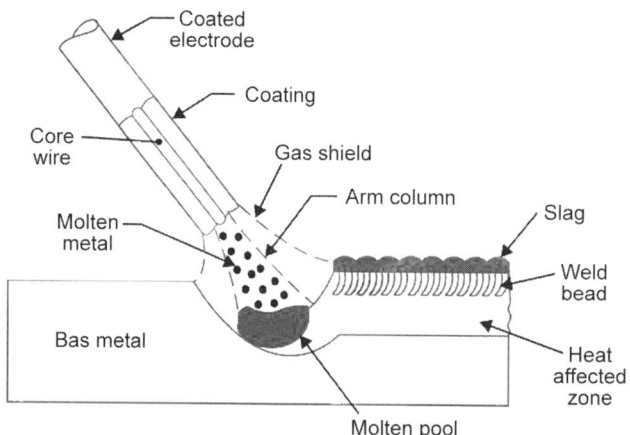

Fig. 5.3: Deposition of weld metal.

1. Due to intense heat, some of the electrode wire metal vaporizes which on hitting relatively cool crater is condensed thereon.
2. At high temperature on the tip of the electrode, carbon monoxide is evolved in the form of small explosions. This explosions project molten metal away from electrode. There is also another thought that dissolved gases arc liberated at electrode tip, expansion of which is the prime force for the transfer of metal across arc.

5.5 FOUR POSITIONS OF ARC WELDING

There are four basic positions in which manual arc welding is done.

1. **Flat position:** It is shown in Fig. 5.4a. Of all the positions, flat position is the simplest, most cost effective and the most used for all shielded arc welding. It provides the strongest weld joints. Weld beads are invariably smooth and free from slag spots. This position is most adaptable for welding of both ferrous and nonferrous metals particularly for cast iron.

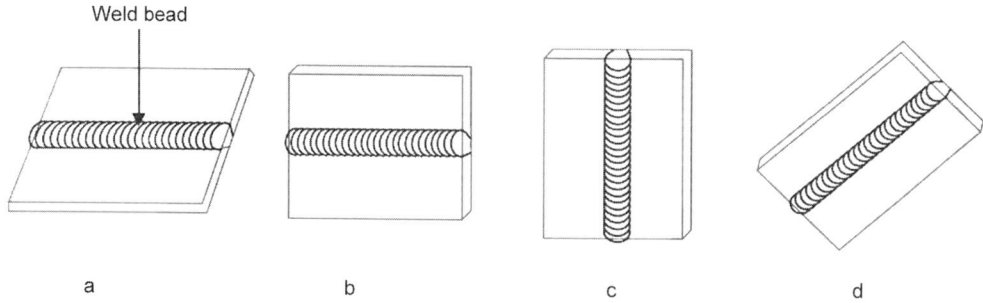

Figs 5.4a to d: (a) Flat position, (b) Horizontal position, (c) Vertical position, (d) Overhead position.

2. **Horizontal position:** It is the second most popular position and is shown in Fig. 5.4b. It requires a short arc length because it helps in preventing the molten puddle of the metal from sagging. However, major errors that happen while welding in horizontal position are undercutting overlapping of the weld zone (Fig. 5.5).

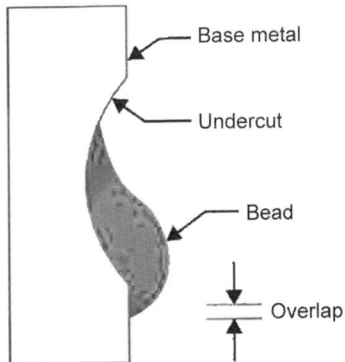

Fig. 5.5: Undercutting the overlapping of the weld zone.

3. **Vertical position:** It is shown in Fig. 5.4c. In this case, the welder can deposit the bead either in the uphill or in the downhill direction. Downhill welding is preferred for thin metals because it is faster than the uphill welding. Uphill welding is suited for thick metals because it produces stronger welds.

4. **Overhead position:** It is shown in Fig. 5.4d. Here, the welder has to be very cautious otherwise he may get burnt by drops of falling metal. This position is thought to be the most hazardous but not the most difficult one.

5.6 BARE METAL ARC WELDING

In metal arc welding, arc is drawn between the work piece and the wire electrode. The wire is melted down into the weld. Bare metal arc welding is not considered to be satisfactory. During welding process, both the globules of molten electrode wire and parent metal are exposed to the oxidizing and nitriding action of oxygen and nitrogen respectively. Oxides reduce the ductility of the weld and nitrides produces embrittlement in the weld deposit. Thus, the weld lacks required strength and ductility. Bare metal arc welding is preferred for unimportant works and for metal deposition on railway wagon wheels, travelling crane wheels, etc. It may be noted that bare electrode welding can be done satisfactory with DC supply only.

5.7 COATED ELECTRODES

For good welds, covered electrodes are always preferred. As per IS: 814-1970, the contact end of the electrode is left bare and clean to a length of 20–30 mm for inserting it into electrode holder (Fig. 5.6).

Lightly coated (or washed) electrodes have a coating layer. This coating usually consists of lime mixed with soluble glass which serves as a binder. The basic purpose of the light coating is to enhance the arc stability, so they are also called the stabilizing coatings. No action is taken to prevent oxidation and no slag is formed on the weld,

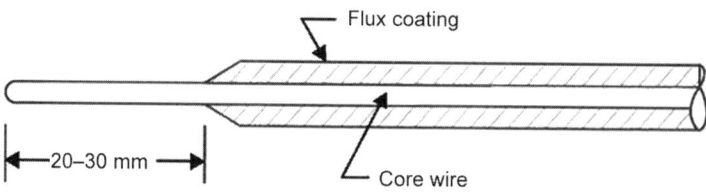

Fig. 5.6: Electrode for metallic arc welding.

nor are the mechanical properties of the weld metal improved. For this reason lightly coated electrodes are only used for welding non-essential jobs.

Heavy coated electrodes, sometimes called shielded-arc electrodes, are used to obtain a weld metal of very high quality, comparable with, and could be even superior to, the parent metal in terms of mechanical properties. Heavy coatings employed are composed of ionising (chalk), deoxidising (aluminium, ferro-managanese, etc.), gas generating (starch), slag forming (kaolin), alloying and binding materials.

i. The physical and metallurgical properties of the weld can be influenced by adding alloying components to the covering.

ii. Basic salts of silicon, magnesium and calcium in the covering form slag which floats on the surface of metal and prevents rapid cooling of the weld. Thus weld does not become brittle.

iii. The weld metal is protected from oxidising action of atmospheric oxygen and nitrifying action of nitrogen of air due to gases formed from the covering material. Therefore, we get quality ductile weld.

iv. In case of AC supply, arc cools at zero current and there is a tendency of deionizing the arc path. Covering gases keep the arc space ionized.

v. During welding the covering extends beyond the core wire. This directs the arc and concentrates the arc stream, reduces thermal losses and causes increase in temperature of electrode tip.

There may also be powder-cored electrodes which have a good portion of the source metal for depositing the joint as powdered iron is mixed with the flux coating. This makes the electrodes much larger on the outside for the same diameter core wire as standard flux-coated electrodes. These electrodes are well suited to down-hand welding and are said to deposit more metal in a given period time than standard flux-covered electrodes.

Both bare and coated electrodes, for manual arc welding, are made in the shape of role up to 12 mm in diameter and 450 mm long. Semiautomatic and automatic welding use electrode wire in coils.

5.8　TYPES OF JOINTS AND APPLICABLE WELDS

Bureau of Indian Standards (BIS) has recommended the following types of joints and the welds applicable to each one of them (Fig. 5.7).

1. Tee joint—with six types of welds.
2. Corner joint—with two types of welds.
3. Edge joint—with one type of weld.

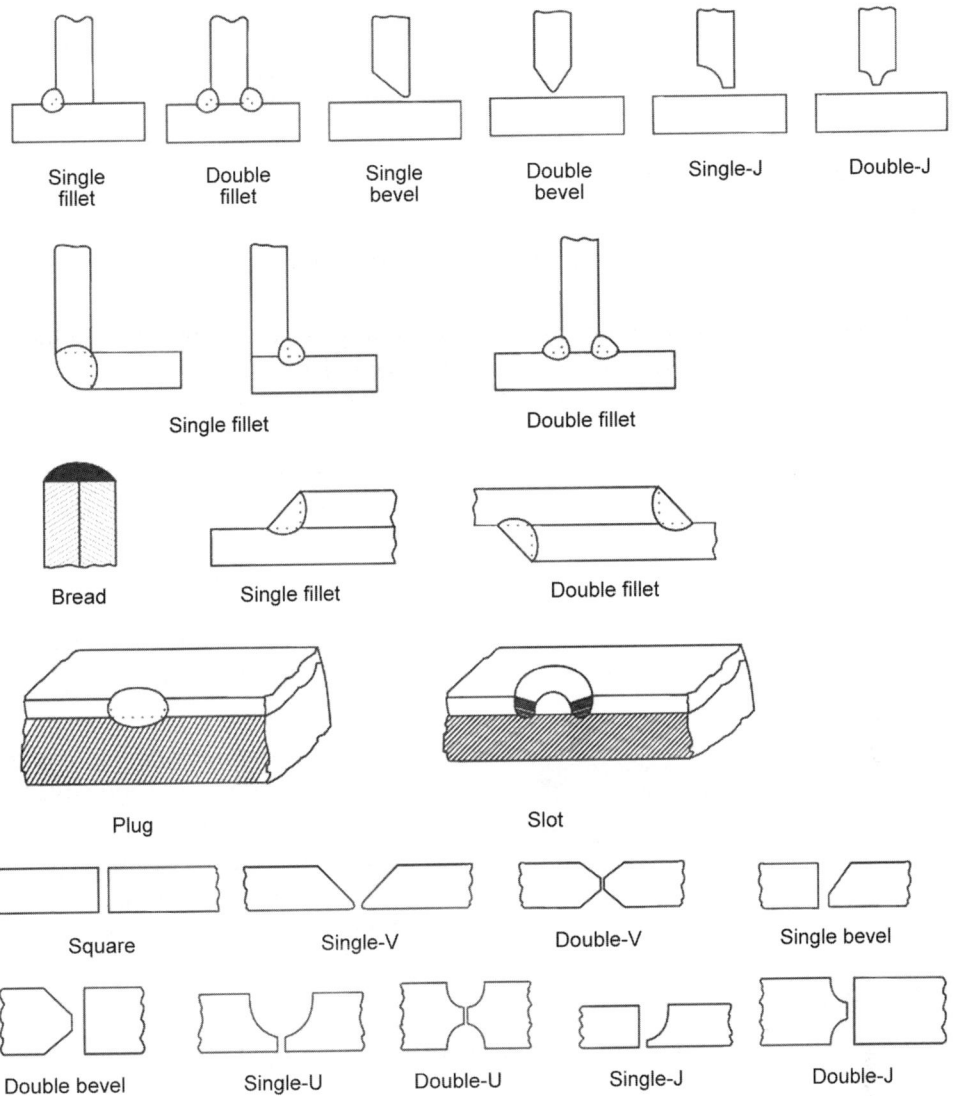

Fig. 5.7: Types of joints and applicable welds.

4. Lap joint—with four types of welds.
5. Butt joint—with nine types of welds.

5.9 REQUIREMENTS OF GOOD WELD

Oxidizing and nitrifying of atmospheric air can be reduced by deploying low voltage gradients in between electrodes. Open circuit voltage required to strike an arc is 50 to 60 V for DC supply and 70 to 100 V for AC supply. DC supply is therefore, considered safer particularly for structural welding work where a shock may cause an operator to lose balance and fall. Size of the electrode is dependent upon the size of the plates to be welded. In order to get good weld it is necessary to obtain absolute fusion of parent metal and electrode metal. Formation of crater of sufficient depth is a confirmation of thorough melting of surface. In Fig. 5.8a bead has been deposited with correct welding current so that the toes of bead smoothly change into parent metal and there is thorough penetration. In Fig. 5.8b bead has been deposited with insufficient current and it does not have proper penetration. This is clear from the toes of bead convexing. In Fig. 5.8c bead has been formed with excess current giving undercutting at the toe of bead. Undercuts are good places for stress concentration and are, therefore, dangerous in case of impact loads. Rough guide for selecting proper welding current is given by formula $I = k \times d$ amps where d is diameter of electrode in mm and K= 40 to 60.

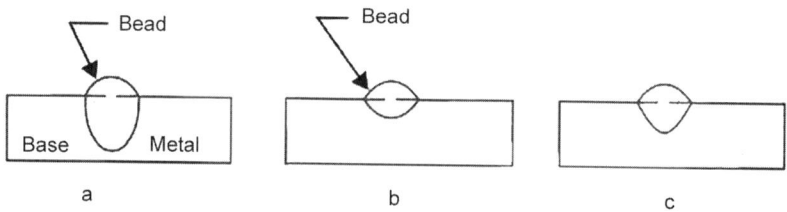

Figs 5.8a to c: (a) Bead has been deposited with correct welding current, (b) Bead has been deposited with insufficient current, (c) Bead has been formed with excess current giving undercutting at the toe of bead.

5.10 ATOMIC HYDROGEN WELDING

It is a non-pressure fusion welding process and the only use of the welder set is to supply heat for the base metal. If additional metal is needed, a filler rod can be melted into the joint. It uses two tungsten electrodes between which an arc column (actually, an arc fan) is maintained by AC supply.

Basic Principle

As shown in Fig. 5.9 an arc column is struck between two tungsten electrodes with an AC power supply. Soon, normal molecular hydrogen (H_2) is passed through this arc column. Due to excessive heat of the arc column, this diatomic hydrogen is dissociated into atomic hydrogen (H). However, atomic hydrogen being very unstable, recombines to form stable molecular hydrogen. Doing so, it releases huge amount of heat at about 3750°C which is used to fuse the metals.

Welding Equipment

The welding equipment primarily consists of the following :
1. **Standard welding machine** (Fig. 5.10) consists of a step-down transformer with tapped secondary powered from normal AC supply. Current requirement ranges from 15 to 150 A.
2. Hydrogen gas supply with a suitable regulator.
3. Atomic hydrogen welding torch having an ON-OFF switch and a trigger for moving the two tungsten electrodes together for striking and maintaining the arc column.

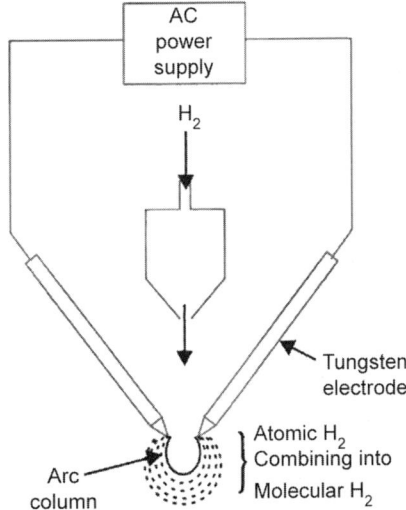

Fig. 5.9: Arc column is struck between two tungsten electrodes.

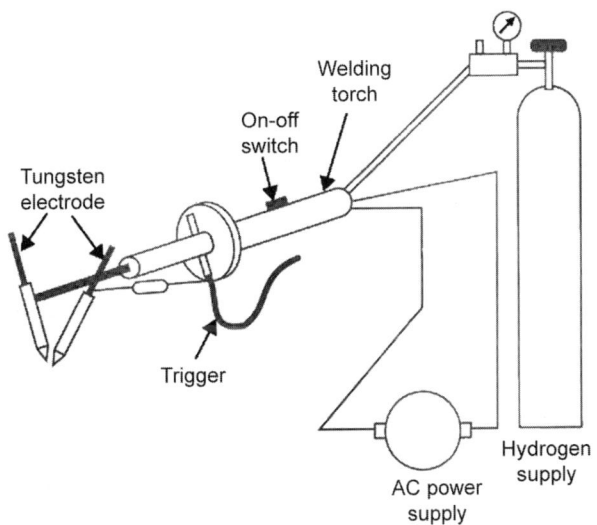

Fig. 5.10: Standard welding machine.

Method of Welding

The torch is held in the right hand with first finger resting lightly on the trigger. The arc is struck either by allowing the two tungsten electrodes to touch and then separate or by drawing the separated electrodes over a carbon block. At the same time, a stream of hydrogen is allowed to pass through the arc. As soon as the arc strikes, an extremely hot flame extends fanwise between the electrodes. When this fan touches the work piece, it melts it down quickly. If filler material is required, it can be added from the rod held in the left hand as in gas welding.

☞ *Advantages*

1. Arc and weld zone are shrouded by burning hydrogen which, being an active reducing agent, it protects them from atmospheric contamination.
2. Can be used for materials too thin for gas welding.
3. Can weld relatively thick sections as well.
4. Gives strong, ductile and stable welds.
5. Can be used for welding of mild steel, alloy steels and stainless steels and aluminium alloys.
6. Can also be used for welding of non-ferrous metals such as nickel, monel, brass, bronze, tungsten and molybdenum, etc.

5.11 SUBMERGED ARC WELDING

In this process, welding is performed under a blanket of granulated flux which protects the weld from all adverse effects of atmospheric gases while a consumable electrode is continuously and mechanically fed into the arc. The arc, the end of the bare metal electrode and the molten weld pool are all submerged under a thick mound of finely-divided granulated powder that contains de-oxidizers, cleansers and other fluxing agents. The fluxing powder is fed from a hopper that is carried on the welding head itself (Fig. 5.11).

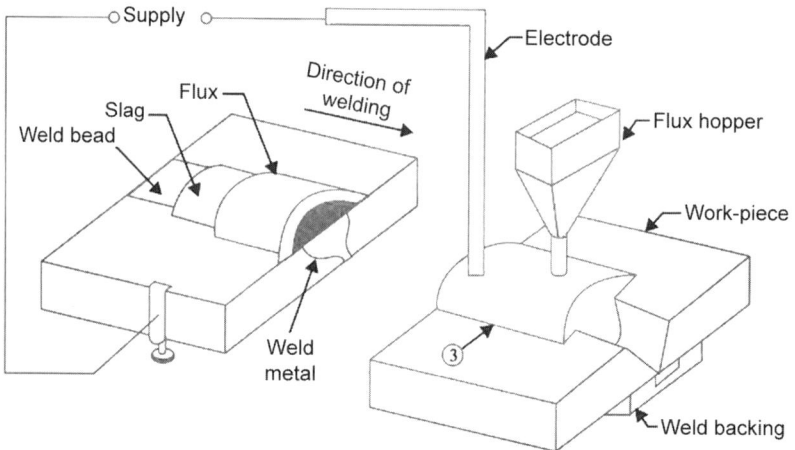

Fig. 5.11: Submerged arc welding.

This hopper spread the powder in a continuous mound ahead of the electrode in the direction of welding. Since arc column is completely submerged under the powder, there is no splatter or smoke and, at the same time, weld is completely shielded from atmospheric contamination. Because of this protection, weld beads are extremely smooth. The flux adjacent to the arc column melts and floats to the top of the molten pool where it solidifies to form slag. This slag is easy to remove. Often it cracks off automatically as it cools. The unused flux is removed and is used repeatedly.

The electrode is either a bare wire or has a slight mist of copper coated over it to prevent oxidation. In automatic or semiautomatic submerged arc welding, wire electrode is fed mechanically through an electrically contacting collet. Though AC power supply may be used, yet DC supply is more used because it assures a simplified and positive control of the welding process. This process requires high current densities about 5 to 6 times of those used in ordinary manual stick electrode welding. As a result, melting rate of the electrode as well as welding speed becomes much higher. Faster welding speed minimizes distortion.

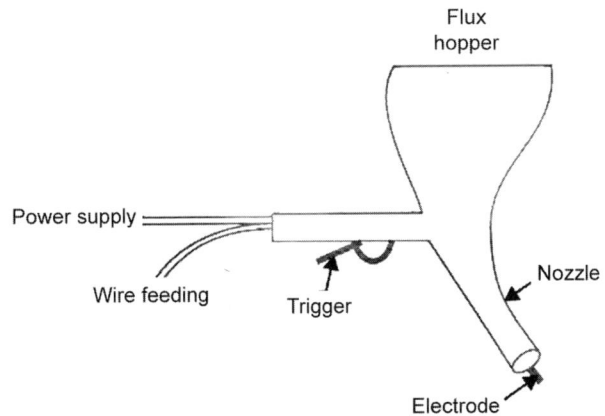

Fig. 5.12: Welding gun.

Submerged arc welding can be done manually where automatic process is not possible such as on curved lines and irregular joints. Such a welding gun is shown in Fig.5.12. Both manual and automatic submerged arc processes are most suited for flat and slightly downhill welding positions.

Application

The submerged arc process is suitable for welding:
1. Low-alloy, high-tensile steels.
2. Low-carbon steels.
3. Nickel, monel and other non-ferrous metals like copper.
4. Medium-carbon steel, heat-resistant steels and corrosion-resistant steels, etc.
5. "Industrial applications" includes fabrication of pipes, boiler pressure vessels, railroad tank cars, structural shapes, etc. which demand welding in a straight line.

Advantages

1. Partly because it is often automated, it is much faster than regular arc welding. Speeds up to 3800 mm/minute are possible on 3 mm thick steel at 100% efficiency.
2. A fairly thick sections can be welded in a single pass without edge preparation.
3. Welds made by this process have high strength and ductility.
4. Deep penetration with quality weld is possible.

5.12 INERT GAS METAL ARC WELDING

In this process, welding is done with bare electrodes but weld zone is shielded from the atmosphere by a gas which is piped to the arc column. Shielding gases used are carbon dioxide, argon, helium, hydrogen and oxygen. No flux is required. Different processes using inert gas metal arc welding are as follows.

   i. Tungsten inert-gas (TIG) Process.
  ii. Metal inert-gas (MIG) Process.
 iii. Metal-active-gas (MAG) process.

5.12.1 Tungsten Inert-Gas (TIG) Process

In this process, non-consumable tungsten electrode is used and filler wire is fed separately. The weld zone is shielded from the atmosphere by the inert gas (argon or helium) which is ducted directly to the weld zone where it surrounds the tungsten and the arc column.

Basic Principle

It is an electric process which uses a bare non-consumable tungsten electrode for striking the arc only (Fig. 5.13). Filler material is added separately. It uses an inert gas to shield the weld puddle from atmospheric contamination. This gas is ducted directly to the weld zone from a gas cylinder.

- The usual TIG welding system consists of the following:
  1. A standard shield arc welding machine complete with cables, etc.
  2. A supply of inert gas complete with hose, regulators, etc.
  3. A source of water supply (in the case of water-cooled torches).
  4. A TIG torch with a control switch to which all the above are connected.

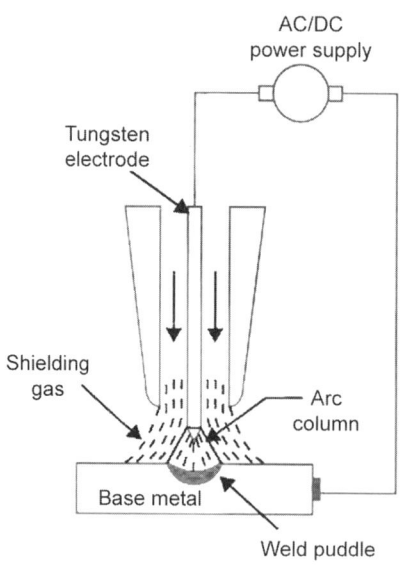

Fig. 5.13: Tungsten inert-gas process.

- The electrodes are made of either pure tungsten or zirconiated or thoriated tungsten. Addition of zironium or thorium (0.001 to 2%) improves electron emission tremendously.

Advantages

1. It provides maximum protection to weld bead from atmospheric contamination.
2. TIG welds are stronger, more ductile and more corrosion-resistant than those of shield metal arc welding.
3. Since no flux is used, there is no flux entrapment in the bead.
4. Since no flux is required, a wider variety of joint designs can be used.
5. No post-weld cleansing is necessary.
6. There is no weld splatter or sparks that could damage the surface of the base metal.
7. It gives relatively fast welding speeds.
8. It is suitable for welding food or medical containers where entrapment of any decaying organic matter could be extremely harmful.
9. It is suitable for all welding positions—the flat, horizontal, vertical and overhead positions.

Application

1. The TIG welding process lends itself ably to the fusion welding of aluminium and its alloys, stainless steel, magnesium alloys, nickel base alloys, copper base allows, carbon steel, and low-alloy steels.
2. TIG welding is also used for the combining of dissimilar metals, hard facing and surfacing of metals.
3. Special industrial applications include manufacture of metal furniture and air conditioning equipment.

5.12.2 Metal Inert-Gas (MIG) Process

It is a refinement of the TIG process. It uses a bare consumable (i.e. fusible) wire electrode which acts as the source for the arc column as well as the supply for the filler material. The weld zone is shielded by argon gas which is ducted directly to the electrode point.

Basic Principle

It is also called inert-gas consumable-electrode process (Fig. 5.14). The fusible wire electrode is driven by the drive wheels. Its function is two-fold: to produce arc column and to provide filler material. This process uses inert gas for shielding the weld zone from atmospheric contamination. Argon is used to weld non-ferrous metals though helium gives better control of porosity and arc stability. This process can deposit large quantities of weld metal at a fast welding speed. The process is easily adaptable to semiautomatic or fully automatic operations.

- The basic MIG welding system consists of the following:

    1. Welding power supply.
    2. Inert gas supply with a regulator and flow meter.

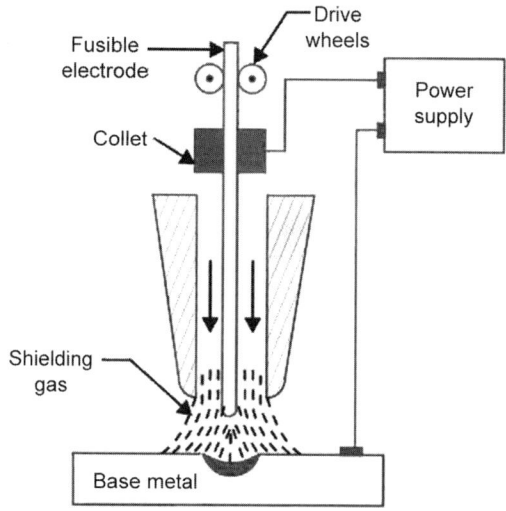

Fig. 5.14: Metal inert-gas process.

3. Wire feed unit containing controls for wire feed, gas flow and the ON/OFF switch for MIG torch.
4. MIG torch.
5. Depending on amperage, a water cooling unit.

Advantages

1. This welding process requires no flux.
2. Gives high metal deposit rates (varying from 2 to 8 kg/h.)
3. Requires no post-welding cleaning.
4. Gives complete protection to weld bead from atmospheric contamination.
5. Adaptable for manual and automatic operations.
6. Can be used for a wide range of metals both ferrous and non-ferrous.
7. Easy to operate (requiring comparatively much less operating skill).
8. This welding process is especially suited for horizontal, vertical and overhead welding positions.

Applications

1. Practically all commercially available metals can be welded by this method.
2. It can be used for deep groove welding of plates and castings, just as the submerged arc process can, but it is more advantageous on light gauge metals where high speeds are possible.

5.12.3 MAG Welding

As stated previously, in MIG welding process, the shielding gas used is mono atomic (argon or helium) and is inert, i.e. chemically inactive and metal transfer takes place by axial pulverization. In MAG (metal-active-gas) process, shielding, gas used is chemically active, i.e. carbon dioxide or its mixture with other gases. Transfer of metal conducted in big drops.

5.13 CARBON ARC WELDING

Carbon arc welding was the first electric welding process invented by a French inventor Auguste de Meritens in 1881. For metal arc welding, size of globule which has to be transported from electrode to the work, limits the size of metal electrode to 10 mm and current of 300 to 400 amps. For heavier welds or where heavy build-up is needed, carbon arc welding is employed with carbon electrode of diameter up to 25 mm and current up to 800 amps. Arc is struck between carbon electrode and work piece and filler electrode may be used for the necessary deposition of metal. To avoid transfer of electrode material into the weld metal, electrode is always kept negative and work as positive. Due to high rate of heat production, carbon arc welding is mainly utilized for copper welding as copper is known to have high thermal conductivity. The basic circuit is shown in Fig. 5.15 and can be used with DC as well as AC supply.

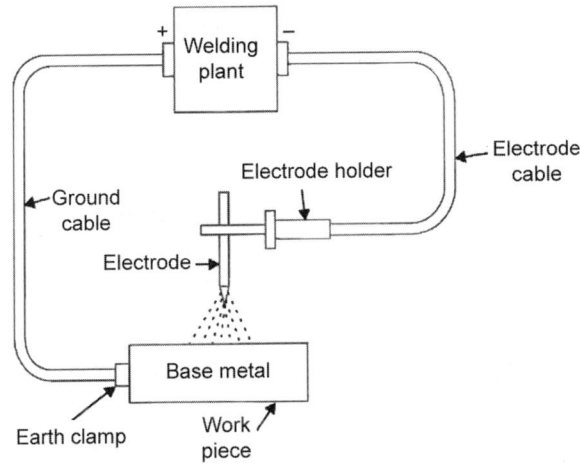

Fig. 5.15: Basic circuit of carbon arc welding.

Applications

1. The joint designs that can be used with carbon arc welding are butt joints, bevel joints, flange joints, lap joints and fillet joints.
2. This process can be easily put into use for automation particularly where amount of weld deposit is large and materials to be fabricated are of simple geometrical shapes such as water tanks.
3. It is suitable for welding galvanised sheets using copper-silicon-manganese alloy filler metal.
4. It is useful for welding thin high-nickel alloys.
5. Monel metal can be easily welded with this process by using a suitable coated filler rod.
6. Stainless steel of thinner gauges is often welded by the carbon-arc process with excellent results.

Advantages

1. The main advantage of this process is that the temperature of the molten pool can be easily controlled by simply varying the arc length.
2. It is easily adaptable to automation.
3. It can be easily adapted to inert gas shielding of the weld.
4. It can be used as an excellent heat source for brazing, braze welding and soldering, etc.

Disadvantages

1. A separate filler rod has to be used if any filler material is required.
2. Since arc serves only as a heat source, it does not transfer any metal to help reinforce the weld joint.
3. The major disadvantage of the carbon-arc process is that blow holes occur due to magnetic arc blow especially when welding near edges of the work piece.

5.14 ELECTRIC SUPPLY FOR ARC WELDING

Both DC and AC are used for electric arc welding. Each having its particular applications.

The electric properties of the arc determines the requirements that should be met. To initiate an arc, a higher voltage is required than to maintain it under normal conditions. For this reason the open-circuit voltage of the power supply source (when there is no arc drawn between the electrode and the work) must be higher than the arc or closed-circuit voltage (when there is an arc drawn between the electrode and the work). With DC the open-circuit voltage must be at least 30 or 35 volts, while with AC it should not be lower than 50 or 55 V. An open arc will be sustained at 18 to 25 volts. Open-circuit voltage usually ranges from 50 to 90 volts.

To initiate an arc, the electrode is touched to the work and is then withdrawn a short distance. On contact, the electrode shorts the welding circuit and an inadvertent short circuit may give rise to an excessive current in the circuit, which may dangerously heat and even burn the insulation. A power supply source for welding should, therefore, be able to limit the state of current, when the arc is initiated, to a small percentage above the desired value.

In metal arc welding the globules of molten electrode metal are carried across the arc, thus changing the resistance of the arc and consequently the arc voltage and current. In fact, the arc resistance is continually varying within broad limits in very short periods of time. The point is that the molten globules of metal pass from the electrode to the plate at the rate of 30 or more per second, continually causing approximate short circuits from the electrode to the work. If the arc is not to go out, the power supply source must be able to quickly vary its voltage with the resistance of the arc.

Of special importance is the ability of a power supply source to restore the voltage quickly after a globule of molten metal has broken away from the electrode. The arc will only restrike at not less than 25 volts. The voltage should, therefore, recover to 25 volts within, 0.05 second.

For manual welding with coated electrodes the static characteristic (curve giving relation between the source voltage and the welding current) is represented by curve 1 in Fig. 5.16. This is a drooping or negative characteristic. OA represents the open-circuit voltage, OB represents the arc or closed circuit voltage, OD represents the arc current and OE represents the short-circuit current. The point C on the curve 1 shows the instant of striking the arc. With this type of volt-ampere characteristic the short-circuit current in the welding circuit is reduced to a limit safe enough for the windings of the welding generator or transformer.

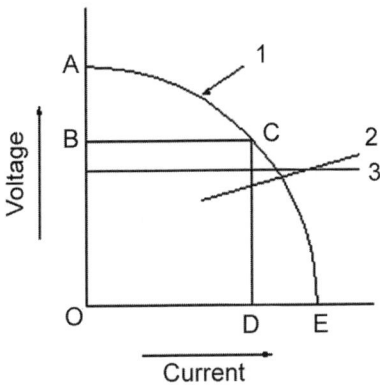

Fig. 5.16: Relation between the source voltage and the welding current.

If the arc has a flat (curve 3) or a rising volt-ampere characteristic (curve 2) such as in shielded-arc welding, automatic welding with large currents the source of power supply should also have a flat or a rising characteristic respectively.

To sum up, the power supply for manual-arc welding should meet the following requirements:

i. The open-circuit voltage must be sufficient for an arc to strike without difficulty, but safe for the operator (below 80 volts).

ii. The short-circuit current must be within limits of safety for generator or transformer windings.

iii. The voltage of the source of power supply should vary rapidly with changes in arc length.

iv. The wattage of the source of power supply should be sufficient to give the desired arc current.

5.15 MACHINES FOR ARC WELDING

Welding is never done directly from the supply mains. Instead, special welding machines are used which provided currents of various characteristics. Use of such machines is essential for the following reasons:

1. To convert AC supply into DC supply when DC welding is desired.

2. To reduce the high supply voltage to a safer and suitable voltage for welding purposes.

3. To provide high current necessary for arc welding without drawing a corresponding high current from the supply mains.

4. To provide suitable voltage/current relationships necessary for arc welding at minimum cost.

There are two general types of arc welding machines:
    a. DC welding machines
        i. motor-generator set
       ii. AC transformers with rectifiers
    b. AC welding machines

5.15.1 DC Welding Machines with Motor Generator Set

Such a welding plant is a self-contained single-operator motor-generator set consisting of a reverse series winding DC generator driven by either a DC or an AC motor (usually 3-phase). The series winding produces a magnetic field which opposes that of the shunt winding. On open-circuit, only shunt field is operative and provides maximum voltage for striking the arc. After the arc has been established, current flows through the series winding and sets up a flux which opposes the flux produced by shunt winding. Due to decreases in the net flux, generator voltage is decreased. With the help of shunt regulator, generator voltage and current values can be adjusted to the desired level. Matters are so arranged that despite changes in arc voltage due to variations in arc length, current remains practically constant. Fig. 5.17 shows the circuit of a DC motor-generator type of welding machine.

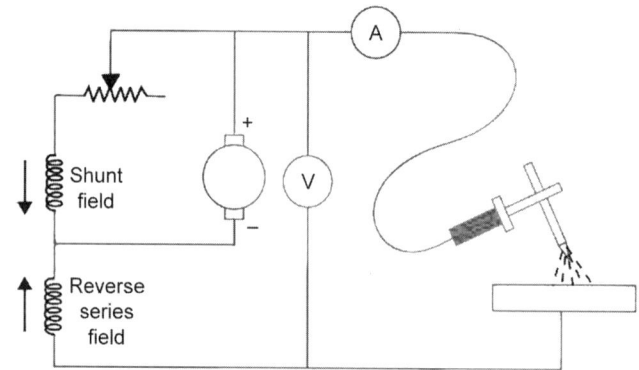

Fig. 5.17: Circuit of a DC motor-generator type welding machine.

Advantages

Such a DC welder has the following advantages:
1. It permits portable operation.
2. It can be used with either straight or reverse polarity.
3. It can be employed on nearly all ferrous and non-ferrous materials.
4. It can use a large variety of stick electrodes.
5. It can be used for all positions of welding.

Disadvantages

1. It has high initial cost.
2. Its maintenance cost is higher.

3. Machine is quite noisy in operation.

4. It suffers from arc blow.

5.15.2 AC Rectified Welding Unit

It consists of a transformer (single-or three-phase) and a rectifier unit as shown in Fig. 5.18. This type of unit has no moving parts and so has long-life. The only moving part is the fan which is used for cooling the transformer. But this fan is not the basic part of the electrical system. Fig. 5.18 shows a single-phase full-wave rectified circuit of the welder. Silicon diodes are used for converting AC into DC. These diodes are hermetically sealed and are almost ageless because they maintain rectifying characteristics indefinitely. Such a transformer-rectifier welder is most adaptable for shield arc welding because it provides both DC and AC polarities. It is very efficient and quiet in operation. These welders are particularly preferred for the welding of (i) pipes in all positions (ii) non-ferrous metals (iii) low-alloy and corrosion—heat and creep-resisting steel (iv) mild steels in thin gauges.

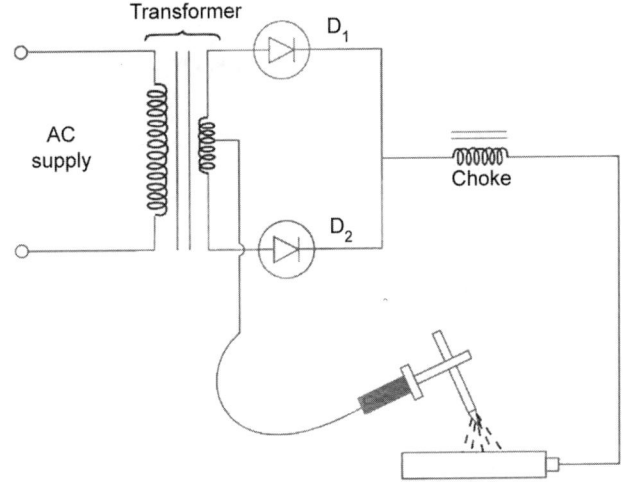

Fig. 5.18: AC rectified welding unit.

5.15.3 AC Welding Machines

As shown in Fig. 5.19, it consists of a step-down transformer with a tapped secondary having an adjustable reactor in series with it for obtaining the drooping V/I characteristics. The secondary is tapped for providing different voltage/current settings.

Advantages

This AC welder which can be operated from either a single-phase or a 3-phase supply has the following advantages:

    i. Low starting cost

    ii. Low operation and maintenance cost

   iii. Low wear

   iv. No arc blow.

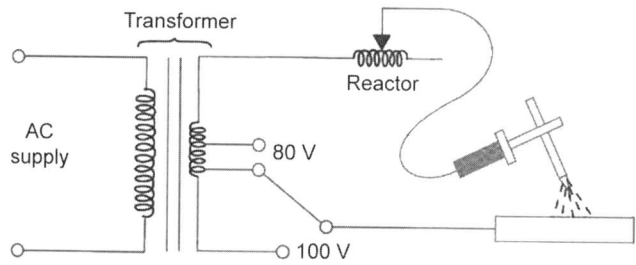

Fig. 5.19: AC welding machine.

Disadvantages

    i. Its polarity cannot be changed

   ii. It is not suitable for welding of cast iron and non-ferrous metals.

      Comparison between DC welding and AC welding is given in Table 5.1

Table 5.1	Comparison between DC and AC arc welding		
Sl. no.	*Particulars*	*DC welding*	*AC welding*
1.	Equipment	Motor-generator set or rectifier is required in case of availability of AC supply, otherwise oil engine-generator set is required.	In case of availability of AC supply only a transformer is required transformer costs less and its maintenance cost is also low.
2.	Prime cost	Two to three times as compared to that of a transformer.	Comparatively low.
3.	Operating efficiency	Low; consequently high cost of electrical energy.	High (about 85%).
4.	No-load voltage	Low (safer)	Frequently too high (dangerous)
5.	No-load requirements	Very high	Low (advantageous)
6.	Power factor	Comparatively higher	Low, requiring capacitors for correction
7.	Arc blow	Pronounced	Not so pronounced with AC (advantageous)
8.	Electrodes	Both hare (non-coated) and thus cheap electrodes can be used	Only coated electrodes-expensive ones
9.	Heating	Uniform	Not so uniform as in case of DC
10.	Connected load (cross-sectional area of conductors and fuses)	Normal	Considerably higher because of low power factor
11.	Welding of non-ferrous metals	Suitable	---
12.	Arc stability	Higher	---

5.16 RESISTANCE WELDING

In resistance welding, heat required for the purpose of weld is produced by the resistance offered to the flow of current at the junction of two metals and is given by following formula.

$$H = 0.24 \int_0^{t_1} i^2 r dt \ Cal$$

where r is the resistance in ohm at the junction of the sheets where weld is required and 'i' is the current in amps. Since junction resistance is usually very small, it is very essential to use very high currents, in the order of 1000 amps and above at low voltage of 1 to 15 V.

Advantages

Some of the advantages of resistance welding are as under:

1. Heat is localized where required.
2. Welding action is rapid.
3. No filler material is needed.
4. Requires comparatively lesser skill.
5. Is suitable for large quantity production.
6. Both similar and dissimilar metals can be welded.
7. Parent metal is not harmed.
8. Difficult shapes and sections can be welded.

Only disadvantages are with regard to high initial as well as maintenance cost.

It is a form of resistance welding in which the two surfaces are joined by spots of fused metal caused by fused metal between suitable electrodes under pressure.

5.17 SPOT WELDING

In spot welding, work pieces which are to be joined, are pressed together by external pressure exerted through electrodes as shown in Fig. 5.20a. And then supply is switched on for a particular period, determined by the size of the work piece. Passage of current will generate heat at the three junction faces, two between electrodes and work and one between two work pieces as shown in Fig. 5.20b. As such heating of junction 1 and 3 is to be avoided. This is achieved by taking following measures.

1. By water cooling the electrodes shown in Fig. 5.21 junction 1 and 3 are cooled.
2. Electrodes are made of materials which have high electrical conductivity as well as thermal conductivity so that heat developed at junctions 1 and 3 is minimum in the first instance. Secondly whatever heat is developed is efficiently conducted away.

Apart from high thermal and electrical conductivity, electrode material should have very high strength to minimize the wear and tear at the tip of the electrode. Electrodes are, therefore, made-up of following material to suit different duties. Table 5.2 shows various types of electrode used in welding.

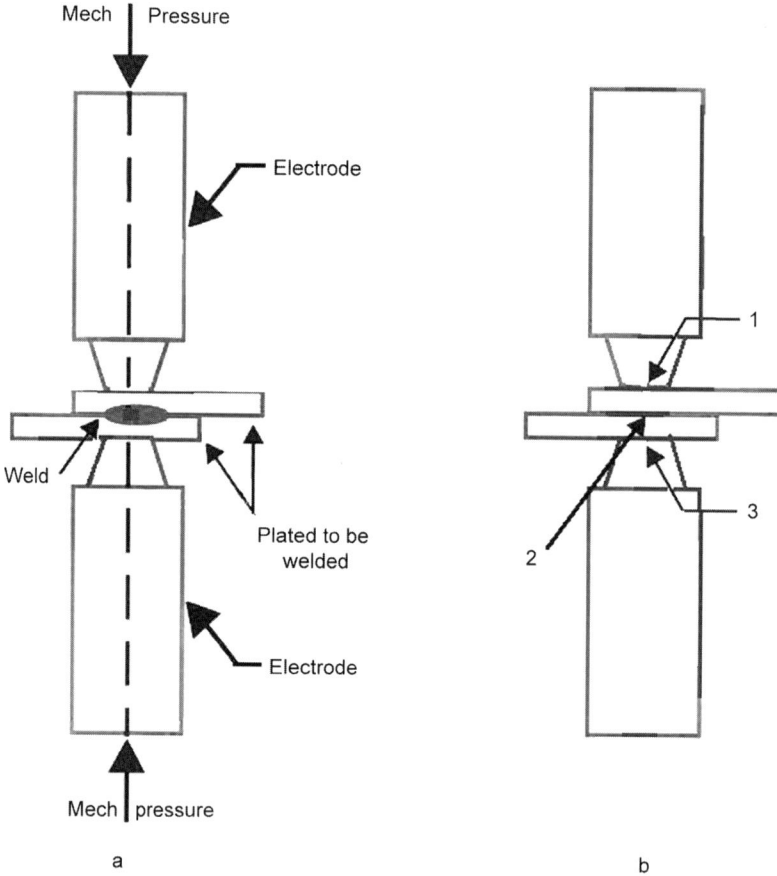

Figs 5.20a and b: (a) Work pieces are pressed together by external pressure exerted through electrodes, (b) Generating heat at the three junction faces.

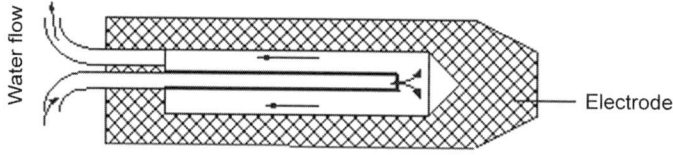

Fig. 5.21: Water cooling process.

Table 5.2	Types of electrode used in welding		
Sl. no.	Material of electrode	Softening temperature	Material to be welded
1.	Hard drawn copper	150°C	Aluminium alloys
2.	Cadmium copper	250°C	Thin MS sheets
3.	Chromium copper	500°C	Steel
4.	Tungsten copper	1000°C	Steel and copper alloys

Tip diameter of the electrode is approximately given by following formula:

$$D = \sqrt{t}$$

where D = tip diameter in inches and
         t = sheet thickness in inches.

Mechanical holding pressure is not less than 10,000 lbs/sq inch. Current density in the body of the spot welding electrodes may be as high as 50,000 Amps/sq inch, while at the tip it may be 3,00,000 Amps/sq. inch. Effective welding current may be less than the transformer secondary current due to shunting effect of adjacent welds as shown in Fig. 5.22.

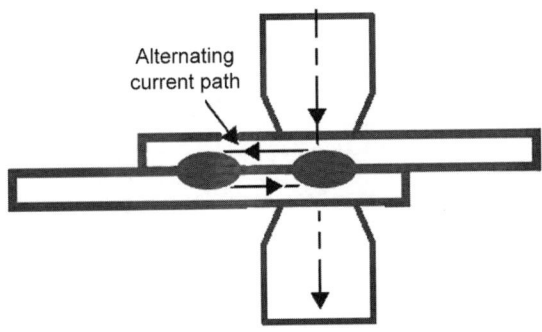

Fig. 5.22: Shunting effect of adjacent welds.

5.18 SEAM WELDING

The seam welder differs from ordinary spot welder only in respect of its electrodes which are of disc or roller shape as shown in Fig. 5.23a. These copper wheels are power driven and rotate whilst gripping the work. The current is so applied through the wheels that the weld spots either overlap as in Fig. 5.23b or are made at regular intervals as in Fig. 5.23c. The continuous or overlapped seam weld is also called stitch weld whereas the other is called roll weld.

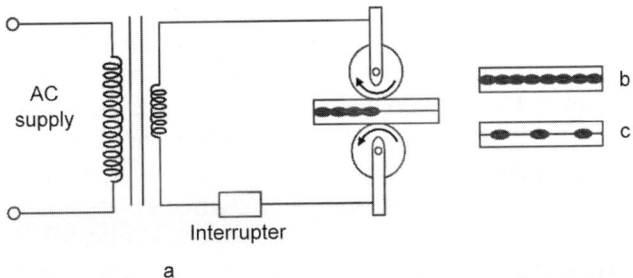

Figs 5.23a to c: (a) Seam welding, (b) Weld spots (overlap), (c) Weld spot (regular intervals).

Seam welding is confined to welding of thin materials ranging in thickness from 2 to 5 mm. It is also restricted to metals having low hardenability rating such as hot-rolled grades of low alloy steels. Stitch welding is commonly used for long water-tight and gas-tight joints. Roll welding is used for simple joints which are not water-tight or gas-tight. Seam welds are usually tested by pillow test.

5.19 PROJECTION WELDING

This is similar to spot welding. However, in this process, large-diameter flat electrodes (also called platens) are used. Instead of bringing the concentration of heat by the shape of electrode, it is the projections or humps on the thinner plate which gives the heat concentration. It is here that welding takes place. Basic arrangement is shown in Fig. 5.24. Thickness of plates to be jointed is limited to 2 × 6 mm in steel, 2 × 4 mm in aluminium, 2 × 2 mm in copper. This reduction in maximum thickness of different metals is due to increase in conductivity.

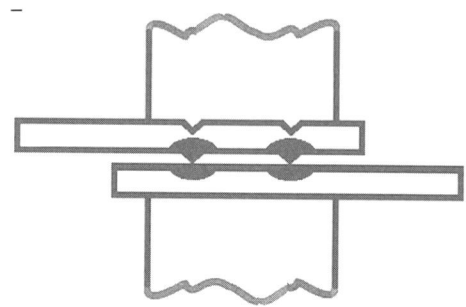

Fig. 5.24: Basic arrangement of projection welding.

☞ *Advantages*

Projection welding has following advantages over spot welding:

    i. More than one weld are done at a time. Therefore more output is obtained.

   ii. Electrode life is increased due to low current density and low pressure.

  iii. Good finished appearance is obtained as surface unindented by the electrodes.

  iv. Welds are automatically located by the position of projections.

5.20 BUTT WELDING

As shown in Fig. 5.25 two rods to be welded together are fixed in clamps and butted squarely against each other. Current flowing through the junction of two butting surfaces produces heat at the joint. When sufficient heat is developed, the pieces are rammed by force to complete the weld. This process is useful where parts have to be joined end-to-end or edge-to-edge. For example, welding pipes, wires and rods. It is also employed for making continuous lengths of chain.

5.20.1 Flash Butt Welding

It is also called by the simple name of flash welding. It is similar to butt welding but with the difference that here current is applied when ends of the two metal pieces are quite close to each other but do not touch intimately. Hence, an arc or flash is set-up between them which supplies the necessary welding heat. As seen, in the process heat is applied before the two parts are pressed together.

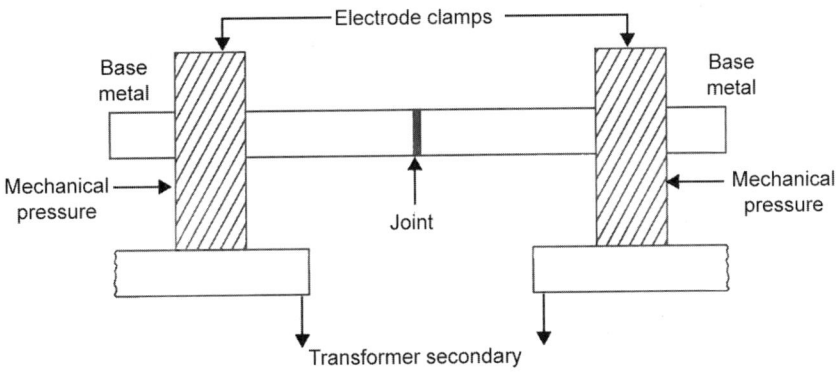

Fig. 5.25: Butt welding.

As shown in Fig. 5.26a, the work pieces to be welded are clamped into specially designed electrodes one of which is fixed whereas the other is movable. After the flash has melted their faces, current is cut off and the movable platen applies the forging pressure to form a fusion weld. As shown in Fig. 5.26b, there is increase in the size of the weld zone because of the pressure which forces the soft ends together.

Advantages

1. Even rough or irregular ends can be flash-welded. There is no need to level them by machining and grinding because all irregularities are burnt away during flashing period.
2. It is much quicker than butt welding.
3. It uses considerably less current than butt welding.
4. One of its major advantages is that dissimilar metals with different welding temperatures can be flash-welded.

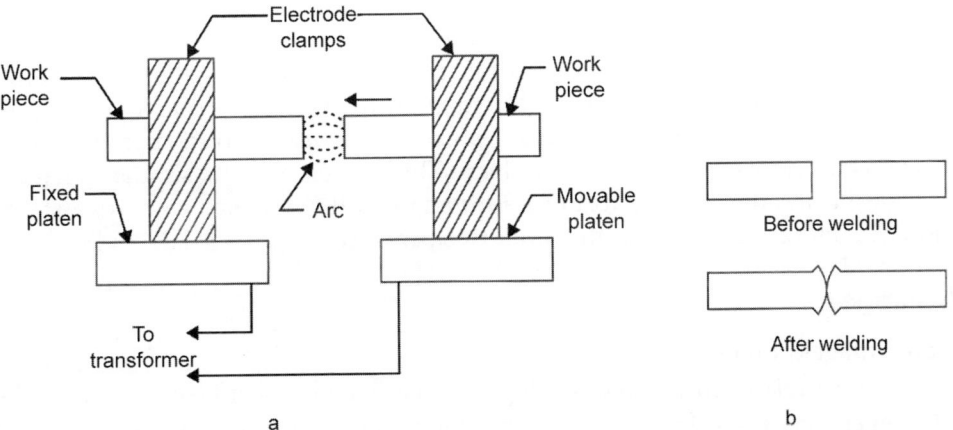

Figs 5.26a and b: (a) Flash butt welding, (b) Work spices (before and after welding).

Applications

1. To manage rods, bars, tubings, sheets and most ferrous metals.
2. In the production of wheel rims for automobiles and bicycles.
3. For welding tubular parts such as automobile break cross-shafts.
4. For welding tube coils for refrigeration plants. etc.

5.20.2 Upset Butt Welding

In this process, no flash is allowed to occur between the two pieces of the metals to be welded. When the two base metals are brought together to a single interface, heavy current is passed between them which heats them up. After their temperature reaches a value of about 950°C, the two pieces of base metal are compressed together more efficiently. This pressing together is called upsetting. This upsetting takes place while current is flowing and continues even after current is switched off. This upsetting action mixes the two metals homogeneously while pushing out many atmospheric impurities. Fig. 5.27 shows a finished upset weld.

Fig. 5.27: Finished upset weld.

5.20.3 Percussion Welding

It is a recent advancement in the field of welding. It is a resistance welding process wherein coalescence is produced simultaneously over the total area of abutting surfaces by heat obtained from an arc produced by a rapid discharge of stored electrical energy. This is a self timing spot welding method. A percussion welding process is shown in Fig. 5.28.

In this process, the pieces or parts of a material to be welded are kept at a distance, one in a stationary holder and the other in a clamp mounted in a slide and backed up against heavy spring pressure. When the movable clamp is slowly released, it moves rapidly carrying with it the pieces to be welded. When the two pieces are about 1.5 mm apart, there is a sudden discharge of electrical energy, creating an intense arc between the two surfaces and heat them to high temperature. As pieces come in contact with each other under heavy pressure the arc is extinguished due to the percussion blow of the two parts and the force between them. This process is quite similar in nature to flash welding and upset welding, but is limited to parts the same geometry and cross-section. It is more complex than the other two processes.

Advantages

The advantages of the process are:

1. There is an extremely shallow depth of heating and the time cycle is very short.
2. This process is so fast (takes about 0.1 s) that there is little heating effect on the material adjacent to the weld.

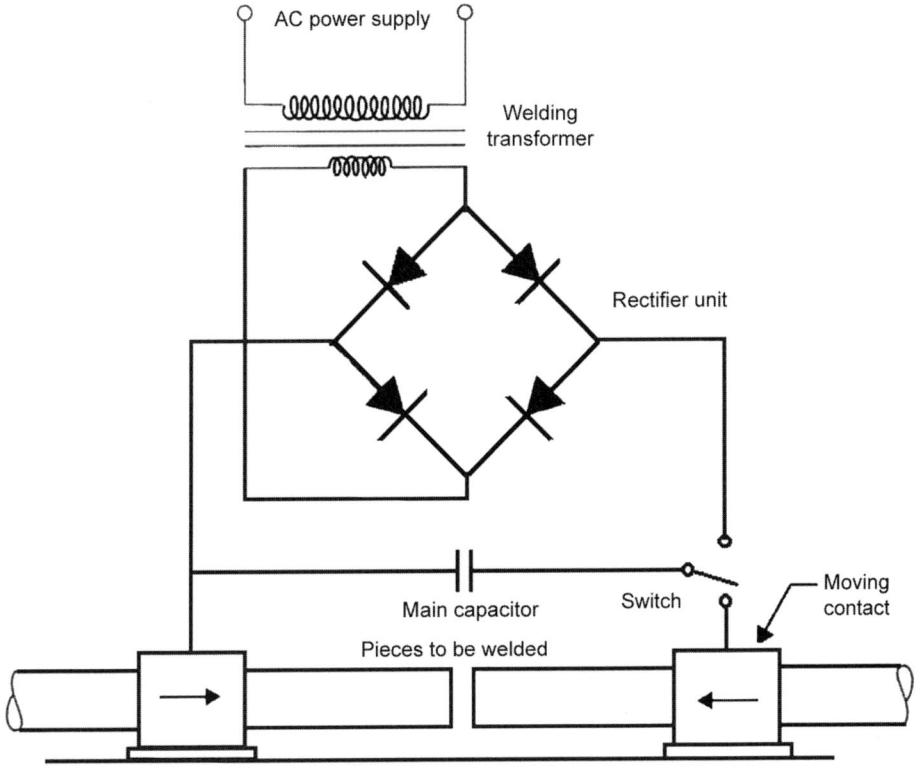

Fig. 5.28: Percussion welding process.

Disadvantages

1. This process is limited to only small areas (up to 3.2 cm²) of nearly regular sections.
2. Thin sheets of equivalent area cannot be joined by this process.

Application

It can be used for welding a large number of dissimilar metals. It is used for very specialized applications and the process is entirely automatic. Metals which can be percussion welded include, copper alloys, aluminium alloys, nickel alloys, low-carbon steel, medium-carbon steel and stainless steels. This process is used for welding satellite tips to tools, copper to aluminium or stainless steel, silver contact tips to copper, cast iron to steel, lead-in-wires on electric lamps and zinc to steel. Gold, silver, copper- tungsten, silver-tungsten and silver-cadmium oxide percussion welded to copper alloys for commonly used assemblies for electric contacts.

5.21 ELECTRON BEAM WELDING

In this process, welding operation is performed in a vacuum chamber with the help of a sharply focused beam of high-velocity electrons (Fig. 5.29). The electrons after being

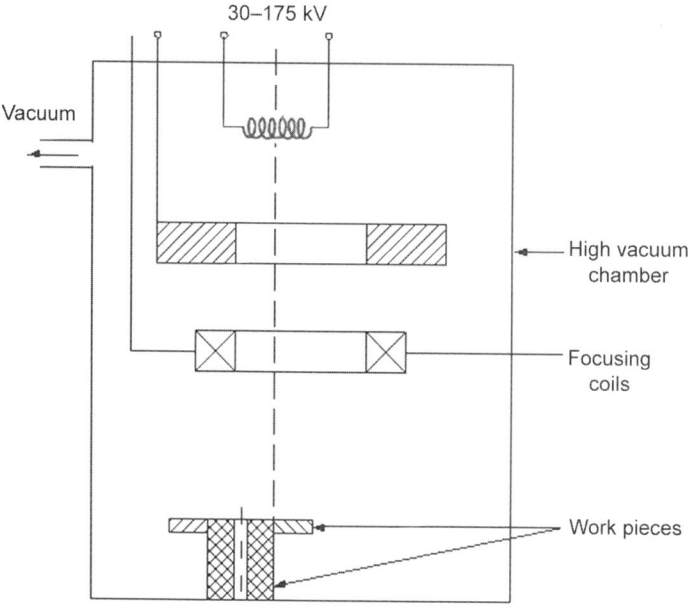

Fig. 5.29: Electron beam welding.

emitted from a suitable electrode are accelerated by the high anode voltage and are then focussed into a fine beam which is finally directed to the work piece. Obviously, this process needs no electrodes. The electron beam produces intense local heat which can melt not only the metal but can even boil it. A properly-focussed electron beam can completly penetrate through the base metal thereby creating a small hole whose walls are molten. As the beam moves along the joint, it melt the material coming in contact with it. The molten metal flows back to the previously-melted hole where it fuses to make a perfect weld for the entire depth of penetration.

Advantages

Electron-beam welding has following advantages:

1. It produces deep penetration with little distortion.
2. Its input power is small as compared to other electrical welding devices.
3. Electron-beam weld is much narrower than the fusion weld.
4. It is especially suitable for reactive metals which become contaminated when exposed to air because this process is carried out in vacuum.
5. It completely eliminates the contamination of the weld zone and the weld bead because operation is performed in a vacuum chamber.
6. It is especially suited to the welding of beryllium which is being widely used in the fabrication of industrial and aerospace components.
7. Its high deposition rate produces welds of excellent quality with only a single pass.
8. It is the only process which can join high temperature metals such as columbium.

At present, its only serious limitations are that it is extremely expensive and is not available in portable form. However, recently a non-vacuum electron-beam welder has been developed.

5.22 ELECTRO SLAG WELDING

It is a metal-arc welding process and may be considered as a further development of submerged-arc welding. This process is used for welding joints of thick sections of ferrous metals in a single pass and without any special joint preparation. Theoretically, there is no upper limit to the thickness of the weld bead. It is usually a vertical uphill process.

It is called electro slag process because heat is generated by passing current through the molten slag which floats over the top of the metal.

As shown in Fig. 5.30, two water-cooled copper shoes (or dams) are placed on either side of the joint to be welded for the purpose of confining the molten metal in the joint area. The electrode is fed into the weld joint almost vertically from special wire guides. There is a mechanical device which raises the shoes and wire-feed mechanism as the weld continues upwards till it is completed. An AC welding machine has 100 percent duty cycle and which can supply currents up to 1000 A if needed.

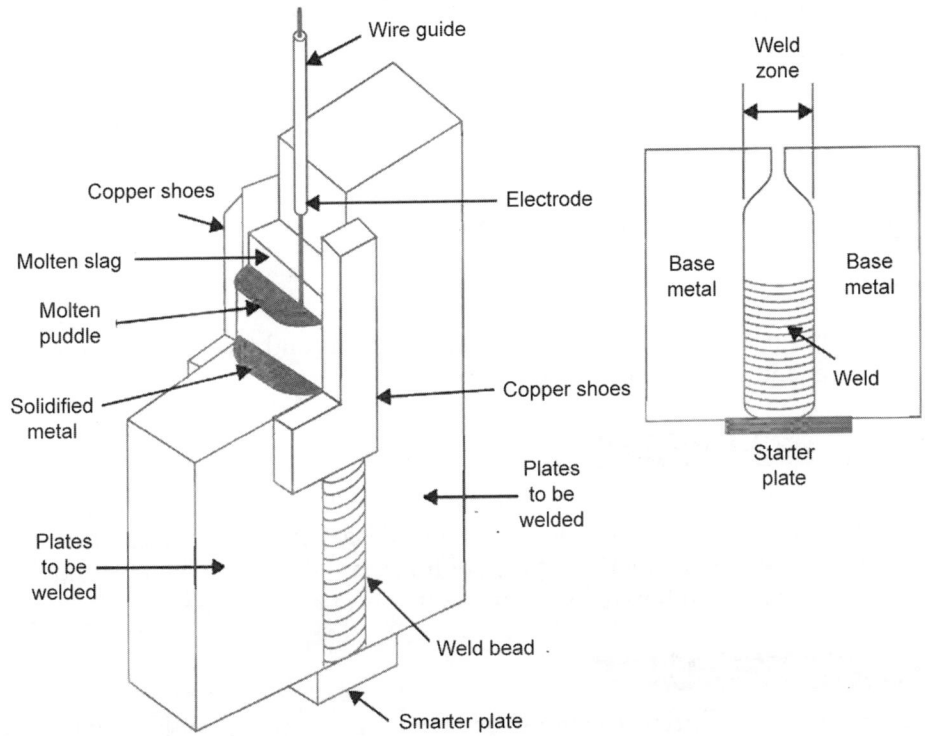

Fig. 5.30: Electro slag welding process.

The electro slag process is initiated just like submerged arc process by starting an electric arc beneath a layer of granular welding flux. When a sufficient thick layer of hot flux or molten slag is formed, arc action stops and from then onwards, current passes from the electrode to work piece through the molten slag. At this point, the process becomes truly electro slag welding. A starting plate is used in order to build up proper depth of conductive slag before molten pool comes in contact with the work pieces.

The heat generated by the resistance to the flow of current through the molten slag is sufficient to melt the edges of the work piece and the filler electrode. The molten base metal and filler metal collect at the bottom of the slag pool forming the weld pool. When weld pool solidifies, weld bead is formed which joins the faces of the base metal. As welding is continued upwards, flux flows to the top in the form of molten slag and cleanses the impurities from the molten metal. A mechanism raises the equipment as the weld is completed in the uphill vertical position.

Advantages

1. It needs no special joint preparation.
2. It does welding in a single pass rather than in costly multiple passes.
3. There is theoretically no maximum thickness of the plate it can weld.
4. There is also no theoretical upper limit to the thickness of the weld bead. Weld beads up to 400 mm thick have been performed with the presently-available equipment.
5. This process requires less electrical power per kg of deposited metal than either the submerged arc welding process or the shield arc process.
6. It has high deposit rate of up to 20 kg of weld metal per hour.
7. It has lower flux consumption.
8. Due to uniform heating of the weld area, distortion and residual stresses are reduced to the minimal amounts.

However, for electro slag welding, it is necessary to have only a square butt joint or a square edge on the plates to be welded.

Applications

It is commonly used in the fabrication of large vessels and tanks. Low-carbon steels produce excellent welding properties with this process.

5.23 ELECTRO GAS WELDING

This process works on the same basic principle as the electro slag process but has certain additional features of submerged arc welding. Unlike electro slag process, the electro gas process uses an inert gas for shielding the weld from oxidation and there is a continuous arc (as in submerged arc process) to heat the weld pool.

5.24 PLASMA ARC WELDING

It consists of a high-current electronic arc which is forced through a small hole in a water-cooled metallic nozzle. The plasma gas itself is used to protect the nozzle from the extreme heat of the arc. The plasma arc is shielded by inert gases like argon and helium which are pumped through an extra passageway within the nozzle of the plasma torch. As seen, plasma arc consists of electronic arc, plasma gas and gases used to shield the jet column. The idea of using the nozzle is to constrict the arc thereby increasing its pressure. Collision of high-energy electrons with gas molecules produces the plasma which is swept through the nozzle and forms the current path between the electrode and the work piece. Plasma jet torches have temperature capability of about 35,000°C.

For stainless steel welding and most other metals, straight polarity tungsten electrodes are used. But for aluminium welding, reverse polarity water-cooled copper electrodes are used.

Plasma arc welding requires DC power supply which could be provided either by a motor-generator set or transformer-rectifier combination. The latter is preferred because it produces better arc stability. The DC supply should have an open-circuit voltage of about 70 V and drooping voltage-ampere characteristics. A high-frequency pilot arc circuit is employed to start the arc (Fig. 5.31).

Welding with plasma arc jet is done by a process called 'keyhole' method. As the plasma jet strikes the surface of the work piece, it burns a hole through it. As the torch progresses along the work-piece, this hole also progresses along with but is filled up by the molten metal as it moves along. Obviously, 100 percent penetration is achieved in this method of welding. Since plasma jet melts a large surface area of the base metal, it produces a weld bead of wine glass design. The shape of the bead can be changed by changing the tip of the nozzle of the torch. Practically, all welding is done mechanically.

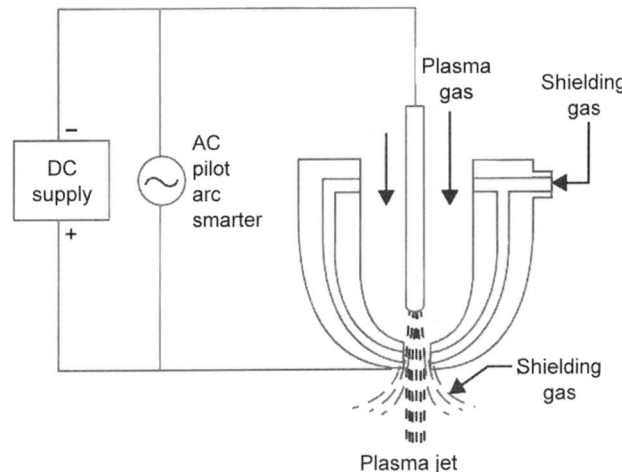

Fig. 5.31: High-frequency pilot arc circuit.

Applications

1. Plasma arc welding process has many aerospace applications.
2. It is used for welding of reactive metals and thin materials.
3. It is capable of welding high-carbon steel, stainless steel, maraging steel, copper and copper alloys, brass alloys, aluminium and titanium.
4. It is also used for metal spraying.
5. It can be modified for metal cutting purposes. It has been used for cutting aluminium, carbon steel, stainless steel and other hard-to-cut steels. It can produce high-quality dross-free aluminium cuts 15 cm deep.

Disadvantages

1. Since it uses more electrical equipment, it has higher electrical hazards.
2. It produces ultraviolet and infrared radiations necessitating the use of tinted lenses.
3. It produces high-pitched noise (100 dB) which makes it necessary for the operator to use ear plugs.

5.25 LASER WELDING

It uses an extremely concentrated beam of coherent monochromatic light, i.e. light of only one colour (or wavelength). It concentrates tremendous amount of energy on a very small area of the workpiece to produce fusion. It uses solid laser (ruby, saphire), gas laser (CO_2) and semiconductor laser. Both the gas laser and solid laser need capacitor storage to store energy for later injection into the flash tube which produces the required laser beam.

The gas laser welding equipment consists of (i) capacitor bank for energy storage (ii) a triggering device (iii) a flash tube that is wrapped with wire (iv) lasing material (v) focussing lens and (vi) a worktable that can rotate in the three X, Y and Z directions. When triggered, the capacitor bank supplies electrical energy to the flash tube through the wire. This energy is then converted into short-duration beam of laser light which is pin-pointed on the workpiece as shown in Fig. 5.32. Fusion takes place immediately and weld is completed fast.

Since duration of laser weld beam is very short (2 ms or so), two basic welding methods have been adopted. In the first method, the workpiece is moved so fast that the entire joint is welded in a single burst of the light. The other method uses a number

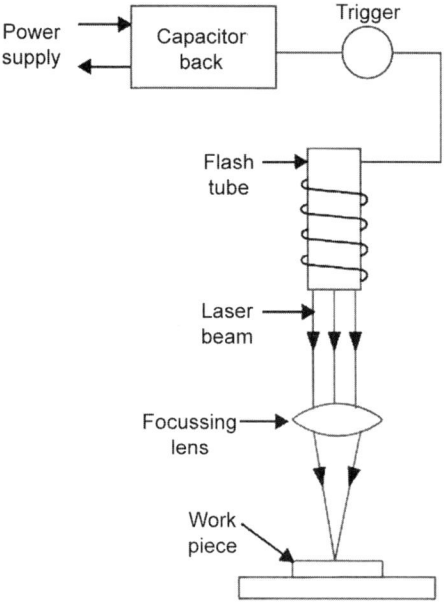

Fig. 5.32: Laser welding.

of pulses one after the other to form the weld joint similar to that formed in electric resistance seam welding.

Laser welding is used in the aircraft and electronic industries for lighter gauge metals.

Advantages

Some of the advantages of laser welding process are as follows:

1. It does not require any electrode.
2. It can make welds with high degree of precision and on materials as thin as 0.025 mm.
3. It does not heat the work piece except at one point. In fact, heat-affected zone is virtually non-existent.
4. Liquidus is reached only at the point of fusion.
5. It can produce glass-to-metal seals as in the construction of klystron tubes.
6. Since laser beam is small in size and quick in action, it keeps the weld zone uncontaminated.
7. It can weld dissimilar metals with widely varying physical properties.
8. It produces minimal thermal distortion and shrinkage because area of heat-affected zone is the minimum possible.
9. It can easily bond refractory materials like molybdenum, titanium and tantalium, etc.

However, the major disadvantage of this process is its slow welding speed. Moreover, it is limited to welding with thin metals only.

5.26 ULTRASONIC WELDING

Ultrasonic welding is a solid-state welding procedure that produces coalescence of material by the application of high-frequency (20–60 kHz) vibratory energy as the work pieces are kept together under pressure. Neither flux nor filler metals are used, electrical current is not allowed to flow through the weld metal as well, and usually no heat is applied. Since vibratory action in ultrasonic welding breaks up and disperses moisture, oxide and other (e.g. insulation) coatings, normally, degreasing may be the only cleaning required before welding (especially with aluminium). Fig. 5.33 shows a ultrasonic welding process.

Welding occurs when the ultrasonic tip or electrode, the energy coupling device, is clamped against the work pieces and is made to oscillate in a plane parallel to the weld interface. The combined clamping pressure and oscillating forces introduce dynamic stresses in the base metal. This creates minute deformations which results in a moderate temperature rise in the base metal at the weld zone. This coupled with the clamping pressure provides for coalescence across the interface to produce the weld. Ultrasonic energy will aid in cleaning the weld area by breaking up oxide films and causing them to be carried away. The vibratory energy that produces the minute deformation comes from a transducer which converts high-frequency alternating electrical energy into mechanical energy. The transducer is coupled to the work by various types of tooling which can range from tips similar to resistance welding tips to resistance roll welding electrode wheels.

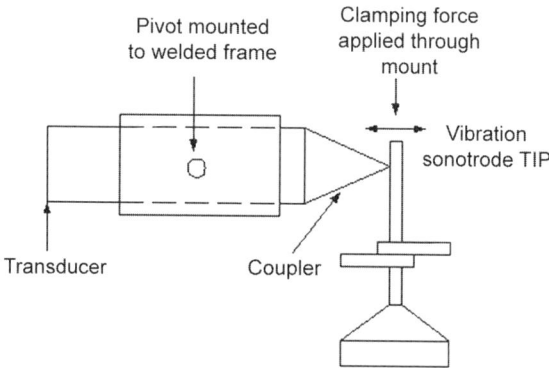

Fig. 5.33: Ultrasonic welding process.

The temperature at the weld is not increased to the melting point and, therefore, there is no nugget similar to resistance welding. Weld strength is almost equal to the strength of the base metal. Most of the ductile metals can be welded together and there are many combinations of dissimilar metals that can be welded. The process is restricted to relatively thin materials normally in the foil or extremely thin gauge thicknesses. The maximum thickness by these processes ultrasonically may vary from 0.38 to 2.5 mm depending upon the metal.

Applications

This process is used extensively in the electronics, aerospace, and instrument industries. It is also used for producing packages and containers and for sealing them. The process can also be employed for joining plastics and is finding wider use in this field than in joining metals.

5.27 POWER SUPPLY FOR RESISTANCE WELDING

AC supply is used for resistance welding because of the ease and covenience with which the required high current at a low voltage can be obtained by means of a transformer. The kVA required for resistance welding, when actually molting a weld, range from a few kVA to as much as 1 MVA. The power factor will be about 0.25 or 0.3 lagging. The power factor is low mainly due to the high ratio of reactance to resistance of the loop formed by the jaws of the welding machine. Such heavy intermittent single-phase loads may cause serious voltage drop difficulties in the supply network. Such problems can be overcome to some extent by connecting capacitors of suitable capacity in parallel with the welding transformer so as to improve the power factor. But with this arrangement the power factor will become leading when welding current is not being drawn. This problem can he avoided by connecting the capacitors in series with the welding transformer to neutralize the reactance drop in the supply circuit.

5.28 MACHINES FOR RESISTANCE WELDING

The machine for resistance welding incorporates a transformer, suitable electrodes for supplying current to the weld and arrangement for controlling the mechanical pressure, and finally, means for controlling the duration of weld current flow. The mechanical pressure may be exerted through levers and clutch by an electric motor or by compressed air. The magnitude of pressure required depends upon the type of work and may vary from a few kg for thin sheets or wires up to a tonne or more for heavy work.

In the older type of welding machines, the electrodes were brought on to the work and the electrical circuit closed by the operation of a pedal. Thus application of pressure and the time duration of current flow used to be controlled by the operator and for this operator needs to be experienced and skilled. The modern practice is to pass heavy currents for shorter time durations (ranging from 10 ms to 100 ms). The equipment used for this purpose may be constant time, current-actuated, or energy-actuated types.

Constant time equipment is employed in high speed production where the work has a consistently clean surface. Constant time equipment may be provided with mechanical control or electrical control. In mechanical control providing up to 300 welds per minute, the device employed is a cam-operated switch, connected in the primary circuit of a welding transformer, driven from the welding machine. For a large number of welds per minute the mechanical arrangement becomes unsuitable because it is not capable of providing consistently accurate timing, due to wear of the cam and operating mechanism, arcing and burning of the contacts and irregularities caused by closing of switch at different instants in the cycle.

An alternative arrangement is to control the timing through grid controlled ignitrons or thyratrons. It is easier to build tubes for a high voltage and small current than with a low voltage and high current. An arrangement using valves in the secondary circuit of a series transformer is shown in Fig. 5.34. When the tubes conduct, the series transformer secondary is almost short-circuited, and whole of the supply

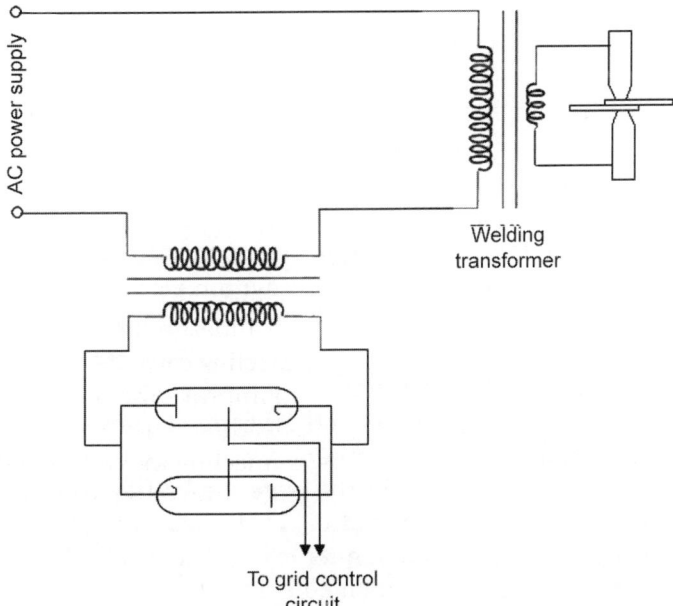

Welding transformer

To grid control circuit

Fig. 5.34: Machines for resistance welding.

voltage is available across the welding transformer primary. But when the tubes are not conducting, the series transformer primary winding offers a high impedance in the circuit of the welding transformer and the current is reduced to a negligible value. Auxiliary valves are employed for controlling the timing of the negative potential applied to the grids of the main tubes.

Constant-time method of control does not yield consistently good results when there may be variations in the conditions under which successive welds are made, due to variations in supply voltage or mechanical pressure, wearing of electrodes, surface irregularities, etc. The energy actuated control, wherein definite amount of energy is supplied to the weld, is used. Constant-time method of control did not prove to be successful, particularly with modern high-speed welding. The energy-actuated control, which permits the current to flow until a predetermined amount of energy has been supplied to the weld, is theoretically an ideal method. However, the control equipment is quite complicated.

5.29 ELECTRONIC WELDING CONTROL

As there are many factors, i.e. current, pressure, heat, time, to be considered, manual control does not yield desired outcome in case of resistance welding. For the efficient control of these factors electronic control welding circuits are used. Under this heading we will consider following:
1. Ignitron contactor.
2. Heat control unit.
3. Weld timer.

5.29.1 Ignitron Contactor

When we have to handle large currents, mechanical switches, if employed to control them, have following drawbacks:
   i. Due to large mass and inertia of mechanical switches, switching operation is very inconvenient. It is next to impossible to achieve switching operation required many times a second.
  ii. Switching operation by mechanical switches cannot have as much precise control as required in certain applications.

To overcome above limitations of mechanical switches, we resort to switching contactors, employing two thyratrons or ignitrons connected back to back.

Theory of ignitrons has been already dealt within § 2 9. We will now discuss its use as a contactor for controlling heavy currents. In Fig. 5.35 switch 'S' controls both the ignitrons.

When terminal 1 is negative and switch is closed, electronic current flows in the path indicated by 1, 3, 4, 5, 6, 7, 8, 9, 10, 11 and 2. Diode 14, prevents electronic current flow from igniter to cathode in tube B. Igniter 6-will fire the tube A and main current will flow in path 1, 3, 4, 5, 11 and 2. In the negative half cycle when terminal 2 is negative, electronic current will flow in path 2, 11, 10, 12, 13, 14, 9, 8, 4, 3 and 1. Diode 7 prevents electronic current flow from igniter to cathode in tube A. Igniter 13 will fire the tube B and main current path will then be 2, 11, 10, 12, 3 and 1. While the current through the main circuit may be very high, current through circuit 8, 9 never exceeds 40 amps. If switch s is opened, no igniter circuit will be working and therefore there is

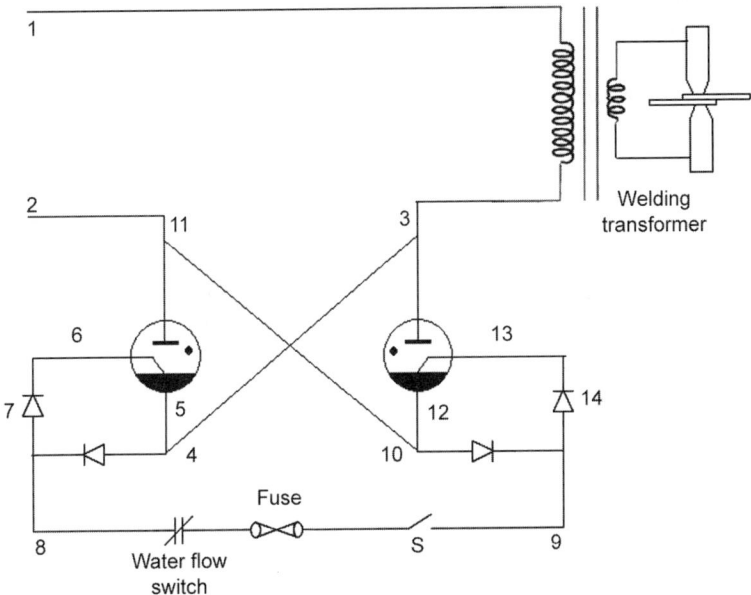

Fig. 5.35: Ignitron contactor

no current flow in main circuit. Since very heavy current flows through the ignitron say 1000 amps. And arc drop be taken constant at 10 V, there will be losses to the tune of 10 kW. As such ignitions are always water cooled. In case of very heavy load, temperature of water becomes too much, thermostat contacts, which are normally closed as shown, will open and igniters will stop conducting.

5.29.2 Heat Control Unit

It is an electronic circuit which helps the delay in the firing of the ignitrons by a definite, predetermined angle in each cycle and operates in conjunction with the line contactor. A typical circuit employed for the heat control is shown in Fig. 5.36. This is essentially

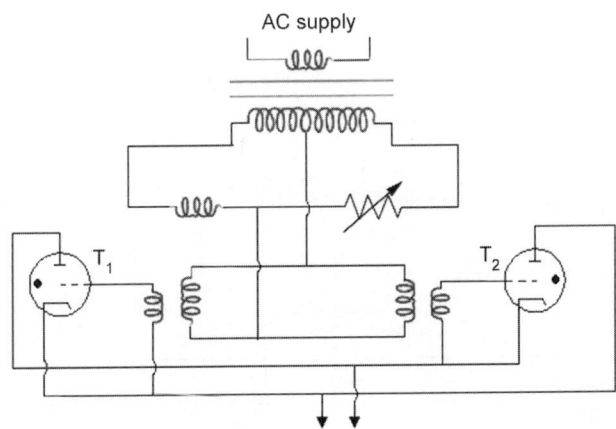

Fig. 5.36: Heat control unit.

a phase shift control circuit and delays the firing of the ignitrons, thus reducing the magnitude of welding current as per requirements.

5.29.3 AC Timer Circuit

Let E be the voltage across condenser before switch S is closed (Fig. 5.37).

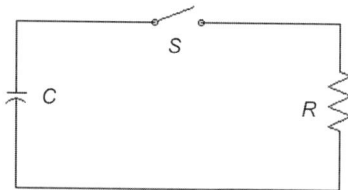

Fig. 5.37: AC timer circuit.

When switch S is closed, condenser will get discharged through resistor R and voltage across the condenser will fall. Greater the capacity of condenser and value of resistor, greater will be the time required for voltage to fall by given amount. It can be proved that voltage falls exponentially and is given by following equation:

$$v = Ve^{-t/CR}$$

$$t = RC \log_e \left(\frac{V}{v}\right)$$

In all timer circuits, therefore, there is provision of charging condenser to a particular value of voltage and time is started by short circuiting switch, till condenser is discharged to a particular value when relay will operate and particular contacts will open or close.

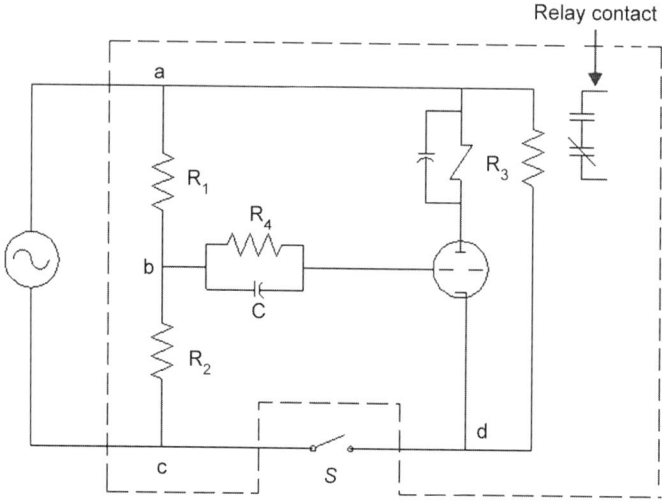

Fig. 5.38: Typical AC timer circuit.

In Fig. 5.38 is shown typical AC timer circuit. When switch `S' is open and terminal 'S' is positive with respect to terminal `b', there will be no electron flow between cathode and grid as cathode is positive with respect to grid and cathode and anode are at Sara, potential. If 'a' is negative with respect to 'b', grid is positive with respect to cathode and anode and electronic current will flow from 'a' to R3, cathode, grid and charge the condenser C to peak value of voltage across 'all'. So long switch `s' remains open, condenser remains charged to this voltage by the grid rectification action. As soon as switch is closed, grid becomes very much negative with respect to cathode and there is no condenser charging current through grid rectification. Condenser will, therefore, start discharging through R_4 and negative bias of grid will gradually decrease depending upon the time constant 'RC' of the discharge circuit. Conduction in the tube will start when grid condenser has discharged sufficiently till grid voltage falls to the critical grid voltage value. Current through relay coil is rectified half-waves. As such to avoid relay terminals' chatter, a condenser is connected across relay coil.

5.30 ENERGY STORAGE WELDING

Resistance welding of metals of very high conductivity, such as aluminium and magnesium, requires high current for small time as otherwise cold surrounding material drains away the heat from the weld spot. Too high demand from line, at the time of weld, will cause dip in the line voltage. This is avoided in energy storage welding circuits. There are basically two such circuits employing electrostatically stored energy and electromagnetically stored energy.

5.30.1 Capacitor Discharge Circuit

As shown in Fig. 5.39, condenser is charged to about 3000 V from grid controlled rectifier. The condenser on being connected to primary of welding transformer by ignitron contactor, will discharge. This will produce high transient current in the secondary to weld the material. Following two points may, however be noted in connection with this method.

1. Rate of charging becomes low as voltage of condenser approaches the voltage of source of supply. Therefore, to achieve high charging rate, 3000 V condenser will require about 5000 to 6000 V DC source.
2. Flux in the welding transformer core should not be present otherwise if there happens to be residual magnetism near saturation, there will be low rate of change of flux linkages in the secondary and, therefore, low heat will be produced.

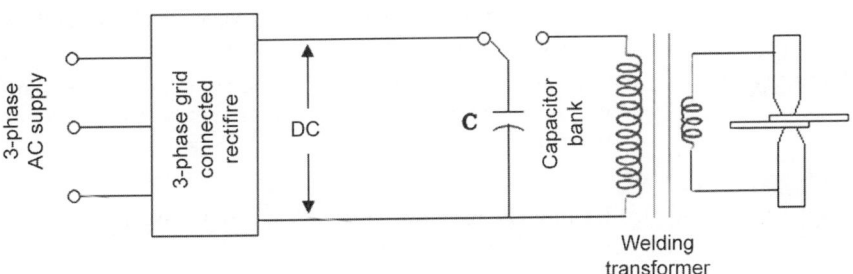

Fig. 5.39: Capacitor discharge circuit.

5.30.2 Magnetic Storage Welding Circuit

In this type of welding, energy stored in magnetic circuit is used in the welding operation. When primary of welding transformer is connected by means of contactor to output side of 3-phase rectifier of about 100 V, magnetizing current will flow to set up magnetic flux in the core depending upon the magnitude of applied DC voltage. Establishment of flux is to be gradual to avoid preheating of metals at the weld joint. This is done by gradual increase in DC output voltage of rectifier. On opening of contactor, DC no longer flows and there is rapid collapse of magnetic field which induces very high currents in the secondary side of transformer to do the welding job. Practical circuit is shown in Fig. 5.40. On closing the switch, magnetizing current flows

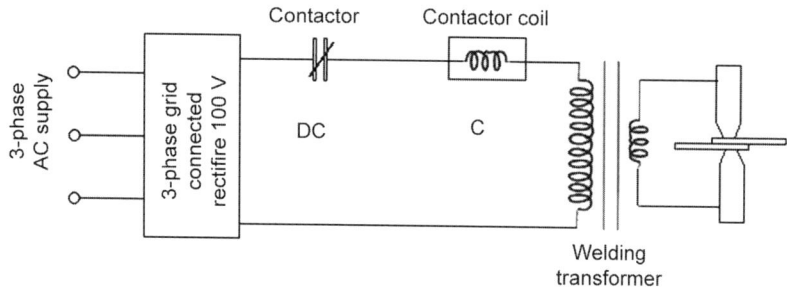

Fig. 5.40: Magnetic storage welding circuit.

Table 5.3	Comparison between resistance and arc welding	
Sl. no.	*Particulars*	*Resistance welding*	*Arc welding*
---	---	---	---
1.	Supply	Usually AC only	AC or DC
2.	Voltage	Very low	The striking voltage is high so requires voltage control
3.	Power factor	Very low	Poor
4.	Additional material requirement	No material is added in any form to get the two pieces joined	Suitable filler metal electrodes are necessary to get proper strength
5.	External pressure	External pressure is required	No external pressure is required hence the equipment is more simple and easy to control
6.	Development of heat	Heat is developed due to flow of current through the contact resistance mainly	Heat is developed due to arc between electrode and the work piece
7.	Temperature	The temperature attained is not very high as in case of arc welding	The temperature of the arc is very high and so likely to damage the work if not properly handled
8.	Applications	It cannot be used for repair work. It is most suitable for mass production	It is not suitable for mass production. It is most suitable for repair work and where more metal is to be deposited

through reactor. This current cannot flow through primary of welding transformer as cathode of ignitron is positive. When required value of magnetizing current flows, coil relay R opens the normally closed contact. This brings condenser C in circuit. Flux of reactor collapses which induces opposite polarity in the reactor coil, i.e. upper end becomes negative and lower end positive. Now ignitron's mercury cathode is negative and it, therefore, fires through diode. Current through primary of welding transformer is established all of a sudden. This will give rise to establishment of flux linkages on the primary side of transformer which in turn induces currents in the secondary for welding purpose. Condenser C also avoids sparking at the terminals of normally closed contacts.

QUESTIONS

1. What is welding?
2. What are different types of electric welding?
3. What do you mean by negative resistance characteristic of an electric arc and what precautions are taken for ill effects due to this?
4. Compare resistance welding and arc welding.
5. What is the technique of weld metal deposition by electric arc?
6. What are the advantages of using coated welding electrodes?
7. What are the qualities of a good weld?
8. What is the advantage of using submerged arc welding?
9. What type of electric supply is suitable for electric arc welding?
10. How storable reactor controls the magnitude of welding current?
11. What do you understand by resistance welding?
12. How do you prevent electrode in spot welding from sticking to the work piece?
13. What properties you would seek for selection of electrode material for spot welding?
14. What is the effect of welding time of resistance welding on the quality of the weld?
15. What are the advantages of projection welding?
16. What are advantages of flash butt welding over simple butt welding?
17. How heat control of the weld is obtained?
18. Give the comparison between AC and DC welding.

6

ELECTRIC TRACTION

6.1 INTRODUCTION

The locomotion in which electric motors are used to provide the driving or tractive force is called electric traction. In DC traction system, mainly series and compound motors are employed whereas single-phase series and three phase induction motors are preferred on AC traction systems. The motors should be built keeping in mind the severe working conditions, they should also comprised of essential security measures to keep itself safe from other working hazards.

6.2 TRACTION SYSTEMS

Broadly all traction systems may be classified into two categories:

a. **Non-electric traction systems:** Electrical energy is not used to run these at any stage. Examples are steam engine drive used in railways and internal-combustion-engine drive used for road transport.

b. **Electric traction systems:** They involve the use of electric energy at some stage or the other. They can be further subdivided into two groups:

1. First group comprises of self-contained vehicles or locomotives. Examples are battery-electric drive and diesel-electric drive, etc.

2. Second group consists of vehicles which receive electric power from a distribution network fed at suitable points from either central power stations or suitably-spaced substations. Examples are railway electric locomotive fed from overhead AC supply and tramways and trolly buses supplied with DC supply.

6.3 STEAM ENGINE DRIVE

Though its use is being limited for various reasons, steam locomotive is still the most widely adopted means of propulsion for railway work. Invariably, the reciprocating engine is employed because:

1. it is very simple.
2. connection between its cylinders and the driving wheels is simple.
3. its speed can be controlled very easily.

However, the steam locomotive suffers from the following disadvantages:

1. Since it is difficult to install a condenser on a locomotive, the steam engine runs non-condensing and, therefore, has a very low thermal efficiency of about 6–8 percent.
2. It has strictly limited overload capacity.
3. It is available for hauling work for about 60% of its working days, the remaining 40% being spent in preparing for service, in maintenance and overhaul.
4. Wear and tear of railway track is very high. Maintenance and replacement of track in regular interval is needed.

6.4 DIESEL-ELECTRIC DRIVE

It is a self-contained motive power unit which employs a diesel engine for direct drive of a DC generator. This generator supplies current to traction motors which are geared to the driving axles. In India, diesel locomotives were introduced in 1945 for shunting service on broad-guage (BG) sections and in 1956 for high-speed main-line operations on metre-guage (MG) sections. It was only in 1958 that Indian Railways went in for extensive main-line dieselisation.

Diesel-electric traction has the following advantages:

1. No modification of existing tracks is required while converting from steam to diesel-electric traction.
2. It provides greater tractive effort as compared to steam engine which results in higher starting acceleration.
3. It is available for hauling for about 90% of its working days.
4. Diesel-electric locomotive is more efficient than a steam locomotive (though less efficient than an electric locomotive).

Disadvantages

1. For same power, diesel-electric locomotive is costlier than either the steam or electric locomotive.
2. Overload capacity is limited because diesel engine is a constant-kW output prime mover.
3. Life of a diesel engine is comparatively shorter.
4. Diesel-electric locomotive is heavier than plain electric locomotive because it carries the main engine, generator and traction motors, etc.
5. Regenerative braking cannot be employed though rheostatic braking can be.

6.5 INTERNAL COMBUSTION ENGINE DRIVE

This is non-electric drive mostly used for road transport. Main advantages of this drive are: low investment and flexibility of routes. Its main disadvantages are: low overload capacity, uneconomical operation at other than normal speed, speed control possible through gearbox only, high maintenance cost, short-life and imported fuel oil required.

6.6   ELECTRIC DRIVE

Here drive is by means of electric motors which are fed from overhead distribution system.

Advantages of Electric Traction

As compared to steam traction, electric traction has the following advantages:

1. **Cleanliness.** Since it does not produce any smoke or corrosive fumes, electric traction is most suited for underground and tube railways. Also, it causes no damage to the buildings and other apparatus due to the absence of smoke and flue gases.

2. **Maintenance cost.** The maintenance cost of an electric locomotive is nearly 50% of that for a steam locomotive. Moreover, the maintenance time is also much less.

3. **Starting time.** An electric locomotive can be started at a moment's notice whereas a steam locomotive requires about two hours to heat up.

4. **High starting torque:** The motors used in electric traction have a very high starting torque. Hence, it is possible to achieve higher accelerations of 1.5 to 2.5 km/h/s as against 0.6 to 0.8 km/h/s in steam traction. As a result, we are able to get the following additional advantages:

   i. High schedule speed

   ii. Increased traffic handling capacity

   iii. Because of (i) and (ii) above, less terminal space is required—a factor of great importance in urban areas.

5. **Braking.** It is possible to use regenerative braking in electric traction system. It leads to the following advantages:

   i. About 80% of the energy taken from the supply during ascent is returned to it during descent.

   ii. Goods traffic on gradients becomes safer and speedier.

   iii. Since mechanical brakes are used to a very small extent, maintenance of brake shoes, wheels, tyres and track rails is considerably reduced because of less wear and tear.

6. **Saving in high grade coal.** Steam locomotives use costly high-grade coal which is not so abundant. But electric locomotives can be fed either from hydroelectric stations or pit-head thermal power stations which use cheap low-grade coal. In this way, high-grade coal can be saved for metallurgical purposes.

7. **Lower centre of gravity.** Since height of an electric locomotive is much less than that of a steam locomotive, its centre of gravity is comparatively low. This fact enables an electric locomotive to negotiate curves at higher speeds quite safely.

8. **Absence of unbalanced forces.** Electric traction has higher coefficient of adhesion since there are no unbalanced forces produced by reciprocating masses as is the case in steam traction. It not only reduces the weight/kW ratio of an electric locomotive but also improves its riding quality in addition to reducing the wear and tear of the track rails.

Disadvantages of Electric Traction

1. Most vital factor against electric traction is high capital outlay on overhead supply system. Therefore, unless heavy traffic is to be handled electric traction becomes uneconomical.
2. Power failure for few minutes can cause dislocation of traffic for hours.
3. Communication lines which also run along with power lines experience interference. Therefore, communication lines have either to be removed away from the track or special expensive cables have to be used for the purpose.
4. Traction is tied up to electrified routes.
5. If provision of negative booster is not made, return current through earth causes lot of corrosion of underground pipe work and interference with telegraph and telephone work.

6.7 BATTERY DRIVE

In this case, the vehicle carries secondary batteries which supply current to DC motors used for driving the vehicle. Such a drive is well-suited for shunting in railway yards, for traction in mines, for local delivery of goods in large towns and large industrial plants. They have low maintenance cost and are free from smoke. However, the scope of such vehicles is limited because of the small capacity of the batteries and the necessity of charging them frequently.

6.8 HYBRID DRIVE

In hybrid drive, locomotive derives power tor part journey from overhead DC system and for other part from batteries. This system proves economical where long tunnels have to be excavated so that neither end of journey is fixed point as cutting face of rock advances so does the dumping end. Extension of overhead, conductors cannot be economically done in small lengths. Employment of diesel traction is precluded on the ground of higher fuel and maintenance cost and fumes. Locomotive takes power from overhead trolley wires for most of the journey but at either ends it is switched on to batteries. These are recharged from trolley wire on return journey. For long tunnels pure battery drive will require large and costly batteries. This system has been used with advantage in the excavation of Sutlej Beas link tunnels.

Hybrid drive also finds its application in mines. Certain sections of main haulage roads either lie within the zone of mining influence or are inadequately ventilated. Mining regulations prohibit the use of trolley wire locomotives. For these sections, locomotive is worked from battery while for the rest of journey, trolley wire feed is used.

6.9 FLYWHEEL DRIVE

In this drive, use is made of KE of flywheel for driving the vehicle. The vehicle is equipped with three phase induction motor which is coupled to the flywheel and DC generator. At each halting station, three phase induction motor is supplied with power for sufficient time to bring the flywheel to full speed. KE of the flywheel on the way between two halting stations drives the DC generator which supplies the traction motors. KE of the flywheel is sufficient to move the vehicle over longest distance between two adjacent halting stations. This system can only be employed in cities where distance between stations is not much. Main advantage of this drive is that no electric supply arrangements have to be made in between halting stations.

6.10 TRAMWAYS

Electric supply is given to tram cars by grooved wheel or bow from overhead conductor at about 600 V DC track forming the return conductor. There are at least two driving axles which enable the tram car to be started from any one end. This incidentally enables series parallel method of speed control to be applied. Speed of tram car can further be increased by weakening the field of motors. Two drum controllers are employed in parallel one at each end of vehicle to control it. They are interlocked in parallel in such a way that only one at a time can control the motors. Motors are provided with ventilating ducts placed in the top half of motor frame in such a way that rain water of track is not splashed into the motors. This system of traction is now losing ground to trolley bus or internal combustion engine omnibus system on account of following reasons:

1. There is lack of flexibility of operation in areas of heavy congestion.
2. There is undesirable effect of track on other road vehicles.
3. Because of heavy expenditure on overhead supply system and track, it becomes economical method of transportation in large cities having dense traffic only.

Life of tram car equipment is much more than that of IC engine omnibus. Tram cars, if provided with magnetic brakes, can be excited by the rheostat braking current. These magnetic brakes are infact electromagnets suspended on springs. These are attracted to the rail track, where they exert sufficient locking force. In this way dual braking effect is obtained. Tram cars are employed for heavy traffic and where streets are congested with private motor cars.

6.11 TROLLEY BUS

Serious drawback of tramway is the lack of manoeuvrability in congested areas and noise. These are overcome by trolley bus system of traction. In this system the vehicle is similar to omnibus having rubber tyres and powered by electric motor. Since there is no track as in case of tramways, there are two overhead trolley wires fed at 600 V DC Since coefficient of adhesion between rubber tyred wheel and road is higher than that in the case of tram car, it is possible in this case to employ single axle drive without running the risk of slip. Motor employed is generally 80 to 150 HP compound wound where speed control can be carried out by inserting resistance in the shunt field or using diverters across series field. In order that driver should have his hands free to steer the vehicle and apply hand brake, master controllers are foot operated. One pedal controls the starting, speed control and regenerative braking if any, and second pedal controls rheostat and compressed air brakes. Employment of regenerative braking for trolley bus or tram car, though used in past is being discouraged because of difficulty of ensuring that supply system is always in position to absorb the energy generated by the motors. In trolley bus system, due to high adhesion between road and rubber tyres, power transmission gear is liable to be damaged if plain rheostat braking is used. To avoid this situation, stabilised rheostat braking is employed. In ordinary rheostat braking applied to shunt motors, field remains connected to the supply and armature after being disconnected from supply is connected across loading rheostat, In stabilized braking, a portion of loading resistance is included in the shunt field circuit.

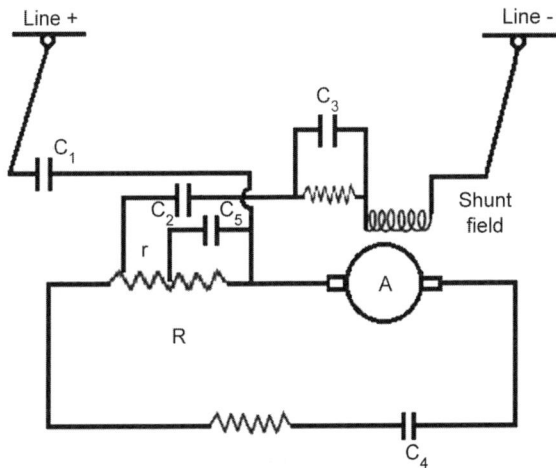

Fig. 6.1: Circuit diagram.

Portion R of loading rheostat Fig. 6.1 carries field as well as armature currents. It will be observed that the voltage applied across the field is equal to supply line voltage less the voltage drop across resistance R. Any tendency for braking current to increase will reduce the voltage applied across shunt field thereby reducing field flux. In this way, any tendency for braking current to increase and thereby overstressing the transmission gear is automatically countered by the fall in field flux. Reverse is also true. Net result of all this arrangement is to maintain the braking current substantially constant over a wide range of speed. On first notch, contactors C_1, C_2 C_3 and C_4 and connected. This puts full resistance R in field circuit. Another rate of braking can be obtained if contactor C_5 is also closed. In that case value of resistance in field circuit is r. If battery is provided on the vehicle, latter can be manoeuvred from the overhead line in case of emergency. If vehicle is halted in the dead section, it can be changed over to battery operation to clear the dead section. Trolley bus is insulated from ground by rubber tyres. It is, therefore, very important to see that no leakage develops which may cause shock to passengers while boarding or alighting if supply system has one pole earthed. Conditions are less dangerous if we use insulated supply system. In the event of leakage, insulation resistance of one supply pole will be in the circuit and current will be very small. To guard against these dangers, trolley buses are once a day insulation tested preferably when coming off from service when they are hot and damp and therefore, they are in their worst condition. Trolley buses have more passenger carrying capacity, higher acceleration and braking retardation than oil engines buses. These are therefore, used for medium traffic density as obtained in inner suburbs. Oil engines buses on the other hand are used for outer suburbs and country side where traffic density is low.

6.12 SYSTEMS OF RAILWAY ELECTRIFICATION

Presently, following four types of track electrification systems are available:
1. Direct current system—600 V, 750 V, 1500 V, 3000 V

2. Single-phase AC system—15–25 kV, $16\frac{2}{3}$, 25 and 50 Hz

3. Three-phase AC system—3000–3500 V at $16\frac{2}{3}$ Hz.

4. Composite system—involving conversion of single-phase AC into 3-phase AC or DC.

6.12.1 Direct Current System

Direct current at 600–750 V is universally employed for tramways in urban areas and for any suburban railways while 1500–3000 V DC is used for main line railways. The current collection is from third rail (or conductor rail) up to 750 V, where large currents are involved and from overhead wire for 1500 and 3000 V, where small currents are involved. Since in majority of cases, track (or running) rails are used as the return conductor, only one conductor rail is required. Both of these contact systems are fed from substations which are spaced 3 to 5 km for heavy suburban traffic and 40–50 km for main lines operating at higher voltages of 1500 to 3000 V. These substations themselves receive power from 110/132 kV, 3-phase network (or grid). At these substations, this high-voltage 3-phase supply is converted into low-voltage 1-phase supply with the help of Scott connected or V-connected 3-phase transformers. Next, this low AC voltage is converted into the required DC voltage by using suitable rectifiers or converters (like rotary converter, mercury arc, metal or semiconductor rectifiers). These substations are usually automatic and are remote controlled. The DC supply so obtained is fed via suitable contact system to the traction motors which are either DC series motors for electric locomotive or compound motors for tramway and trolley buses where regenerative braking is desired.

It may be noted that for heavy suburban service, low voltage DC system is undoubtedly superior to 1-phase AC system due to the following reasons:

1. DC motors are better suited for frequent and rapid acceleration of heavy trains than AC motors.
2. DC train equipment is lighter, less costly and more efficient than similar AC equipment.
3. when operating under similar service conditions, DC train consumes less energy than a 1-phase AC train.
4. the conductor rail for DC distribution system is less costly, both initially and in maintenance than the high-voltage overhead AC distribution system.
5. DC system causes no electrical interference with overhead communication lines.

The only disadvantage of DC system is the necessity of locating AC/DC conversion substations at relatively short distances apart.

6.12.2 Single-phase Low Frequency AC System

In this system, AC voltages from 11 to 15 kV at $16\frac{2}{3}$ or 25 Hz are used. If supply is from a generating station exclusively meant for the traction system, there is no difficulty in getting the electric supply of $16\frac{2}{3}$ or 25 Hz. If, however, electric supply is taken from the high voltage transmission lines at 50 Hz, then in addition to step-down transformer, the substation is provided with a frequency converter. The frequency converter equipment consists of a 3-phase synchronous motor which drives a 1-phase alternator having or 25 Hz frequency. The 15 kV $16\frac{2}{3}$ or 25 Hz supply is fed to the electric locomotor via a single overhead wire (running rail providing the return path). A step-down transformer carried by the locomotive reduces the 15-kV voltage to 300–400 V for feeding the AC series motors. Speed regulation of AC series motors is achieved by applying variable voltage from the tapped secondary of the above transformer. Low-frequency AC supply is used because apart from improving the commutation properties of AC motors, it increases their efficiency and power factor. Moreover, at low frequency, line reactance is less so that line impedance drop and hence line voltage drop is reduced. Because of this reduced line drop, it is feasible to space the substations 50 to 80 km apart. Another advantage of employing low frequency is that it reduces telephonic interference.

6.12.3 Three-phase Low Frequency System

It uses 3-phase induction motors which work on a 3.3 kV, $16\frac{2}{3}$ Hz supply. Substations receive power at a very high voltage from 3-phase transmission lines at the usual industrial frequency of 50 Hz. This high voltage is stepped down to 3.3 kV by transformers whereas frequency is reduced from 50 to $16\frac{2}{3}$ Hz by frequency converters installed at the substations. Obviously, this system employs two overhead contact wires, the track rail forming the third phase (of course, this leads to insulation difficulties at the junctions). Induction motors used in the system are quite simple and robust and give trouble-free operation. They possess the merits of high efficiency and of operating as a generator when driven at speeds above the synchronous speed. Hence, they have the property of automatic regenerative braking during the descent on gradients. However, it may be noted that despite all its advantages, this system has not found much favour and has, in fact, become obsolete because of its certain inherent limitations given below:

1. The overhead contact wire system becomes complicated at crossings and junctions.
2. Constant-speed characteristics of induction motors are not suitable for traction work.
3. Induction motors have speed/torque characteristics similar to DC shunt motors. Hence, they are not suitable for parallel operation because, even with little difference in rotational speeds caused by unequal diameters of the wheels, motors will becomes loaded very unevenly.

6.12.4 Composite System

We have seen above that no system of electrification is good in all respects. Composite system combines any two of above systems with a view to incorporate good points of each system. At present there are two such composite systems in vogue:
   a. Kendo system or single phase to 3-phase system.
   b. Single phase to DC system.

6.12.4.1. Kando System

In this system, single-phase 16 kV, 50 Hz supply from the substation is picked up by the locomotive through the single overhead contact wire. It is then converted into 3-phase AC supply at the same frequency by means of phase converter equipment carried on the locomotives. This 3-phase supply is then fed to the 3-phase induction motors. As seen, the complicated overhead two contact wire arrangement of ordinary 3-phase system is replaced by a single wire system. By using silicon controlled rectifier as inverter, it is possible to get variable-frequency 3-phase supply at 1/2 to 9 Hz frequency. At this low frequency, 3-phase motors develop high starting torque without taking excessive current. In view of the above, Kando system is likely to be developed further.

6.12.4.2. Single Phase to DC System

1. This system combines the advantages of high-voltage AC distribution at industrial frequency with
2. the DC series motors traction. It employs overhead 25 kV, 50 Hz supply which is stepped down by the
3. transformer installed in the locomotive itself. The low-voltage AC supply is then converted into DC
4. supply by the rectifier which is also carried on the locomotive. This DC supply is finally fed to DC
5. series traction motor fitted between the wheels. The system of traction employing 25 kV, 50 Hz,
6. 1-phase AC supply has been adopted for all future track electrification in India.

6.13 ADVANTAGES OF 25 KV, 50 HZ AC SYSTEM

Advantages of this system of track electrification over other systems particularly the DC system are as under:

1. **Light overhead catenary:** Since voltage is high (25 kV), line current for a given traction demand is less. Hence, cross section of the overhead conductors is reduced. Since these small-sized conductors are light, supporting structures and foundations are also light and simple. Of course, high voltage needs higher insulation which increases the cost of overhead equipment (OHE) but the reduction in the size of conductors has an overriding effect.
2. **Less number of substations:** Since in the 25 kV system, line current is less, line voltage drop which is mainly due to the resistance of the line is correspondingly less. It improves the voltage regulation of the line which fact makes larger spacing of 50–80 km between substations possible as against 5–15 km with 1500 V DC system and 15–30 km with 3000 V DC sysem. Since the required number of substations along the track is considerably reduced, it leads to substantial saving in the capital expenditure on track electrification.
3. **Flexibility in the location of substations:** Larger spacing of substations leads to greater flexibility in the selection of site for their proper location. These substations can be located near the national high-voltage grid which, in our country, fortunately runs close to the main railway routes. The substations are

fed from this grid thereby saving the railway administration lot of expenditure for erecting special transmission lines for their substations. On the other hand, in view of closer spacing of DC substations and their far away location, railway administration has to erect its own transmission lines for taking feed from the national grid to the substations which consequently increases the initial cost of electrification.

4. **Simplicity of substation design:** In AC systems, the substations are simple in design and layout because they do not have to install and maintain rotary converters or rectifiers as in DC systems. They only consist of static transformers along with their associated switch gear and take their power directly from the high-voltage national grid running over the length and breadth of our country. Since such substations are remotely controlled, they have few attending personnel or even may be unattended.

5. **Lower cost of fixed installations:** The cost of fixed installations is much less for 25 kV AC system as compared to DC system. In fact, cost is in ascending order for 25 kV AC, 3000 V DC and 1500 V DC systems. Consequently, traffic densities for which these systems are economical are also in the ascending order.

6. **Higher coefficient of adhesion:** The straight DC locomotive has a coefficient of adhesion of about 27% whereas its value for AC rectifier locomotive is nearly 45%. For this reason, a lighter AC locomotive can haul the same load as a heavier straight DC locomotive. Consequently, AC locomotives are capable of achieving higher speeds in coping with heavier traffic.

Table 6.1	Comparison between pure AC and DC systems		
Sl. no.	Aspects	DC series motors	AC series motors
1.	Development of starting as well as running torque for the same size	More	Less
2.	Number of speeds	Limited number of speeds (except by Chopper method)	Many speeds (by tap changing method)
3.	Cost, weight, efficiency for the same HP	Less costly, lighter, more efficient	More costly, heavier, less efficient comparatively
4.	Maintenance required	Less	More
5.	Regenerative braking	More efficient and less complications	Less efficient and more complications comparatively
6.	Negative boosting required	Less elaborate	More elaborate (due to high impedance of track)
7.	Number of substations required for a given track kilometerage	More in number	Less in number
8.	Overhead distribution system	Heavier and more costly comparatively	Lighter and less costly
9.	Interference with communication lines	Low	High
10.	Rail conductor system of track electrification	Possible with DC system	Not possible

7. **Higher starting efficiency:** An AC locomotive has higher starting efficiency than a straight DC locomotive. In DC locomotive supply voltage at starting is reduced by means of ohmic resistors but by on-load primary or secondary tap-changer in AC locomotives.

Disadvantages of 25 kV AC System

1. Single-phase AC system produces both current and voltage unbalancing effect on the supply.
2. It produces interference in telecommunication circuits. Fortunately, it is possible at least to minimize both these undesirable effects.

6.14 COMPARISON BETWEEN DC AND AC SYSTEMS OF RAILWAY ELECTRIFICATION FROM THE POINT OF VIEW OF MAIN LINE AND SUBURBAN LINE RAILWAY SERVICE

A. Main Line Railway Service

Following are the most important requirements of main line railway service:

1. Minimum cost of overhead structure.
2. Higher maximum speed.
3. Acceleration and retardation are not so important for main line service as in suburban service.

For main line railway service single-phase AC system is preferred due to following advantages:

i. Due to the use of high voltage, the cost of distribution system employing conductor of smaller section to carry the low current required is reduced.
ii. Less number of substations required.
iii. The initial, maintenance and operating costs of AC substations are less comparatively.

B. Suburban Line Railway Service

Following are the most important requirements of suburban line railway service:

1. Rapid acceleration and retardation (since frequent starting and stopping is required).
2. The working of motors should not be affected by the fluctuations in voltage.
3. There should be no interference to the communication network running along the track on account of power system adopted.

All the above requirements are fulfilled by the DC system and is invariably used for suburban service.

The advantages and disadvantages of DC system are listed below:

Advantages

i. For exerting same torque, current required in DC system is less than that in AC system.
ii. Less energy consumption, in comparison to AC system.
iii. The DC locomotive and motor coach equipment are lighter in weight, cheaper in initial as well as in maintenance cost and more efficient.

Disadvantages

    i. Less efficiency of the substations.

    ii. In order to keep the voltage of return rail within limits, additional equipment such as negative booster, etc. is required.

    iii. Number of substations required is more.

    iv. Substations are more costly as converting machinery is required.

The above mentioned disadvantages are partly eliminated by using composite system-single-phase AC to DC system.

6.15 TRACTION MECHANICS

In this article we will mainly concern ourselves with the applied mechanics side of the traction.

6.15.1 Units used in Traction Mechanics

Units of distance speed, etc. used in the mechanics of train are altogether different, from those used in solving the problems of ordinary applied mechanics. Various such units used and their notations are given as follows:

    D = Distance in kilometres (km). V = Speed in kilometre per hour (kmph)

    α or β =acceleration retardation in kilometre per hour per sec (Kmphps)

    W = Weight in tonnes. T = time in seconds. F = Tractive effort or force in Newton's.

    T = Torque in Newton metres.

6.15.2 Types of Services

There are three types of passenger services offered by the railways:

1. **City or urban service.** In this case, there are frequent stops, the distance between stops being nearly 1 km or less. Hence, high acceleration and retardation are essential to achieve moderately high schedule speed between the stations.

2. **Suburban service.** In this case, the distance between stops averages from 3 to 5 km over a distance of 25 to 30 km from the city terminus. Here, also, high rates of acceleration and retardation are necessary.

3. **Main line service.** It involves operation over long routes where stops are infrequent. Here, operating speed is high and accelerating and braking periods are relatively unimportant. On goods traffic side also, there are three types of services:

    i. Main-line freight service.

    ii. Local or pick-up freight service.

    iii. Shunting service.

6.15.3 Speed–Time Curve

Typical speed/time curve for electric trains operating on passenger services is shown in Fig. 6.2. It may be divided into the following five parts:

1. **Constant acceleration period (0 to t_1):** It is also called notching-up or starting period because during this period, starting resistance of the motors is gradually cut out so that the motor current (and hence, tractive effort) is maintained nearly

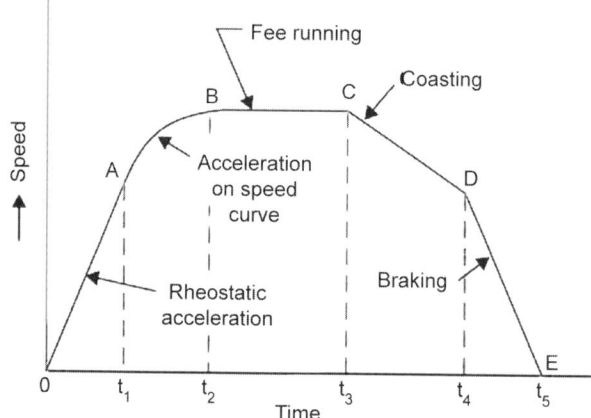

Fig. 6.2: Typical speed–time curve for main line service,

constant which produces constant acceleration alternatively called 'rheostatic acceleration' or 'acceleration while notching'.

2. **Acceleration on speed curve (t_1 to t_2):** This acceleration commences after the starting resistance has been all cut out at point t_1 and full supply voltage has been applied to the motors. During this period, the motor current and torque decrease as train speed increases. Hence, acceleration gradually decreases till torque developed by motors exactly balances that due to resistance to the train motion. The shape of the portion AB of the speed/time curve depends primarily on the torque/speed characteristics of the traction motors.

3. **Free-running period (t_2 to t_3):** The train continues to run at the speed reached at point t_2. It is represented by portion BC in Fig. 6.2 and is a constant-speed period which occurs on level tracks.

4. **Coasting (t_3 to t_4):** Power to the motors is cut off at point t_3 so that the train runs under its momentum, the speed gradually falling due to friction, windage, etc. (portion CD). During this period, retardation remains practically constant. Coasting is desirable because it utilizes some of the kinetic energy of the train which would, otherwise, be wasted during braking. Hence, it helps to reduce the energy consumption of the train.

5. **Braking (t_4 to t_5):** At point t_4, brakes are applied and the train is brought to rest at point t_5. It may be noted that coasting and braking are governed by train resistance and allowable retardation respectively.

6.15.4 Simplified Speed–Time Curve

In order to make actual speed–time diagram amenable to calculations, simplified speed–time curves are taken in such a way as to cause least error and at the same time calculations are made easy.

In Fig. 6.3 OABC is the actual speed–time curve. The basis of constructing simplified speed–time curves to keep both acceleration and retardation values same and area under actual and simplified curves (i.e. distance between stations) also same. In case of simplified trapezoidal speed–time curve OA'B'C, speed curve running and

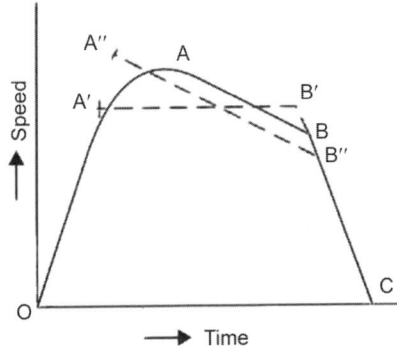

Fig. 6.3: Actual speed–time curve.

coasting periods are replaced by constant speed period. On the other hand, in case of simplified quadrilateral speed–time curve OA"B"C, speed curve running and coasting periods are extended. Trapezoidal speed–time curve gives closer approximation of the conditions of main line service where long distances are involved and quadrilateral speed–time curve for urban and suburban services.

(a) Trapezoidal speed–time curve. Fig. 6.4 shows simplified trapezoidal speed–time curve. If V_m is maximum speed attained,

Time of acceleration $t_1 = \dfrac{V_m}{\alpha}$ seconds and

Time of retardation $t_3 = \dfrac{V_m}{\beta}$ seconds.

Total distance D = Area OABC

= OAE + AEB + BEC

= OAE + CBD

= $\dfrac{1}{2}$ AD.OE + $\dfrac{1}{2}$ BE.DC

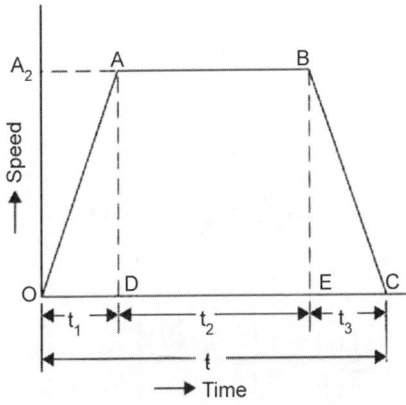

Fig. 6.4: Simplified trapezoidal speed–time curve.

$$= \frac{1}{2} V_m \left(\frac{T - t_3}{3600} \right) + \frac{1}{2} V_m \left(\frac{T - t_1}{3600} \right)$$

$$= \frac{V_m}{7200} [2T - t_3 - t_1]$$

$$= \frac{V_m}{7200} \left[2T - \frac{V_m}{\beta} - \frac{V_m}{\alpha} \right]$$

$$= \frac{V_m}{7200} \left[2T - V_m \left(\frac{1}{\beta} + \frac{1}{\alpha} \right) \right]$$

If

$$\frac{\alpha + \beta}{2\alpha\beta} = K$$

$$D = \frac{V_m}{3600} [T - V_m K] \qquad (6.1)$$

or $\qquad K V_m^2 - V_m T + 3600D = 0$

or $\qquad V_m = \dfrac{T \pm \sqrt{T^2 - 14400\ KD}}{2K}$

Positive sign gives very much high value of V_m which is not possible in practice. Hence negative sign is adopted.

$$V_m = \frac{T \pm \sqrt{T^2 - 14400\ KD}}{2K} \qquad (6.2)$$

From equation (6.1) we get,

$$D = \frac{V_m T}{3600} - \frac{V_m^2 K}{3600}$$

$\therefore \qquad K = \dfrac{3600}{V_m^2} \left[\dfrac{V_m T}{3600} - D \right]$

$$= \frac{3600}{V_m^2} \left[\frac{V_m}{3600D} - 1 \right]$$
$$\qquad\qquad\qquad\qquad T$$

$$\frac{1}{2} \left[\frac{1}{\alpha} + \frac{1}{\beta} \right] = \frac{3600D}{V_m^2} \left[\frac{V_m}{V_a} - 1 \right]$$

where V_a is the average speed.

$$\therefore \qquad \frac{1}{\alpha} + \frac{1}{\beta} = \frac{7200D}{V_m^2}\left[\frac{V_m}{V_a} - 1\right] \qquad (6.3)$$

(a) Quadrilateral speed–time curve. In Fig. 6.5

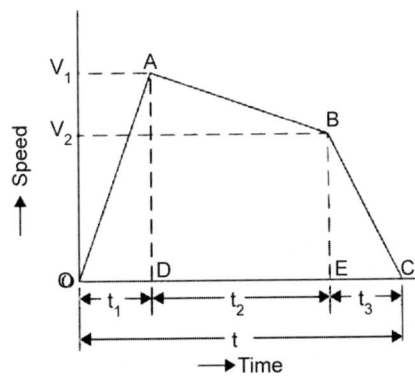

Fig. 6.5: Quadrilateral speed–time curve.

Total distance D = Area OABC

= OAE + AEB + BEC

= OAE + CBD

$$= \frac{1}{2}\,AD.OE + \frac{1}{2}\,BE.DC$$

$$= \frac{1}{2}V_1\left(\frac{T-t_3}{3600}\right) + \frac{1}{2}V_2\left(\frac{T-t_1}{3600}\right)$$

$$= \frac{1}{7200}\left[T(V_1 + V_2) - (V_1 t_3 + V_2 t_1)\right]$$

$$= \frac{1}{7200}\left[T(V_1 + V_2) - \left(V_1\frac{V_2}{\beta} + V_2\frac{V_1}{\alpha}\right)\right]$$

$$= \frac{1}{7200}\left[T(V_1 + V_2) - V_1 V_2\left(\frac{1}{\alpha} + \frac{1}{\beta}\right)\right] \qquad (6.4)$$

$$V_2 = V_1 - \beta_0 t_2$$

$$= V_1 - \beta_0\left(T - t_1 - t_3\right)$$

$$= V_1 - \beta_0\left(T - \frac{V_1}{\alpha} - \frac{V_2}{\beta}\right)$$

$$\left(V_2 - \frac{\beta_0}{\beta}V_2\right) = V_1 - \beta_0\left(T - \frac{V_1}{\alpha}\right) \qquad (6.5)$$

Solving equation no (6.4) and (6.5), we get value of D, V_1, or V_2.

6.15.5 Average Speed and Schedule Speed

While considering train movement, the following three speeds are of importance:

    1. Crest speed. It is the maximum speed (V_m) attained by a train during the run.

    2. Average speed $= \dfrac{\text{Distance between the stops}}{\text{Actual time of run}}$ (6.6)

    In this case, only running time is considered but not the stop time.

    3. Schedule speed $= \dfrac{\text{Distance between the stops}}{\text{Actual time of run} + \text{stop time}}$ (6.7)

    Factors affecting schedule speed. Following are the factors which affect the schedule speed of a train engaged in a given service. Obviously, schedule speed can be obtained from average speed by including the duration of stops. For a given distance between stations, higher values of acceleration and retardation will mean lesser running time and, consequently, higher schedule speed. Similarly, for a given distance between stations and for fixed values of acceleration and retardation, higher crest speed will result in higher schedule speed. For the same value of average speed, increase in duration of stops decreases the schedule speed.

6.16 TRAIN MOVEMENT

The movement of trains and their energy consumption can be conveniently studied by means of speed/time and speed/distance curves. As their names indicate, former gives speed of the train at various times after the start of the run and the later gives speed at various distances from the starting point. Out of the two, speed/time curve is more important because

    1. its slope gives acceleration or retardation as the case may be.
    2. area between it and the horizontal (i.e. time) axis represents the distance travelled.
    3. energy required for propulsion can be calculated if resistance to the motion of train is known.

6.17 TRACTIVE EFFORT

Tractive effort is the force developed by the traction unit at the wheel rims for moving the traction unit and its train. Draw bar pull is the force exerted by the traction unit through the draw bar for moving the train. Thus draw bar pull is less than the tractive effort by the force required to move the traction unit. Tractive effort exerted by the traction unit has to perform the following functions.

    a. To give necessary linear and angular acceleration to the train mass.
    b. To overcome the gravity component of the weight of the train.
    c. To overcome the wind and frictional resistance of the train.
    d. To overcome curve resistance.
    a. Tractive effort for acceleration. Force is required to give linear acceleration to the train and is given by

$$F_a = 1000\ W\,\frac{\alpha \times 1000}{3600}$$

$$277.8 \, W \, \alpha \text{ newton} \tag{6.8}$$

When the speed of the train is being changed, it behaves as a mass greater than its dead weight. This is due to angular speed variation of its rotating parts. If linear acceleration is f metre/sec^2 angular acceleration of rotating parts having radius of gyration r metres will be $\dfrac{f}{r}$. Now if I is the moment of inertia of rotating masses, and F the force in newton applied at the rim of the wheel, then torque applied (F.r) will be given by

$$F.r = I\frac{f}{r} \tag{6.9}$$

$$F = \frac{I}{r^2} f$$

Magnitude of force as per equation 6.9 will be in addition to the force required to give linear acceleration and equivalent additional mass of rotating masses is $\dfrac{I}{r^2}$ kg. This additional mass is best expressed as a percentage of the dead weight of train the value of which lies between 8 to 15%. We will now introduce the concept of effective weight of the train 'W_e' which takes into account not only motion of translation of train but also the motion of rotation of wheels, axles, armatures of motors and gears. Equation (6.8) now becomes

$$F_a = 277.8 \, W_e \, \alpha \text{ newton} \tag{6.10.1}$$

$$= 28.3 \text{ kg} \tag{6.10.2}$$

Tractive effort to give acceleration of 1 Kmphps to 1tonne is given as:

$F_a = 28.3 \times (1.1 \, W) \times 1$. Assuming $W_e = 1.1 \, W$

$$= 31.1 \, W \text{ kg} \tag{6.11}$$

$$\therefore \qquad \frac{\text{Tractive effort}}{\text{train weight}} = \frac{31.1}{100} \times 100 \tag{6.11.1}$$

$$= 3.11\%$$

From equation 6.11.1. It is observed that the tractive effort required to give an acceleration of 1 Kmphps is 3.11% of the weight of the train which is equivalent to 5% for an acceleration of 1 mpbps.

Usual values of acceleration adopted for different services are:

Goods	.21 to .29 Kphps
Parcel	.35 to .53 Kphps
Passenger	.53 to 1.08 Kphps
Suburban	1.08 to 2.16 Kphps
Metropolitan	2.16 to 3.16 Kphps

b. **Tractive effort required to balance the gravitational pull:** When train is on an upgradient, gravity component of dead weight of train parallel to the track (W sin θ) (Fig. 6.6a) will be responsible for the train to come down. In order to prevent this tractive effort has to be applied in upwards direction whose magnitude is given by equation 6.12.

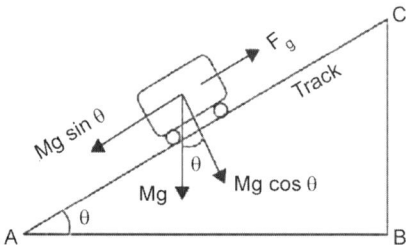

Fig. 6.6a: Phasor diagram of different forces.

$F_g = 1000 \text{ W} \sin\theta \times 981 \text{ Newtons}$ (i)

In railway practice, gradient is expressed as rise in metres in a track resistance of 1000 metres and is denoted by letter G%

∴ $G = \sin\theta \times 1000$ (ii)

Substituting (ii) in (i) we get,

$$F_g = 1000 \text{ W.}\frac{G}{1000} \times 9.81$$

$$= 9.81 \text{ W.G Newtons}$$

$$F_g = \text{W.G kg}$$ (6.12)

Location of gradient holds crucial role on the overloading which can be allowed on the traction motors. Gradient does not create any problem as momentum of the train takes it up the steep gradient. On the other hand if ruling gradient happens to be in the middle of a rising gradient, the length of gradient becomes very important in determining the temperature rise of the motor. Another factor which affects the overloading of motor is the location of signal with respect to the gradient. If stopping signal is at the foot of a rising gradient, speed pick-up will be tow and motors are likely to be overloaded.

c. **Tractive effort to overcome train resistance:** Train resistance means all the forces which hinder the motion of the train on a level track. These forces can be categorized as those forces which are internal to the rolling stock such as friction at journals, axle guides, bogie pivots, buffers, etc., and those forces which are external to the rolling stock such as friction between wheels and rails, flange friction, resistance as a result of temporary deflection of track and aerodynamic drag. Flange friction rises with oscillations of the coach and is affected by side wind pressure, whereas track resistance depends upon the strength of the track and nature of the ballast. Aerodynamic drag consists of pressure drag and frictional drag. Former may be due to ends and the latter due to the length of the train. End resistance comprises of head resistance and tail resistance due to suction of air at the rear. Both of the end resistances depend upon the area perpendicular to motion, the shape of front and rear and wind velocity. Since head resistance is about ten times the suction resistance shape of the front carriage is very important in reducing the wind resistance. Wind resistance due to the length of train is due to air friction on sides and to underside of the train.

This is sometimes termed skin friction. The pressure drag is found to be far less than the frictional drag for long trains. It is also found that endeavour to diminish the pressure drag is not so effective for reduction of total drag. Aerodynamic drag of Tokaido line (JNR) was determined by the formula

$$F = (0.20 + .00451)\frac{P}{2}av^2 \text{kg} \tag{6.13}$$

Where l is length of train in metres, 'a' is cross sectional area of train in sq. metres, P air density in kg $\frac{sec^2}{m^4}$. And v is speed of the train in m/sec. All the internal and external resistances, excluding wind resistance, are termed mechanical resistance. Although different components of mechanical resistance behave in a different way with the inch ease in speed, it is reasonably correct assumption to consider me char leaf resistance as a whole to remain constant specially at high speeds and is proportional to the weight of the train. Wind resistance on the other hand is considered to vary with the square of the speed of the train. Total train resistance F_r, in the case of electric train hauled by electric locomotive is represented by equation (6.14)

$$F_r = a + bv + Cv^2 \qquad\qquad = (6.14)$$

where v = train speed in Kmph

a = constant representing mechanical resistance such as journal friction, track resistance, etc.

b = constant to cover several factors like flange resistance and internal frictions of train.

c = coefficient covering air resistance.

However total train resistance is also given as follows

F_r = 9.81 r. W Newton

= r.W. kg. (6.15)

where W is the weight of the train including locomotive and r is the train resistance in kg/tonne.

Starting friction. Starting friction which is also known as 'striation' is more than the running friction. This depends heavily upon the starting conditions of the train. For instance starting resistance of the train will be more if inter vehicle couplers are in tension which happens in the case of upgradient. On the other hand, enter vehicle couplers will be in compression and starting resistance will be low for train standing on downgradient. It is due to this reason that signals should, as far as possible, not be placed on upgradients. Even on level tracks it is a better practice to mote back locomotive by few metres so as to release inter-vehicle tension. This makes the starting friction low as it has been observed that the starting resistance is due to first few vehicles. Once train starts moving, resistance drops and remaining of vehicles are easily hauled by locomotive. Formulae adopted by SNCF are given below:

(I) for gradients up to 7%.

r = (Gradient per thousand + 4.5) kg. up to 10 tonne axle load

= (Gradient per thousand + 4.2) kg. above 10 tonne axle load (6.16)

i. For gradient above 7%

$r = (1.25 \times$ gradient per thousand $+ 2.75)$ kg. Up to 10 tonne axle load

$= (1.25$ gradient per thousand $+ 2.45)$ kg. Above 10 tonne axle load   (6.17)

d. **Tractive effort to overcome curve resistance.** Curve resistance is due to the friction at the wheel flanges. More the radius of curvature less will be the curve resistance. Curve resistance is given by the following empirical formula and is usually added to the track resistance.

$$F_e = \frac{700}{R} W \text{ kg.} \tag{6.18}$$

where R is radius of curvature in metres. In railway practice sharpness of the curve is expressed more conveniently in degrees of curvature $(C°)$ than by radius of curvature. $1°$ curve is defined as that which in 100 ft. (or metres) turns through $1/360$th of a complete circle or in a track of 100 meters it turns through $3.27°$. $C°$ curvature will have metres as radius of curvature.

Substituting this for value of R in equation (6.18) we get

$$F_e = \frac{700 \times 3.27}{5730} C.W. \text{ kg}$$

$= 0.4$ C.W. kg.   (6.19)

Curve resistance is usually taken to be the part of train resistance and is combined with it. Total tractive effort in that case is given as

$F_t = F_a + F_g + (F_r + F_c)$

$= 28.3 \, W_e \alpha \pm WG + W(r + 0.4C)$ kg.

$= 277.8 \, W\alpha \pm 9.81 \, WG + 9.81 \, W(r + 0.4C)$ Newtons   (6.20)

Sometimes in traction mechanics, it is convenient to express curve resistance in equivalent gradient resistance given by equation (6.21) and added to the actual per thousand gradient.

Equivalent gradient per thousand $= 0.4C$ (6.21)

Expression for total tractive effort, now, becomes

$F_t = 28.3 \, W_e \alpha + (0.4C \pm G)W + W.r.kg$

$= 277.8 \, W_e \alpha + 9.81(0.4C \pm G)W \pm 9.81 \, W.r$ *Newtons*   (6.22)

Positive sign is to be taken for train movement upgradient and negative sign for train movement downgradient.

6.18 TRACTIVE EFFORT-SPEED CHARACTERISTIC

Tractive effort-speed characteristic are family of curves showing the relationship between tractive efforts exerted by the locomotive at various speeds for different notch positions of the controller. One such curve for 75.2 tonne MT locomotive is shown in Fig. 6.6b where line AB represents the limit of adhesion. Taking an average coefficient of adhesion of 33%, we get starting tractive effort from 75.2 tonne locomotive as 25 tonne. This drops as speed increases according to equation 6.37. In order that slip should not take place, tractive effort at any time should not exceed this adhesion limit. To ensure this line CD is drawn tangential to this adhesion curve which gives starting tractive effort corresponding to point C as 23.6 tonnes. About 6% of the tractive effort is required to give acceleration to the rotating masses and this is taken into account by drawing line EF parallel to CD giving reduced tractive effort of 22.1 tonne at starting.

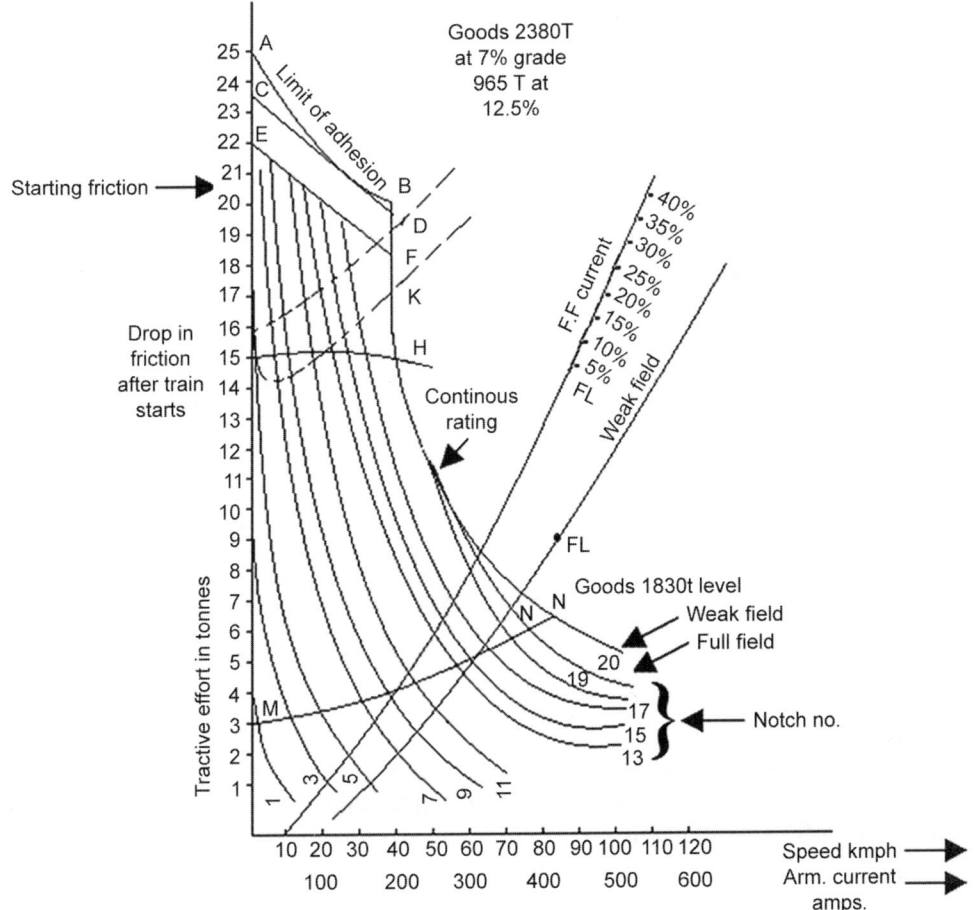

Fig. 6.6b: Tractive effort-speed characteristic.

Therefore, final limits of tractive effort during acceleration period is line EF. Line IK gives the train resistance for 965 tonne train on 12.5% gradient. Drop in starting friction may be noted in line IK. Inter section of train resistance curve and tractive effort curve gives the balancing speed for 965 tonne train on gradient of 12.5% is given by point K and that of 1830 tonne goods train on level track by point N. Line GH is a continuous rating line. If starting torque developed is above this line, it overloads the motor and this period during staring in kept small. Allowable maximum tractive effort during staring period is given by line CD.

6.19 POWER OF THE TRACTION MOTOR

Power is the rate of doing work and is given by

$$P = F_t \times \text{velocity} \tag{6.23}$$

Now consider an instant at point D in Fig. 6.4. Just after D speed remains constant but tractive effort required is less as no acceleration of train take place. Just before D, tractive effort required is maximum and also speed is approaching maximum value. Therefore, at instant D power output required from the driving axles to propel the train is maximum and is given by

$$P_{max} = F_t \frac{V_m 1000}{3600} \text{Watts}$$

$$= 0.278 \text{ watts.} \tag{6.24}$$

where F_t is expressed in Newton's.
If tractive effort is expressed in kg.

$$P_{max} = \frac{0.278 \times 9081 \, F_t V_m}{735.5} \text{metric HP}$$

$$= \frac{F_t \times V_m}{270} \text{metric HP} \tag{6.25}$$

If η be the efficiency of transmission gear, maximum power output of motors

$$= \frac{0.278 \, F_t \times V_m}{\eta} \text{Watts when } F_t \text{ is in Newtons} \tag{6.26}$$

$$= \frac{F_t \times V_m}{270 \, \eta} \text{metric HP when is in kg.} \tag{6.27}$$

Fig. 6.7 shows the relationship between the HP developed by a 75.2 tonne MT locomotive at various speeds. It will be observed that BHP. developed falls for speeds beyond 60 kmph because of the fall in the tractive effort.

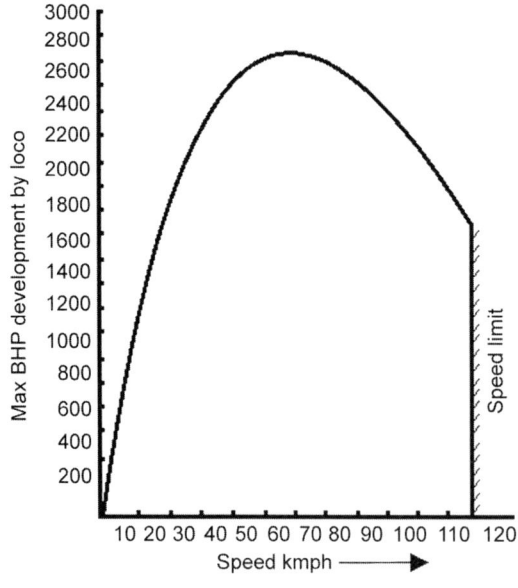

Fig. 6.7: Relationship between HP *vs* speed.

It is interesting to distinguish between the overloading of motor due to excessive HP developed and overloading of motor due to the armature current. Former depends upon the tractive effort and speed while the latter depends upon the armature current of motors. For example, in Fig. 6.6b at the speed of 10 kmph, tractive effort corresponding to line EF is 21 tonne and armature current for full excitation as 610 amperes which is 34% overloading. On the other hand from Fig. 6.7, BPH developed at 10 kmph is only about 850 as against full-load BHP of 2800.

6.20 SPECIFIC ENERGY CONSUMPTION

It is the energy consumed in watt hours per tonne kilometrage of train. We will first find out specific energy output of driving wheels, when this is divided by overall efficiency of transmission gear and motor, we will get specific energy consumption.

Total energy output of driving axles is spent as follows:

a. to accelerate the train.
b. to overcome gradient.
c. to overcome train resistance including curve resistance.
a. Energy output of driving axles to accelerate the train (E_a).
   We assume trapezoidal speed–time curve of Fig. 6.4
   $E_a = F_a \times$ distance OAD

$$= 277.8\, W_e\, \alpha \left[\frac{1}{2} \frac{V_m \times 1000}{3600} \times \frac{V_m}{\alpha} \right] \text{watt seconds}$$

$$= 277.8\, W_e \alpha \left[\frac{1}{2} \frac{V_m \times 1000}{3600} \times \frac{V_m}{\alpha} \right] \frac{1}{3600} \text{ Watt hours}$$

$$= 0.01072\, V_m^2 W_0 \text{ watt hours.} \tag{6.28}$$

b. Energy output of driving axles to overcome gradient (E_g).
   $E_g = F_g \times D_1$
   Where D_1 the distance is over which power remains on and its maximum value is equal to area OABE.
   $E_g = 9.81\, WG \times 1000\, D_1$ joules or watt-sec.

$$= \frac{9.81 \times 1000}{3600} WGD_1 \text{ Whrs}$$

$$= 2.725\, WGD_1 \text{ Whrs} \tag{6.29}$$

c. Energy output of driving axles to overcome friction (E_r).
   $E_r = F_r \times D_1$ Joules

$$= \frac{W.r.D_1\, 1000}{3600} \text{Whrs}$$

$$= 0.2778\, W.r. \text{ Whrs} \tag{6.30}$$

Total energy output of driving axles
$$E = 0.01072\, V_m^2 W_0 \pm 2.725\, WGD_1 + 0.2778\, W.r.D_1 \text{ Whrs} \tag{6.31}$$

$$\text{Specific energy output} = \frac{E}{W.D}$$

$$= \frac{0.01072 \, V_m^2}{D} \frac{W_e}{W} \pm 2.725 \, G\frac{D_1}{D} + 0.2778 \, r\frac{D_1}{D} \text{Whrs / tonne km} \tag{6.32}$$

$$\text{Specific energy consumption} = \frac{\begin{array}{c}\text{Specific energy output} \\ \text{at driving wheels}\end{array}}{\begin{array}{c}\text{Overall efficiency of} \\ \text{transmission gear and motor}\end{array}} \tag{6.33}$$

6.20.1 Factors Affective Specific Energy Consumption

Study of equations 6.32 and 6.33 will suggest the factors affecting specific energy consumption as follows:

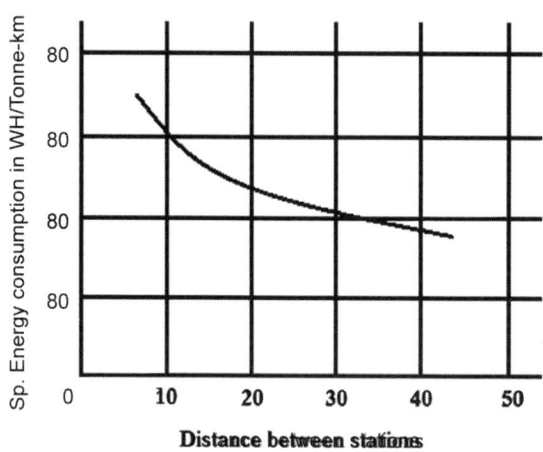

Fig. 6.8: Characteristics of energy consumption over distance.

a. **Distance between stops.** The greater the distance between stops, lesser will be the specific energy consumption Fig. 6.8. Typical values of specific energy consumption are:

Suburban service 50 to 75 w-hrs. per km.

Main line service 18 to 31 w-hrs. per tonne km.

b. **Retardation and acceleration values.** For given schedule speed, greater the value of acceleration and retardation, more will be the period of coasting and, therefore, less the period during which power is on. Hence D_1 will be small and, therefore, specific energy consumption will accordingly be less.

c. **Gradient.** Steep gradient will naturally involve more energy consumption even though regenerative braking is employed.

d. **Train resistance.** This depends upon the nature of the route, speed of the train and shape of front and rear of the train. More the train resistance, greater will be the specific energy consumption.

e. **Type of train equipment.** Overall efficiency for a given specific energy output at axles, will determine the specific energy consumption. Greater the overall efficiency less will be the specific energy consumption.

6.21 MECHANICS OF TRAIN MOVEMENT

How traction motor produces tractive effort at the rim of driving wheel is schematically shown in Fig. 6.9. Motor drives pinion which meshes with gear wheel keyed to the driving axle.

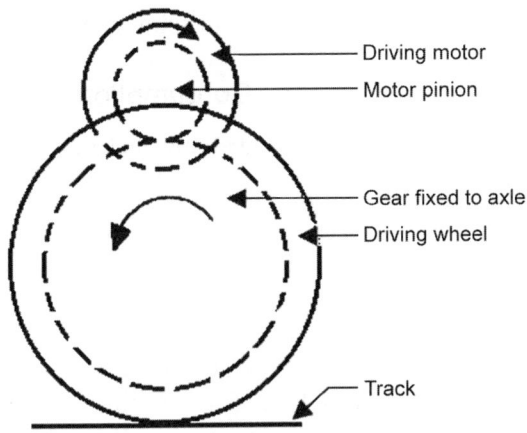

Fig. 6.9: Essential of driving mechanism in an electric vehicle.

If T = The torque exerted by motor.

F = The tractive effort exerted by driving wheel.

γ = Overall gear ratio.

$$= \frac{\text{speed of the motor pinion in rpm}}{\text{speed of the driving axle in rpm}}$$

$$= \frac{N_1}{N_2}$$

η = Efficiency of transmission of power from motor to driving axle.

r_w = Radius of driving wheel.

Then we have output at driving wheel = η input at motor pinion.

$\therefore F r_w N_2 = \eta T N_1$

$$F = \eta \frac{T}{r_w} \frac{N_1}{N_2}$$

$$= \eta \frac{T}{r_w} \gamma \qquad (6.34)$$

6.22 COEFFICIENT OF ADHESION

Adhesion between two bodies is due to interlocking of the irregularities of their surfaces when in contact. The adhesive weight of a train is equal to the total weight to be carried on the driving wheels. Equation 6.34 suggests that tractive effort at the driving wheels can be increased by increasing the torque exerted by the motor. This is possible up to a certain limit after which any increase in the motor torque does not increase the tractive effort but causes driving wheels slip. It has been found that the maximum value of tractive effort at which driving wheels will not slip, depends upon the dead weight over the driving axle.

$$\therefore \qquad\qquad F \propto W$$

Or
$$F = \mu_a W \qquad\qquad (6.35)$$

If F is expressed in Newton's and W in tonnes, equation 8.35 becomes.

$$F = 9.81 \qquad\qquad (6.36)$$

Where μ_a is called the coefficient of adhesion. In order, therefore, to increase the tractive effort of a locomotive, it is not enough to increase the HP of the traction motors along but at the same time weight on the driving wheels has to be increased. Railway standards provide certain maximum axle load. As such adhesive weight can only be increased by increasing the number of driving axles. In steam locomotive this is done by using coupling rod to connect driving wheels.

Adhesion also plays an important role in braking. If braking effort happens to be more than the adhesive weight of the vehicle, skidding will result. Therefore, maximum braking effort is possible only under maximum adhesion condition which have to be provided in the design of the rolling stock. When skidding occurs as a result of failure of adhesion for which many possibilities exist as a result of reduction of the contact area between the wheels tread and rail while negotiating a curve or passing over points and crossing. The braking retardation is suddenly reduced and this results in dangerous conditions jeopardizing safely. As such for braking, while every effort is made to utilize the braking capabilities to maximum, but at same time, care is taken not to exceed the adhesion limit.

Maximum braking force is limited by the capability and the rate of dissipation of kinetic energy. On downgrades since braking distance are longer, reliability of safe braking is ensured by limiting the speeds. Even though braking force in not applied to the adhesion limit, high adhesion capabilities give necessary insurance against the risk of wheel skidding, possibility for which exists more due to the improved brake performance of composite brake blocks.

As coefficient of friction is more at starting than at braking so also coefficient of adhesion at any speed will be less than that at lower speed or at starting. This is due to the cyclic variation of axle loads caused due to the acceleration and retardation to which unsprang masses are subjected as they have to follow track irregularities. Relationship between the coefficient of adhesion is given by equation 6.37 used by French National Railways or by equation 8.38 given by Curtis and Kniffler.

$$\mu_{ar} = \mu_{as} \frac{8 + 0.1\,V}{8 + 0.2\,V} \qquad\qquad (6.37)$$

Or
$$\mu_{ar} = \frac{7.5}{V + 44} + 0.16 = {} + 0.16 \qquad\qquad (6.38)$$

where μ_{ar} = coefficient of adhesion at running
μ_{as} = coefficient of adhesion at starting.
V = speed of locomotive in kmph.

This explains why adhesion limit as shown in Fig. 6.6b reduces as speed increase.

6.22.1 Factors Affecting Slip

Adhesion is a result of meshing of rough patches that occur on even the most highly polished bodies, and inter atomic forces between the protrusions of these rough patches which results in a sort of welding between them. In practice instead of pure steel to steel contact between rail and wheels, there are foreign bodies on the surfaces such as films of organic matters, metallic oxides or even layers of absorbed gases which have a detrimental effect on the adhesion. Various factors, influencing adhesion can be classified as due to:

1. The track
2. Mechanical parts
3. Electrical parts

6.22.1.1 Influence of Track on the Coefficient of Adhesion

Wheel rail condition affects the coefficient of adhesion to a great extent. Depending upon whether the rail is completely wet, partially wet, thoroughly dry or oily, the rail wheel adhesion varies. Presence of oil, dew, etc. reduces adhesion, whereas rust, sand, dust or dry rails improve adhesion. It is due to this reason that the loss of adhesion takes place in the morning hours which improves in the afternoon. It has also been observed that adhesion value before first slip occurs is less than that after first slip has occurred. This is due to the fact that the rail table gets cleaned and dried with first slip of the wheel thus increasing the adhesion on subsequent slips. In a test, coefficient of adhesion improved from 0.25 to 0.33 when rails were chemically treated with stone W 20. With sodium met silicate treatment of rail, coefficient of adhesion improved from 0.17 to 0.25. This is because colloidal contaminants responsible for the reduction of adhesion are removed by chemical treatment. Sand treatment although detrimental to rail and wheel and bringing track circuit difficulties. It is the cheapest device to bring about temporary improvement in the rail adhesion. The sand should be dry and mobile and accurately directed to the rail wheel contact area with velocity of ejection more than the velocity of cross wind. Electric sparking of rails and wheels destroys foreign bodies in their surfaces and remove the gases absorbed by the upper layers of metals. This improves the adhesion. Unevenness of rail wheel contact surfaces brings about reduction in the adhesion. This may be due to worn out rails, loose and uneven track packing's, warp in the wheel rims, difference in wheel diameters, irregular wheel tread profile, etc. Variation in the cross and longitudinal levels of track also reduce the adhesion. At points and crossing, the area of contact between rail and wheel reduces. This tends to increase the possibility of wheel slip as a result of reduction in adhesion.

6.22.1.2 Influence of Mechanical Parts on Coefficient of Adhesion

Wheel rail adhesion reduces owing to following causes:

a. When a bogie, whose axles have to remain parallel, is negotiating a curve, movement of axle is necessarily accompanied by sliding, no matter what the speed be. This sliding is acute when wheel base of axles is large and degree of curvature is more. This, therefore, results in the reduction in adhesion at curves.

b. Tractive effort produced at the rim of the wheels appears at the drawbar after passing through connections between wheel axles to bogie and then to body.

Greater the height of these links above rails, greater will be their moment. This will reduce the weight on front axles of bogie and increase the same on rear axles. This is technically called weight transfer. This results in less adhesion at front axles which may slip. Weight transfer is also caused by the turning moment exerted by the motor. For instance in the case of nose will produce downward force on the axle. This results in an unequal distribution of axle loads which results in reduced adhesion for locomotive as a whole. By employing what is technically called low traction, the effort of weight transfer can be reduced.

c. Under the effect of vertical shock, extent of surface of contact between wheel and rail depends upon the elastic reaction of suspension. Better the elastic suspension and damping arrangement in the bogie, more will be the coefficient of adhesion. Guiding of axles by means of slides is, therefore, in-advisable because this causes friction between axle boxes and their guides.

d. On the run, track irregularities can cause vertical acceleration of 0.2 to 0.4g particularly under resonant condition. Under the condition of vertical acceleration, greater the proportion of sprung weight, less will be the variation of rail pressure and, therefore, better the coefficient of adhesion. Nose suspended motors are, therefore, no good and flexibility of transmission system is desirable.

e. Greater the moment of inertia of revolving masses achieved by coupling the axles by means of coupling rod or coupling gear wheel in mono motor bogie, greater will be the adhesion and less will be the slip.

f. For unequal static loading of axles, tendency of slipping or skidding is pronounced in the case of least loaded axles.

g. Bogie pitching oscillations cause appreciable variation of axle loads. This reduces the adhesion.

h. Sudden increase of tractive effort or braking force affects adhesion in as much as it changes the weight transfer effect.

6.22.1.3 Influence of Electrical Parts

It is due to the part played by electrical parts in affecting that it has been possible to have light AC electrical locomotive for the same tractive effort produced as compared to DC electric locomotive. While DC locomotive is capable of hauling 1000 tonne trailing load on a gradient of 10 in 1000, it's AC counterpart of equal weight can haul a trailing load of 1600 tonne under same condition. This is 60% improved performance. We will now analysis and discuss these factors more closely.

a. *Shape of tractive effort-speed curve.* In case of DC locomotive, motors are started by inserting resistance in series with the armature of motors. Insertion of resistance makes the torque speed curve flatter as shown in Fig. 6.10a.

In case of AC locomotives, reduced voltage at starting is applied by step down transformer with the help of on-load tap changer and since no resistance is connected in series with the armature, speed torque curve is steeper as compared to that of DC locomotive Fig. 6.10b. Slope of the speed torque curve has an important effect on the slip and thereby on the adhesion. In case of AC locomotive, on occurrence of slight slip, torque developed reduces by larger amount because of the steepness of the curve. This reduction in torque is so much that new tractive effort is below the adhesion limit and this register grip on the track. In other words wheel will be reattached to the rail before wheel

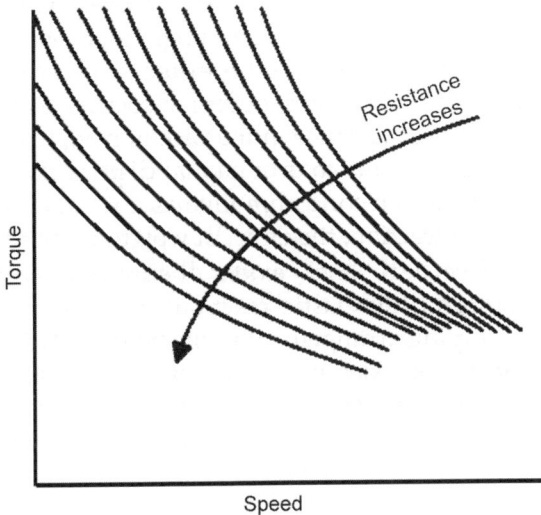

Fig. 6.10a: Torque-speed curve.

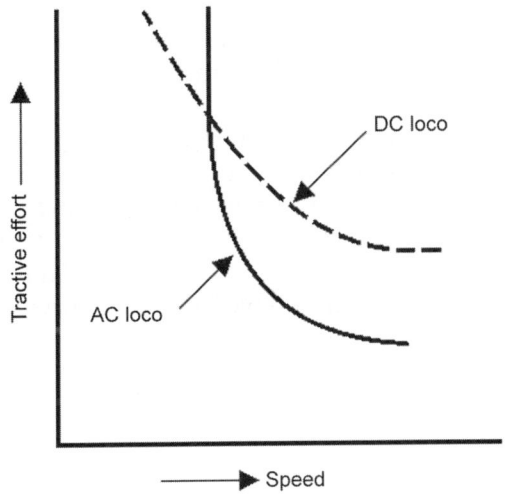

Fig. 6.10b: Tractive effort *vs* speed curve.

slip becomes perceptible. Therefore, probability of wheel slip being reabsorbed is more in case of AC locomotive. On the other hand, in case of DC locomotive, since speed torque is more flat, consequent to the occurrence of initial slip, there will not be as much reduction of tractive effort developed as in case of AC locomotive with the result that tractive effort now produced may still be above adhesion limit. As such changes of grip on the rails are low and consequently probability of wheel slip being absorbed is low.

b. **Method of control.** In case of Bo-Bo DC locomotive there are two pairs of axle driven series motors, each pair being connected permanently in series. At starting both the pairs of motors are in series and resistance is connected in series which is gradually cut out. By means of series parallel control, these two pairs of motors are then connected in parallel after passing through transition notches. We will now understand the sequence of events which takes place when slip occurs on any one axle. Before slip takes place since speed of two motors in series is same, their back emf are equal and, therefore applied voltage across each motor is same. As soon as slip takes place due to (say) imperfection of track, speed of motor driving slipping axle will rise. This will increase the back emf of this motor. As a result of this proportion of the applied voltage across slipping motor will increase and that across non-slipping motor will decrease. Direct result of this situation is that the speed of slipping motor will go on increasing and that of non-slipping motor will go on decreasing. This reduces the adhesion. In case of AC locomotive, all the motors are connected in parallel and independently fed. Slipping of any one motor has no effect on the variation of applied voltage on other motor. This therefore, does not impair the adhesion. In case of series fed motors, equalizing connection shown dotted Fig. 6.24b between two groups of series connected motors restricts the tendency of wheel racing.

Loss of adhesion is also prevented by shunting motor armature with resistance which diverts a part of armature current of slipping motor and reduces its torque developed.

c. **Effects of variation of tractive effort at notch positions.** It has been observed that greater the sudden variation of tractive effort between starting notches, less will be the coefficient of adhesion. Starting of traction motors either by rheostat method, series parallel method or by tap changing method is inferior to the method where applied voltage is smoothly and continuously changed such as by means of grid control in exaction or ignitron or by gate control in the case of silicon controlled rectifier. Resistance to Fig. 6.11 will make this amply clear where curve no. 1 represents adhesion limit, curve no. 2 gives tractive effort at start with continuously variable voltage, and curve no. 3 gives the actual tractive effort during the motor notch. Curve no. 3 is so manipulated that its

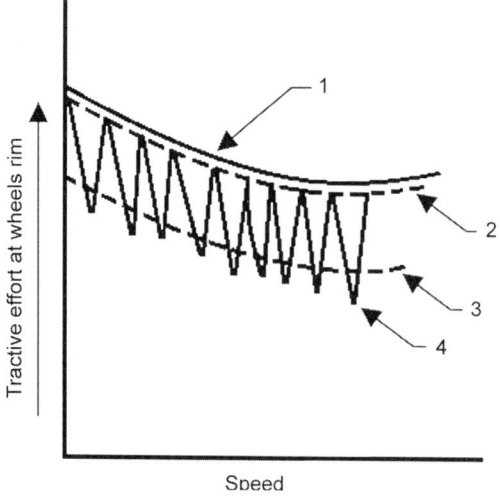

Fig. 6.11: Tractive effort speed characteristic.

crests never cross above curve no. 2. The mean tractive effort during starting by notch method is given by curve no. 4 which is very much below curve no. 2. This explains how starting gear of the motor affects the adhesion. This further explains the adverse effect which variation of supplied voltage on the adhesion of the locomotives.

6.23 ELECTRIC TRACTION SYSTEMS—POWER SUPPLY

6.23.1 Transmission Lines to Substations

The transmission lines to substations depend on the location of the source of power supply. When the power station is situated at one extreme end, it is then probably the worst location. At the receiving station, the correct voltage must be made available.

In duplicate transmission lines arrangement, two separate transmission lines are run, one on each side of the railway track. Under normal operating conditions these lines operate together but when some fault develops in one line, the power can be transmitted by the other. For a double line, the reactance voltage drop is jIX, since each line carries a current I. In the event of fault occurring in one line, the other has to carry a current equal to 2I and therefore the reactance voltage drop becomes j(2IX) and resistance drop equals 2IR (where R is the resistance of one line), then total drop becomes [2IR + j(2IX)]. If both the lines operate in parallel then the voltage drop is equal to [IR + (jIX)]. A duplicate transmission line feeding three substations is shown in Fig. 6.12.

On the occurrence of a fault the line circuit breakers open the faulty section and the supply is continued through other line. In the event of a fault occurring over a particular section, the voltage drop in the other line of this section is doubled since it has to carry double the current.

For every single-unit substation there are four line circuit breakers and one transformer circuit breaker; since the cost of high voltage equipment increases very

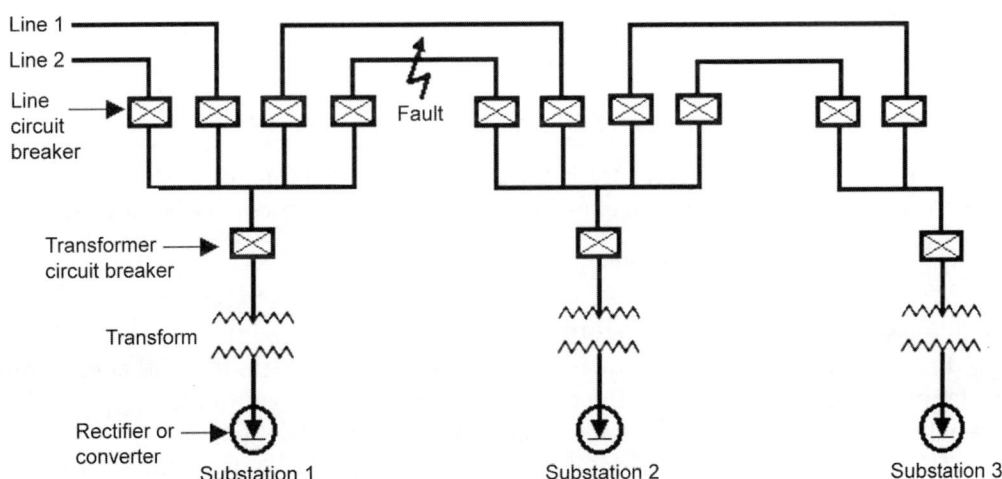

Fig. 6.12: Duplicate transmission line.

rapidly with higher voltages, it is evident that a transmission line of voltage 66 to 132 kV must be tapped as seldom as possible.

Where supply for traction work is available from national grid, the arrangement shown in Fig. 6.13 may be adopted.

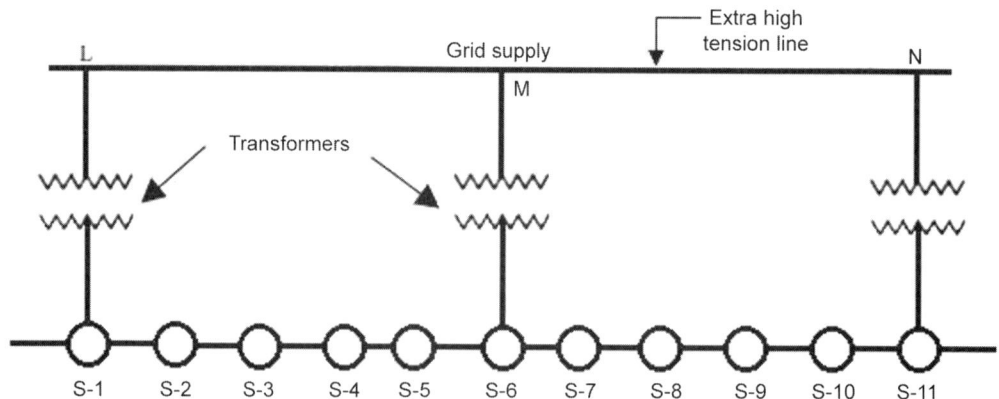

Fig. 6.13: Alternative method of power supply.

The extra high tension supply is tapped at suitable points L, M and N and the power is converted to a lesser voltage at the transformer stations L, M and N. A single high tension line may be taken to the substations from the transformer stations.

An alternative proposal will be to have no transmission line between substations S-3 and S-4 and S-8 and S-9. Duplicate transmissions lines can then be run between (S-1, S-2, S-3), (S-4, S-5, S-6) and (S-7, S-8) and (S-9, S-10, S-11). The transformer station at L will feed substations: S-1, S-2, S-3; that at M: S-4, S-5, S-6, S-7, S-8 and that at N: S-9, S-10, S-11.

6.23.2 Substations

The type of substations depends upon:
    i. The nature of the primary supply
   ii. The system of electrification.

The substations for single-phase system and DC substations are discussed below:

☞ *Substations for Single-phase System*

When the power supply is only for the purpose of feeding a traction system on single-phase as is the case will most European railways, the generators may be single-phase, $16\frac{2}{3}$ Hz machines. The substations may be supplied at 66–132 kV.

When thc supply is available from a national grid, it has to be converted to suit traction requirements, using transformers and frequency changers.

In case of AC traction system at standard 50 Hz frequency, the grid substations are tapped from the 66–132 kV busbars to obtain 25 kV single-phase supply either through Scott-connected transformer or straight transformers.

The spacing of the substations is based on the permissible voltage drop expressed as percentage of the normal voltage. It should not exceed 20% and 40% under normal and abnormal conditions respectively.

The substation spacing on 25 kV is about 1.5 times the spacing on 3000 V DC system.

☞ DC Substations

The DC substations equipped with mercury arc rectifiers, as compared to rotary converter, have much higher all-day efficiency, reliable operation acceptability to remote control. Since the rectifier is non-reversible, suitable arrangements need to be made to receive the energy return during regenerative braking. The rectifier working as inverter feeds back power into the AC system; however, where such an arrangement is not employed, the energy may be dissipated in a loading resistor connected to the DC busbars.

6.23.3 Feeding and Distribution System on AC Traction

The calculations of voltage drop in a distributor for single-phase AC system, are made as follows Refer to Fig. 6.14 which shows the phasor diagram for current and voltage.

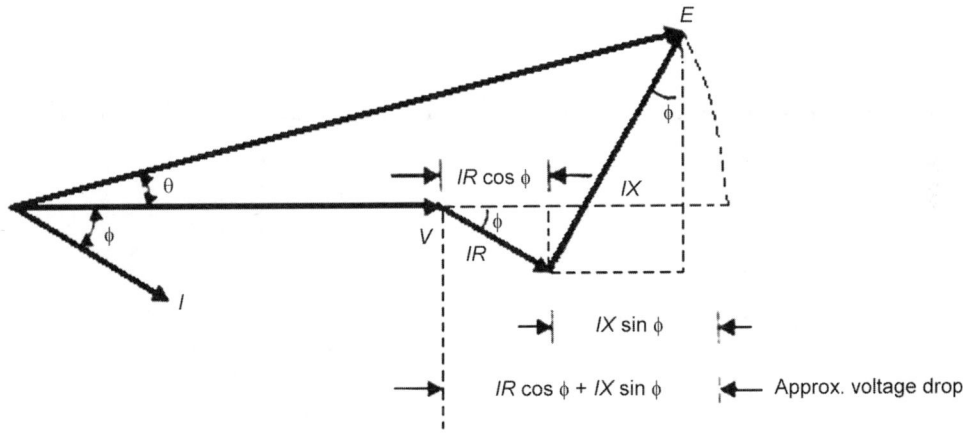

Fig. 6.14: Voltage drop in a distributor.

Let, R = Resistance in ohms per km of trolley wire and rail return,
         X = The reactance in ohms for the same,
   $\cos \phi$ = The power factor of the system,
         V = Voltage at the receiving end,
         E = Voltage at the sending end, and
         I = The current flowing in the feeder/distributor.

Then, since θ is small, the voltage drop in the feeder/distributor is approximately given as:

$$IR \cos \phi + IX \sin \phi = : I(R \cos \phi + X \sin \phi)$$

For conductor rails the value of resistance at $16\frac{2}{3}$ Hz is 3.3 times the actual DC value on account of "skin effect" and 7.8 times the DC value at 50 Hz. The inductance is about 1.55 mH per km of the route.

6.23.4 Feeding and Distribution System for DC Tramways

The tramways need to conform the following Indian regulations:
1. The voltage at the trolley is not to exceed 550 V and at the sending-end stations 650 V.
2. The trolley wire shall be divided into sections not exceeding 1.61 km in length.
3. The potential difference between any two points of the rail return shall not exceed 7 volts.

6.23.5 Electrolysis by Currents through Earth

When the track is used as return conductor, the currents flow through the rail as well as through the earth, as shown in Fig. 6.15. As these currents spread out into the earth, they follow low-resistance paths provided by gas mains, water pipes, cable sheaths, etc. these currents enter such conductors and leave them causing corrosion due to electrolytic action. Such effects can be minimised by using the following methods (i) By providing a return path of very low resistance (by providing good bonding and using insulated negative feeder); (ii) By discouraging the entry of currents into the pipes by incorporating insulating joints.

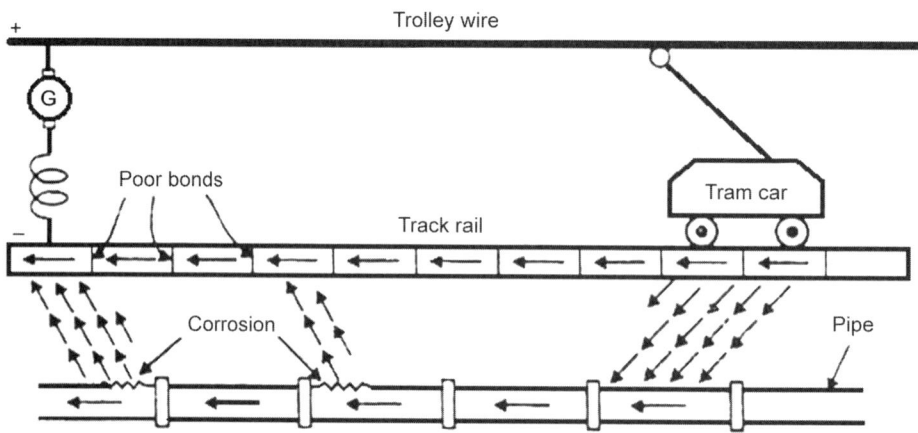

Fig. 6.15: Electrolysis by currents through earth.

6.23.6 Negative Boosters

Negative Boosters are employed to conform to the regulation that the potential difference between any two points of the rail return shall not exceed 7 V.

Figure 6.16 shows the negative boost control.

Two boosters, positive and negative, are used which are mechanically coupled together driven by a DC motor. The positive booster is connected to the trolley wire (near the generating station) and negative booster (separately excited) is connected to the track rail. The 'positive booster' adds voltage to the line while the 'negative booster' lowers the potential of the point it is connected to.

As we go along trolley wire away from the generating station/substation, the potential drop increases and the voltage of the trolley wire falls, since the current returns via the track rail, points away from the generating station acquire high potentials.

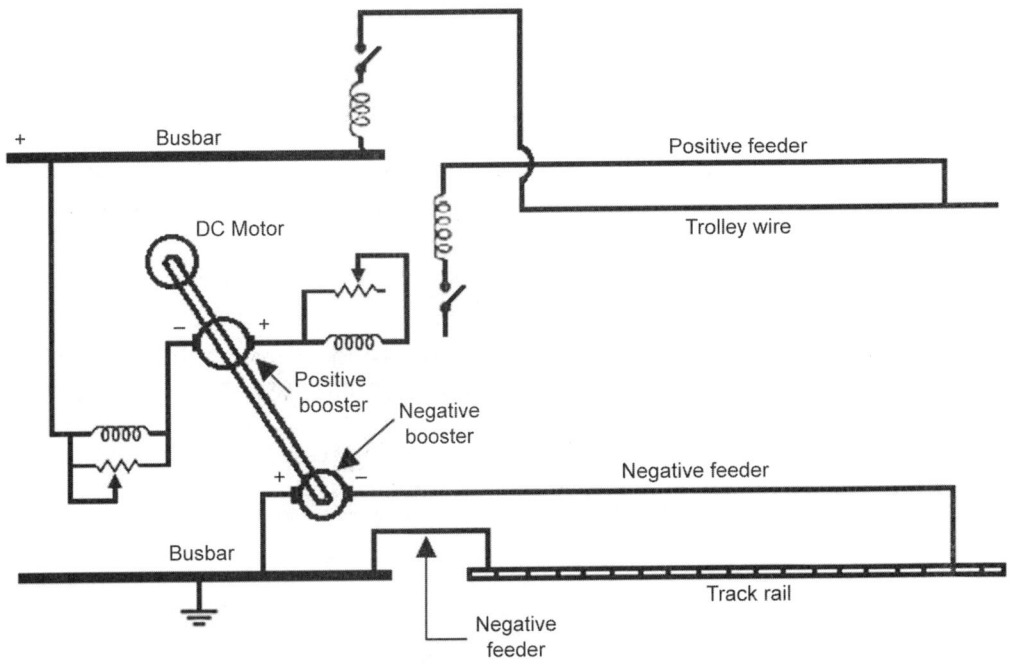

Fig. 6.16: Negetive booster control.

This potential is brought down by the negative boost provided by the negative booster.

◈ When the load is sufficiently far away from the generating station, the trolley wire is fed by the positive booster. The current in the positive booster provides the excitation for the negative booster. The feeder current as it flows through the booster maintains the voltage of the trolley wire within limits. The potential at the corresponding point on the track is reduced by the negative booster since its voltage is regulated by the same feeder current. Thus, the track is maintained nearly at earth potential.

6.24 BLOCK DIAGRAM OF AC ELECTRIC LOCOMOTIVE

The various components of an AC locomotive running on single-phase 25 kV, 50 Hz AC supply are numbered in Fig. 6.17.

As seen, power at 25 kV is taken via a pantograph from the overhead contact wire and fed to the step-down transformer in the locomotive. The low AC voltage so obtained is converted into pulsating DC voltage by means of the rectifier. The pulsations in the DC voltage are then removed by the smoothing choke before it is fed to DC series traction motors which are mounted between the wheels.

The function of circuit breakers is to immediately disconnect the locomotive from the overhead supply in case of any fault in its electrical system. The on-load tap-changer is used to change the voltage across the motors and hence regulate their speed.

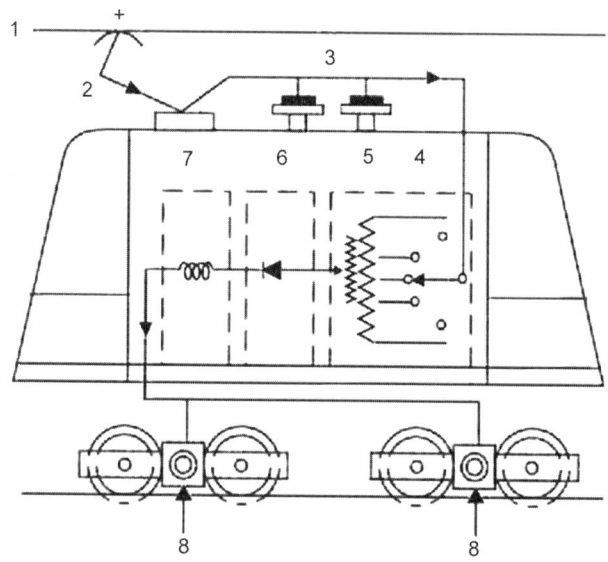

Fig. 6.17: Block diagram of an AC locomotive.

1. OH contact wire
2. pantograph
3. circuit breakers
4. on-load tap-changers

5. transformer
6. rectifier
7. smoothing choke
8. DC traction motors.

6.25 OVERHEAD EQUIPMENT (OHE)

The first and foremost function in electric traction is to keep the traction unit fed with the energy that it needs. For this purpose, broadly speaking, there are two systems of current collection, namely from third rail and from overhead wire. Current collection from overhead wire is far superior to that from the third rail. This is because both theoretically and experimentally current collection is more difficult from a rigid body than from an elastic one. Further the insulation of the third rail at high voltages used on single phase AC traction would also be impracticable proposition and endanger the safety of the personnel. Hence power collection from an overhead wire by means of pantograph is exclusively adopted in the present practice of high speed surface traction.

Simplest type of overhead equipment consists of a single contact wire supported either by bracket or by overhead span. This maximum distance between the two consecutive supports with this system is restricted to 30 m. Because there is appreciable sag of wire between supports, this limits the speed up to maximum of 30 kmph. The single contact wire system therefore, finds application for tram cars or in complicated yards and terminal stations where simplicity of layout is desired and speeds are low.

Since speeds employed in railway traction are more than those of tramway, we cannot employ same magnitude of sag. Pantograph makes sliding contact with trolley or contact wire. As the speed of the train increase, pantograph may tend to leave the contact of trolley wire due to inertia action which may result in acting thereby spoiling the pantograph. It, therefore, becomes essential that trolley wire should be maintained

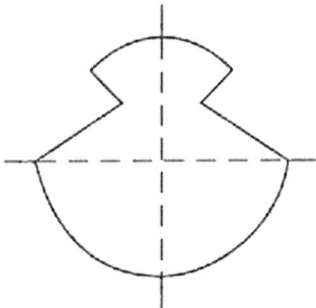

Fig. 6.18: Trolley-wire section.

level without abrupt changes in the height between the supporting structures. If a single wire system is to be suitable for pantograph collection and no excessive tension in the wire is to be produced, span has to be limited to a very small value. To avoid such short spans, we have to use single catenary system which consists of one catenary or messenger wire of steel or stranded cadmium copper with high sag and the other contact or trolley wire supported from messenger wire by means of droppers clipped to both the messenger and contact wires. The droppers made of solid copper of usual cross section of 5 sq. mm. are spaced closely along the contact wire at 9 m distance from one another, the lengths are so adjusted that although there is a sag in the messenger wire, contact wire itself is practically level. By means of insulated tension members attached to the supporting structure known as "pull offs" position of contact wire over the track is maintained. These "pull offs" only affect the horizontal position but do not affect the vertical height of the contact wire. To enable the attachment of pull offs and droppers to the contact wire without interfering with the running of pantograph along it, two grooves are cut on the opposite sides of contact wire as shown in Fig. 6.18.

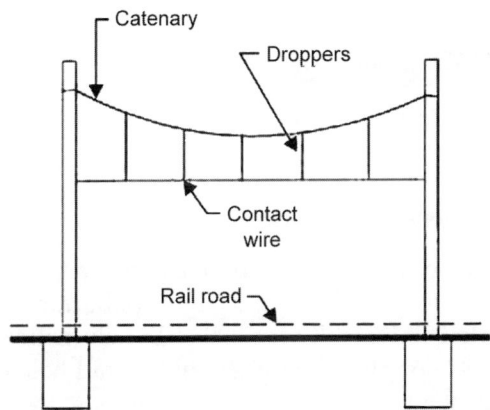

Fig. 6.19: Typical single catenary system.

Typical single catenary system as employed on 25 kV AC traction is shown in Fig. 6.19. OHE is supported by the bracket assemble of swiveling type attached suitably to the traction masts. The bracket assembly is provided with insulators near the support to avoid fumes and smoke ejected from steam locomotives setting on these insulators.

The bracket assembly is made-up of mast fittings, solid core insulators, telescopic stay arm, and high tensile bracket tubes, register arm, steady arm, register arm dropper and associated fittings. Bracket assembly Fig. 6.20 permits adjustment of equipment to cater for slewing or raising of track during maintenance. This type of construction, employing bracket assembly, offers mechanical and electrical independence to each track which is helpful in maintenance. Any irregularity or damage or maladjustment of the OHE of one track does not impair or in anyway affect the performance of the other. Besides this, it is easy for erection. Concrete foundation, which is cast already, is having central core hole in which mast is subsequently erected and grouted. This helps in achieving high speed of erection of OHE.

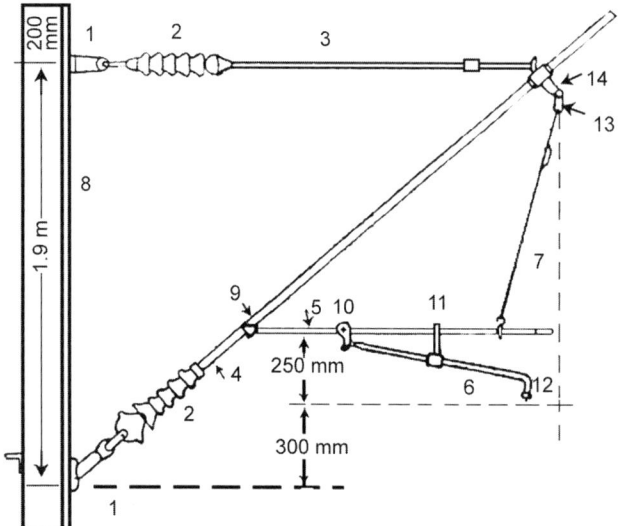

1. Mast fitting
2. Solid core insulator
3. Telescopic stay arm
4. Bracket tube
5. Register arm
6. Steady arm
7. Register arm dropper
8. Mast
9. Register arm hook
10. Steady arm clamp
11. Anti wind clamp
12. Contact wire swivel clip
13. Suspension clamp
14. Catenary supension bracket

Fig. 6.20: Bracket assembly.

6.26 CURRENT COLLECTION SYSTEM

There are mainly two current collection systems namely current collection from conductor rail and current collection from OHE.

6.26.1 Conductor Rail System

The conductor rails may be divided into three classes depending on the position of the contact surface which may be located at the top, bottom or side of the rail. The top contact rail is adopted universally for 600 V DC electrification. The side contact rail is used for 1200 V DC supply. The under contact rail has the advantage of being protected from snow, sleet and ice.

Fig.6.21a shows the case when electric supply is collected from the top of an insulated conductor rail C (of special high-conductivity steel) running parallel to the track at a distance of 0.3 to 0.4 m from the running rail (R) which forms the return path. L is the insulator and W is the wooden protection used at stations and crossings.

The current is collected from top surface of the rail by flat steel shoes (200 mm × 75 mm), the necessary contact pressure being obtained by gravity. Since it is not always

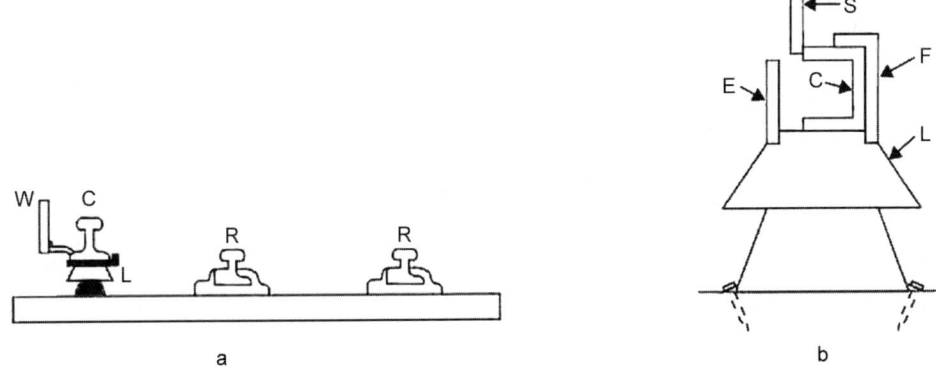

Figs 6.21a and b: (a) Conductor rail system, (b) Conductor rail with side running contact.

possible to provide conductor rail on the same side of the track, shoes are provided on both sides of the locomotive or train. Moreover two shoes are provided on each side in order to avoid current interruption at points and crossings where there are gaps in the running rail.

Fig. 6.21b shows the side contact rail and the method of the mounting. The conductor rail (C) rests upon a wooden block recessed into the top of the por-celain insulator L. Current is collected by steel shoes (S) which are kept pressed on the contact rail by springs. E and F are the guards which rest upon ledges on the insulator.

6.26.2 Current Collection Gear for OHE

The most essential requirement of a collector is that it should keep continuous contact with trolley wire at all speeds. Three types of gear are in common use:

1. trolley collector
2. bow collector
3. pantograph collector.

To ensure even pressure on OHE, the gear equipment must, be flexible in order to follow variations in the sag of the contact wire. Also, reasonable precautions must be taken to prevent the collector from leaving the overhead wire at points and crossings.

6.26.2.1 Trolley Collector

This collector is employed on tramways and trolley buses and is mounted on the roof of the vehicle. Contact with the OH wire is made by means of either a grooved wheel or a sliding shoe carried at the end of a light trolley pole attached to the top of the vehicle and held in contact with OH wire by means of a spring. The pole is hinged to a swivelling base so that it may be reversed for reverse running thereby making it unnecessary for the trolley wire to be accurately maintained above the centre of the track. Trolley collectors always operate in the trailing position. The trolley collector is suitable for low speeds up to 32 km/h beyond which there is a risk of its jumping off the OH contact wire particularly at points and crossing.

6.26.2.2 Bow Collector

It can be used for higher speeds. As shown in Fig. 6.22, it consists of two roof-mounted trolley poles at the ends of which is placed a light metal strip (or bow) about one metre long for current collection. The collection strip is purposely made of soft material (copper, aluminium or carbon) in order that most of the wear may occur on it rather than on the trolley wire. The bow collector also operates in the trailing position. Hence, it requires provision of either duplicate bows or an arrangement for reversing the bow for running in the reverse direction. Bow collector is not suitable for railway work where speeds up to 120 km/h and currents up to 3000 A are encountered. It is so because the inertia of the bow collector is too high to ensure satisfactory current collection.

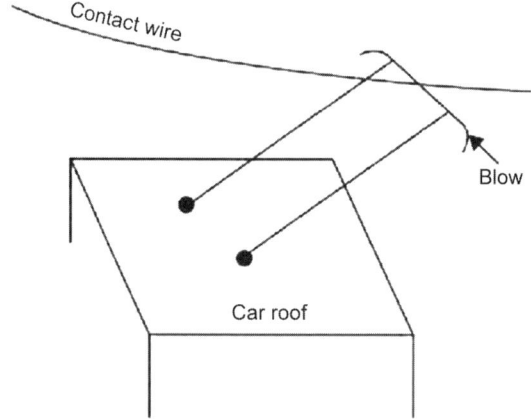

Fig. 6.22: The blow collector.

6.26.2.3 Pantograph Collector

Its function is to maintain link between overhead contact wire and power circuit of the electric locomotive at different speeds under all wind conditions and stiffness of OHE. It means that positive pressure has to be maintained at all times to avoid loss of contact and sparking but the pressure must be as low as possible in order to minimize wear of OH contact wire.

A 'diamond' type single-pan pantograph is shown in Fig. 6.23. It consists of a pentagonal framework of high-tensile alloy-steel tubing. The contact portion consists of a pressed steel pan fitted with renewable copper wearing strips which are forced against the OH contact wire by the upward action of pantograph springs. The pantograph can be raised or lowered from cabin by air cylinders.

6.27 TRACTION MOTOR CONNECTIONS

In traction work, number of motors employed is always more than one. Traction motors can be grouped in anyone of the following ways—series, series parallel and Parallel as shown In Figs 6.24a to c respectively. Main advantages of series and series parallel grouping are the economy which can be achieved in the control gear and also

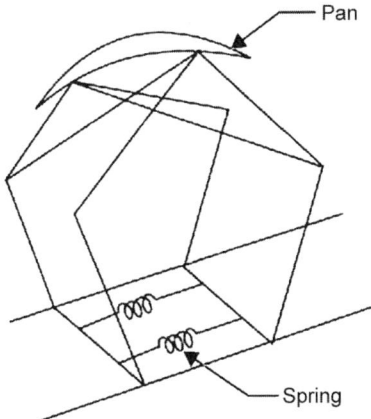

Fig. 6.23: Pantograph collector.

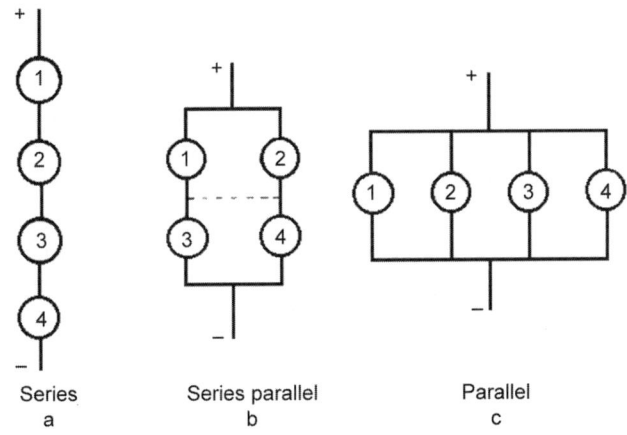

Figs 6.24a to c: Traction motor connection.

keeping the transformer secondary current to a lower value. However, these grouping of motors give rise to more tendency for wheel slip. In series parallel grouping it is desirable to have equaliser connection shown dotted. In parallel grouping there is less tendency of wheel slip and as such locomotives and motor coaches, where high tractive effort is to be produced, invariably employ this type of motor grouping.

6.28 SMOOTHING REACTORS

The DC output of a rectifier especially from single phase supply is pulsating and contains ripples. The magnitude of the ripples which is the alternating component at 100 cycles has to be limited to 20 to 30% from 40 to 50%. These AC harmonics, besides producing heating in the magnetic circuit of the motor, produce commutation troubles. Smoothing reactor is introduced in between output terminals of the rectifier and DC motors and its main job is to smoothen out the ripples in the DC output of the rectifiers. It is made of rectangular copper conductor wound edge wise over laminated core. Spacers are put in between turns to help cooling. Reactor is of blower air cooled type mounted usually on the under frame.

6.29 DESIRABLE CHARACTERISTICS OF TRACTION MOTORS

Traction motors are expected to have both electrical and mechanical characteristics, as far as possible, to suit the stringent service conditions of traction work which are discussed as follows:

1. **Suitable speed torque characteristic.** Article 6.18 clearly shows that tractive effort required is maximum during starting period t_1, Fig. 6.6b. During constant speed running period t_2, even though train is moving at high speed, its tractive effort requirement is not high as it only requires to overcome track resistance and gravity component. Since tractive effort at the tread of the driving wheel is always proportional to the torque of the armature shaft of motor, the dynamics of train requires motors to develop very high starting torque which should fall off at high speeds.

2. **Parallel running.** In traction work usually more than one motor is required. It is, therefore, mandatory that these motors should be operated in parallel for which they must have suitable speed-torque and current-torque characteristics. For small difference in the rotational speeds of various motors wide differences in the torques developed and currents drawn by various motors are not desirable, i.e. speed-torque and speed-current curves should not be flat.

3. **Voltage fluctuations.** In traction work, due to heavy current inrush at starting, voltage fluctuation of supply line is a normal feature. Traction motors should be capable of withstanding voltage fluctuations without any effect on their performance.

4. **Temporary interruption of supply.** Temporary interruption in the supply to the motors occurs when crossovers and section insulators are crossed with controller on. Motor should be in a position to withstand temporary interruptions in the supply without very much undue inrush of current.

5. **Overload capacity.** Motors are subjected to very arduous duty. Large current rush may produce large armature reaction magnetic flux along the brush axis. It may be noted that voltage between adjacent commutator segments is dependent upon the flux distribution in the air gap. Under the conditions of weak field, un-neutralized armature reaction and slow growth of commutating field flux, high voltage between adjacent commutator segments may be developed near brush axis which may lead to the formation of arc between brush and the segment. This arc may not only be maintained but extended from segment to segment by the high voltage existing between segments as the commutator rotates. Such conditions are technically called flash over which have to be avoided at all costs. Motor selected should be capable of taking heavy loads without flash over.

6. **Self relieving property.** In case motor has speed torque characteristic such that torque is inversely proportional to speed, their product will always be constant. Therefore, HP which is proportional to the product of torque and speed will also remain constant. This gives motor a self-protective property against excessive overloading.

7. **Amenability to speed control.** Motor should be amenable to simple speed control methods.

8. **Amenability to electric braking.** Motor should be amenable to easy and simple methods of rheostat and or regenerative braking.

9. **Limitation of weight and size.** In case of all transport vehicles, every kg. weight requires expenditure of energy. Therefore, high power/weight ratio is aimed at in the design of traction motor. With a view to increase the pay load capacity of the vehicle every care is exercised to reduce the weight of the vehicle. This in achieved by employing high speed motors, upper speed limit being fixed by the excessive centrifugal stresses. For given HP of a motor, its physical size depends upon the type of insulation employed. Traction motors are now wound with class H insulation as against class B insulation used earlier.

10. **Robustness.** Motors are subjected to service vibrations, as such they should be robust enough to withstand such service conditions. Motors should further be provided with mechanical protection so as to prevent ingress of dirt, water, mud, etc. into it.

11. **High efficiency.** Motor should have high mechanical and electrical efficiency.

6.30 TRACTION MOTORS

None of the motors can have all the desirable operating characteristics as described above. Some of the motors which find application in traction are:

    a. DC series motors supplied with straight DC or rectified DC.

    b. AC series motors single phase.

    c. Repulsion motors.

    d. Three phase induction motors.

    e. Linear induction motors.

6.30.1 Suitability of Series Motor for Traction Duty

In this article we will analyse the suitability of series motor for traction purpose in view of the requirements laid down in § 6.29.

1. DC series motor develops high torque at low speeds and low torque at high speeds. This is exactly the requirement of traction units. High starting torque developed is needed to accelerate the vehicle.

2. Speed torque and speed current characteristics of series motor are steep. Therefore, for a given difference in rotational speed of various motors running in parallel caused by unequal wear of different driving wheels, torque developed and current taken by each motor will not differ very much.

3. As a rule, for motors where field flux varies as the armature current, as in the case of series motor, torque corresponding to given current will be unaffected by the variations in the line voltage. Voltage variations will, however, affect the speed and efficiency of the motor.

4. Series speed torque curve leads to stability in control. This is because if driving torque happens to be more than resisting torque, speed of the motor will increase. Increase in speed will lead to falling off in the torque so that stable situation is reached very rapidly.

5. Commutating properties of series motor are also very good as increase in the armature current as a result of heavy load torque results in the decrease of armature speed. This reduces the magnitude of emf induced in the coils undergoing commutation which helps in achieving sparkles current collection.

6. In case of series motors, up to the point of magnetic saturation, torque developed by motor varies as the square or armature current. Therefore, for a given percentage increase in the torque to be developed, the drain on electric current in case of series motor will be comparatively low. Thus series motor of a given rating is capable of withstanding heavier load torque.

7. HP $\propto T \times N$ But in series motor

$\therefore$
$$HP \propto T \frac{V}{I_a} \text{ but } T \propto I_a^2$$

$\therefore$
$$HP \propto \sqrt{T}.V \tag{6.39}$$

Equation 6.39 indicates that series motor to some extent has self-relieving property. For a given increase in torque there is no increase in HP to the corresponding degree.

8. Speed of series motor can be controlled by various methods as discussed in Chapter 2.

9. Series motor is basically not suitable for adopting it for regenerative braking as it suffers from electrical instability unless certain measures are taken.

10. Since the time constant of series motor field is low, field flux will die in short time with the result that counter emf of motor ceases with temporary interruption of supply. Initial rush of current on temporary interruption of supply in the case of series motor is, therefore, more.

Due to low weight and high starting torque developed, DC series motors are capable of producing high rates of acceleration. As such DC series motors are better suited for urban and suburban services.

In 1500 V DC system, whilst motors are insulated for 1500 V, they may operate at 1500 V or at 50 V by connecting two motors permanently in series. In 3000 V deck system motors are insulated for 3000 V but operated at 1500 V by connecting two in series.

Drawbacks of series motors arise first from commutator which restricts speed, current and voltage. There is the risk of flash over and the brush gear requires considerable maintenance. Moreover because of the windings on the rotor there are limitations to peripheral speed arising from centrifugal force even if restrictions due to commutator could be overcome.

6.30.2 Series Motor Using Undulating DC

When supply to motor is given after rectifying AC it draws current which is undulating. This undulating DC can be split up into a straight deck producing useful torque and an alternating component of non sinusoidal frequency 2f. This alternating component creates certain number of phenomenon which have an unfavourable effect on the proper working of the motor such as commutation, heating etc. The effect of the alternating component gets reduced in so far as the value of alternating current component is low in comparison with the value of DC component.

The method by which only the magnitude of alternating component is reduced and not the DC component is to increase the inductance of the circuit. Since the inductance of motor is comparatively weak, it become necessary to add fairly large smoothing choke Fig. 6.25 in the supply circuit. The size of the choke should be such as to prevent any appreciable saturation throughout the working range of the motor.

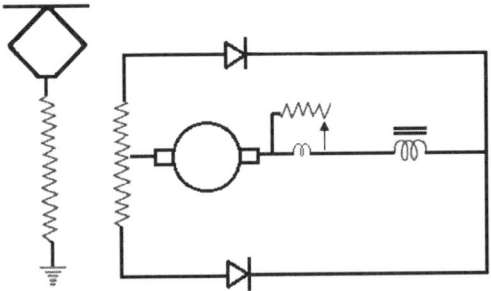

Fig. 6.25: Use of large smoothing choke.

Alternating component produces following effects on the performance of the motor.

1. Lot of eddy current loss and hysteresis loss takes place in the magnetic circuit.
2. Alternating component produces transformer emf in the coils undergoing commutation. This, therefore, gives rise to bad sparking.

Both of the above ill effects can therefore, be warded off by fitting resistance of about ten times the field resistance across the field Fig. 6.25. Because main field has high resistance, the current through this inductance consists mainly of harmonics. This resistance is sometimes called ripple divert resistance. Provision of this ripple divert resistance across field helps to a great extent reduce the loss of rating of motor. Another precaution taken by many manufacturers is to employ laminated magnetic circuit throughout as with AC motors.

Since current, flux and shaft torque are of undulating nature, there exists tendency of starting vibrations. Special provision has, therefore, to be made for damping the vibrations of collecting brushes.

Another precaution taken against the effects of undulating flux is the insulation of the motor bearings to prevent circulating eddy currents flowing through them. This reduces the chances bearing from being spoiled as a result of corrosion due to circulating currents. Further large air gap has to provide in the path of stray flux capable of inducing emf in the ball race of the bearings.

6.30.3 Suitability of Shunt Motor for Traction Duty

Shunt motor is not considered suitable for traction purpose. Various points for and against are given below:

1. Its speed torque characteristic does not match with the requirements of traction. It is a constant speed motor.
2. Speed torque and speed current characteristics motor arc flat. As such they are not suitable for parallel running.
3. Since field flux is affected by the applied voltage torque, voltage fluctuations of line have considerable effect on the magnitude or voltage developed.
4. With the increase of load, speed decreases only very little Motor, therefore, besides getting overloaded develops commutation troubles.

5. Since field flux in case of shunt motor remains constant, torque developed varies with armature current. Therefore, for a given increase in torque developed, current drawn by shunt motor will be more as compared to that by series motor.

6. Since speed of shunt motor is constant, HP of shunt motor is directly proportional to the torque developed. Therefore, for a given increase in torque developed, there will be corresponding increase in HP unlike in series motors.

7. Since time constant of shunt field is high, temporary interruption in supply will not make all the field to die out. Therefore, there will be some back emf in the armature on the resumption of supply. As such in-rush of current will not be much.

6.30.4 Single-phase Series Motors

These motors have various characteristics just similar to those of DC series motor. Maximum voltage for these motors in limited to 400 volts. For given HP these motors have 1.5 to 2 times size than corresponding DC motors. Moreover starting torque of AC series motors is lower than that of DC motors due to poor PF at starting. AC single phase series motors, therefore, are not suitable to impart high acceleration. As such they are not used for suburban services. However, single phase motors are extensively used for main line services. These motors have better performance on reduced frequency supply say 16.66 or 25 CPS. However, in case of industrial frequency motors, it becomes necessary to lower the flux density employed. This makes speed torque characteristic steeper and as a consequence of this, motor needs not only fewer voltage steps required for starting but the division of load between different motors is less affected by we and tear of wheel tyres.

6.30.5 Three-phase Induction Motor

For traction purposes, robustness, trouble free operation, less maintenance, high voltage operation consequently needing reduced amount of current and automatic regeneration are the main advantages of 3 phase induction motor. Flat speed torque characteristic, constant speed operation and complicated overhead feeding system are few of main disadvantages of three phase induction motor. Three phase induction motors for traction purpose as such have now become obsolete.

6.30.6 Linear Motor

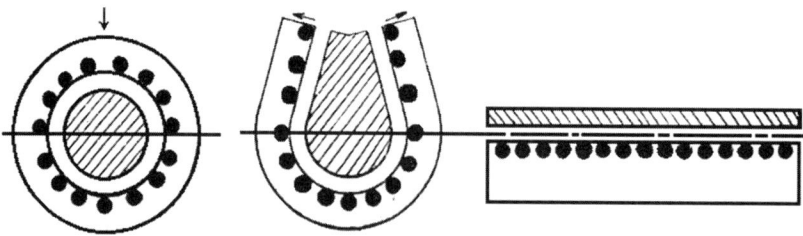

Fig. 6.26: Deriving linear motor from the conventional induction motor.

Linear motor has same principle of operation as rotary induction motor. Whenever there occurs relative movement between the field and the short circuited conductors, currents are induced in them which give rise to electromagnetic forces. Under the influence of these forces, according to lenz's law, the conductors try to more in such a way as to eliminate the currents. When the movement of the field is rotary about an axis, the movement of conductors is also rotary and this is a conventional rotary induction motor. On the other hand when movement of the field is rectilinear, motion of conductors will also be linear and this is the linear motor.

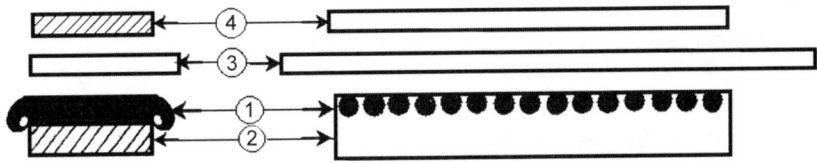

Fig. 6.27: Linear motor construction.

Constructionally linear motor can be derived from the conventional induction motor by cutting the latter on plane through its axis and then opening it out at right angle to that plane progressively as shown in Fig. 6.26. The main drawback of this construction was that there was magnetic pull between the rotor and stator. In conventional induction motor, induced rotor currents flow only through the outer skin of the rotor. If we consider rotor as made-up of a cup of conducting material having fixed magnetic core and now if this motor is cut and opened out, we would get linear motor construction as shown in Fig. 6.27.

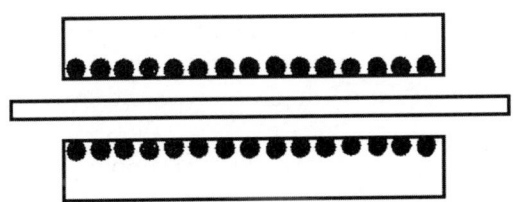

Fig. 6.28: Use of second field winding in linear motor.

Now if yoke and primary magnetic circuit including field windings are fixed to the moving vehicle, we will be avoiding magnetic pull between the moving parts and the fixed parts. The magnetic pull to be overcome is now acting between two members which are mounted on vehicle. The yoke will further be helping in completing the flux path. Since the field winding is provided on one side of the reaction rail, it would require fixing over long length of the vehicle. This will bring problem of long air gap to negotiate a track of given curvature. Therefore, further improvement was brought by replacing the yoke with second field winding as shown in Fig. 6.28. The two field windings carry currents in such a way that opposite poles are produced on them face to face. The main advantage of this construction is that current loading can be double or in other words stator losses may be reduced to one fourth in relation to those of equivalent conventional rotary machine. Moreover due to shorter field lengths, less air gap will be required.

As shown in Fig. 6.29b polyphase currents fed to the primary produce travelling magnetic field whose speed of travel is $V_p = 2PF$ moving backwards. As a result of the interaction of this field with the short circuit currents in the reaction rail Fig. 6.29d,

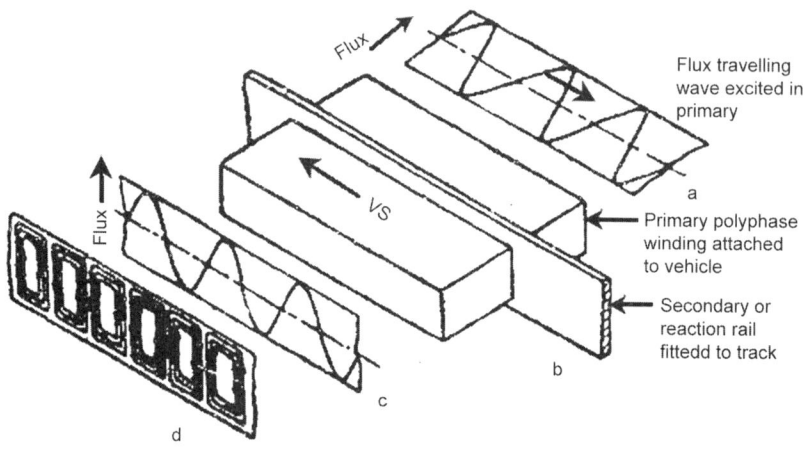

Figs 6.29a to d: Production of travelling magnetic field in polyphase current fed linear induction motor.

force is produced which moves primary winding in forward direction. This reduces the relative speed of travel of the flux with respect to reaction rail. If the speed of vehicle is equal to that of magnetic field; latter would be stationary when viewed from reaction rail. This is corresponding to the synchronous speed of induction motor. If the vehicle is propelled faster than this speed, the direction of force would be reversed and a form of regenerative braking based on the principle of induction generation will come into being. Tractate effort will be the function of slip, i.e. $V_p - V_s$. Control of tractive effort can be effected by so adjusting the frequency of supply given to the motor that a small amount of positive slip is available during motoring and small amount of negative slip during coasting.

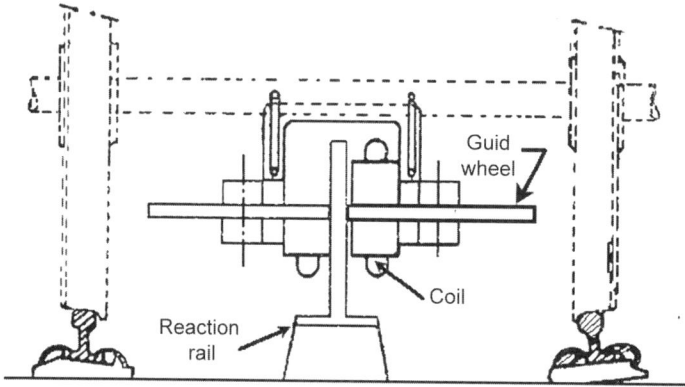

Fig. 6.30: Preferred arrangement.

The arrangement which is most often preferred is the one in which rotor is fixed along the centre line of the track and is embraced by two stator windings Fig. 6.30. The stator is attached to the vehicle so as to have as small air gap as possible. For this purpose stator is guided by means of guide wheels running on the reaction rail. It, therefore, becomes necessary to have some flexible connection between the stator and vehicle. It is desirable that the reaction plate should be of some nonmagnetic metal such as aluminium so as to avoid transverse forces and in particular the risk of adhesion between the plate and the stator. If that happens the latter would act as a magnetic brake shoe.

Tractive effort control of linear motor can be carried out by varying both frequency and voltage simultaneously so that induction density remains constant. This is the only possible most efficient method for use with motors of high power rating especially if they have to operate for long periods at variable speed.

☞ Advantages of Linear Motor

We will summarize the advantages of linear motor as:
1. Simplicity of machine as compared to the rotating motors.
2. Low initial cost.
3. Low maintenance cost in view of absence of rotating parts.
4. No limitations of tractive effort due to adhesion between the wheel and the rail.
5. Overheating of the rotor being eliminated since motor moves continuously over cool rotor plate leaving behind heated rotor portion.
6. No limitation of maximum speed due to centrifugal forces.
7. Tractive effort is independent of speed. In rotary motor tractive effort depends upon the coefficient of adhesion which is very low at high speeds.
8. We can dispense with current collection gear if we fix stator on the track and rotor plate is carried on the vehicle.
9. Power to weight ratio is better than for rotary motor. This is because rotor is not carried on the traction unit. This is also partly due to the use of double field system and partly due to the rotor being continually replaced with fresh cool metal so that very high rotor: losses can be dissipated without increasing the size of the machine.

☞ Disadvantages of Linear Motor

1. Capital cost of reaction rail fixed along the centre line of the track is very high.
2. Adequate clearances at points and crossings may be difficult to maintain due to axle movements on curves, etc.
3. Large air gap and nonmagnetic reaction rail require more magnetizing current and, therefore, power factor and efficiency are low.
4. Provision of three phase collector system along the track is not only costly but it also creates complications.
5. Due to end effect and edge effect, utilization of the motor is inferior.
6. In case of linear motor employing single field system and ferromagnetic reaction rail, lot of attractive force will be exerted between the stator and rotor.

☞ Scope of Application of Linear Motor

It can be used on trolley cars for the internal transport in workshops. It can also be used as booster accelerator for moving heavy trains from rest or up the inclines or on curves or as a propulsion unit in marshalling yards in place of shunting locomotives. For instance in sections where there are heavy gradients for short stretches it may be economical to have a linear motor whose stator is fixed only at up gradient spots. This arrangement would be cheaper than using diesel or steam locomotive as booster.

Linear motor has superiority over conventional motor for speeds over 200 mph. Linear induction motor forms excellent source of motive force suspended trains where conventional motor which depends for torque conversion to linear tractive force upon the adhesive weight on driving wheels, fails.

6.31 TRACTION MOTOR CONTROL

Control of traction motor envisages following operation. To supply without taking excessive:

1. Connecting the motor current at starting.
2. Providing smooth acceleration without sudden avoid damage to couplings.
3. Adjusting the speed according to the type of service and route conditions.

We will now take up various control methods for motors supplied from straight DC supply rectified AC supply and pure AC supply.

6.31.1 Control of DC Traction Motors

The starting current of motor is limited to its normal rated current by starter during starting.

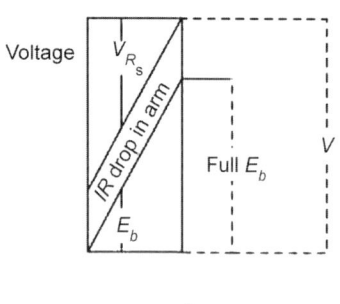
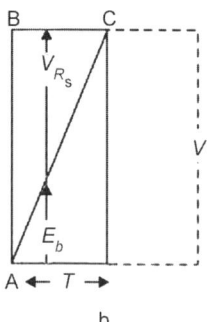

Figs 6.31a and b: (a) Voltage during starting of DC motor, (b) Voltage during starting neglecting IR drop.

At the instant of switching on the motor, back emf $E_b = 0$

$V = IR +$ Voltage drop across R_S $\hspace{4cm}$ (6.40)

At any other instant during starting

$$V = IR + \text{Voltage drop across } R_S + E_b$$

At the end of accelerating period, when total R_S is cut-off

$$V = IR + E_b$$

If T is the time in sec. for starting and neglecting IR drop, total energy supplied =

VIT watt-sec. From Fig. 6.31 Energy wasted in R_S = Area of triangle ABC $\times$ I = $\dfrac{1}{2}$ =.

TVI watt-sec. = $\dfrac{1}{2}$. VIT watt sec. But total energy supplied = VIT watt sec.

∴ Half the energy is wasted in starting

∴ $\eta_{\text{starting}} = 50\%$.

6.31.2 Series Parallel Control

With a 2 motor equipment ½ the normal voltage will be applied to each motor at starting as shown in Fig. 6.32 (series connection) and they will run up to approximate ½ speed, at which instant they are switched on to parallel and full voltage is applied to each motor. R_S is gradually cutout, with motors in series connection and then reinserted when the motors are connected in parallel, and again gradually cut-out.

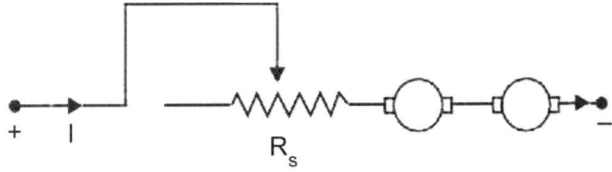

Fig. 6.32: Motors in series.

In traction work, 2 or more similar motors are employed. Consider 2 series motors started by series parallel method, which results in saving of energy.

a. **Series operation.** The 2 motors, are started in series with the help of R_S. The current during starting is limited to normal rated current 'I' per motor. During series operation, current 'I' is drawn from supply. At the instant of starting OA = AB = IR drop in each motor. OK = Supply voltage 'V' The back emfs of 2 motors jointly develop along OM as shown in Fig. 6.34a. At point E, supply voltage V = Back emfs of 2 motors + IR drops of 2 motor. Any point on the line BC represents the sum of back emfs of 2 motors + *IR* drops of 2 motors + Voltage across resistance Rs of 2 motors.

OE = time taken for series running.

At pt 'E' at the end of series running period, each motor has developed a back emf

$$= \frac{V}{2} - IR$$

EL = ED – LD

b. **Parallel operation.** The motors are switched on in parallel at the instant 'E', with R_S reinserted as shown in Fig. 6.33. Current drawn is 2I from supply. Back

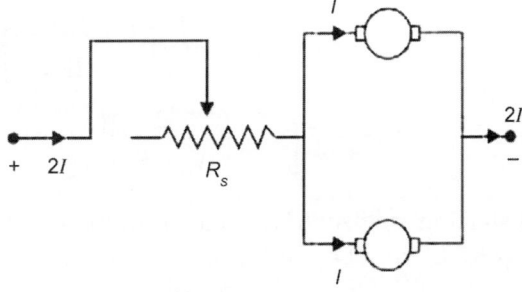

Fig. 6.33: Motors in parallel.

emf across each motor = EL. So the back emf now develops along LG. At point 'H' when the motors are in full parallel, (R_S = 0 and both the motors are running at rated speed)

Supply voltage = V = HF = HG + GF

= Normal Back emf of each motor + IR drop in each motor.

6.31.2.1　To find t_s, t_p and η of Starting

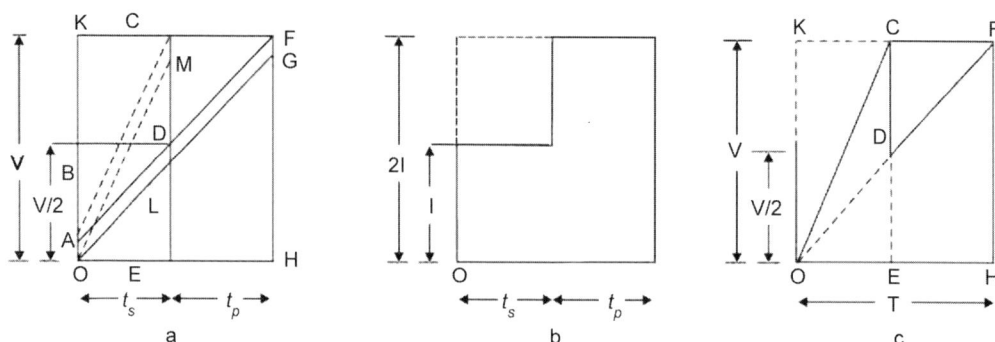

Figs 6.34a to c: (a) Voltage built-up in series parallel starting, (b) Variation of current in series parallel starting, (c) Efficiency of starting by series parallel method.

The values of time t_s during which the motors remain in series and t_p during which they are in parallel can be determined from Figs 6.34a to c. From Fig. 6.34a, triangles OLE and OGH are similar

∴
$$\frac{OE}{OH} = \frac{LE}{GH}$$

$$\frac{t_s}{t} = \frac{DE - DL}{FH - FG}$$

$$= \frac{\dfrac{V}{2} - IR}{V - IR}$$

∴
$$t_s = \frac{1}{2}\left(\frac{V - 2IR}{V - IR}\right)T$$

$$t_p = T - t_s = T - \left\{\frac{1}{2}\left(\frac{V - 2IR}{V - IR}\right)T\right\}$$

$$t_p = T\left\{1 - \frac{1}{2}\left(\frac{V - 2IR}{V - IR}\right)\right\}$$

To calculate η of starting, neglect IR drop in armature circuit.

This modifies Figs 6.34a to c. 'D' is midpoint of CE and back emf develops along DF in parallel combination. KC = CF, i.e. time for series combination = time for parallel combination.

i.e. $t_s = t_p = t$ and average starting current $= I$ per motor.

Energy lost in R_s = Area under triangle OKC + Area under triangle CDF

$$= \left(\frac{1}{2}VI\right) \times t + \left(\frac{1}{2}\frac{V}{2}2I\right) \times t = VIt$$

But total energy supplied $= IVt \quad + \quad 2IVt$

$$\qquad\qquad\qquad\qquad \text{(Series)} \qquad \text{(Parallel)}$$

$$\qquad\qquad\qquad\qquad = 3VIt$$

$\therefore$ η of starting $= \dfrac{3VIt - VIt}{3VIT} \times 100$

$$= \frac{2}{3} \times 100$$

$$= 66.67\%$$

$\therefore$ η is increased by 16.66% as compared to pervious case. If there are 4 motors then $\eta_{starting}$ = 73%. So there is saving of energy lost in R_s, during starting period as compared with starting by both motors in parallel.

6.31.2.2 Series Parallel Control by Shunt Transition Method

The various stages involved in this method of series–parallel control are shown in Fig. 6.35 in steps 1, 2, 3, 4 the motors are in series and are accelerated by cutting out the R_s in steps. In step 4, motors are in full series. During transition from series to parallel, R_s is reinserted in circuit–step 5. One of the motors is bypassed -step 6 and disconnected from main circuit–step 7. It is then connected in parallel with other motor–step 8, giving 1st parallel position. R_s is again cut-out in steps completely and the motors are placed in full parallel.

The main difficulty with series parallel control is to obtain a satisfactory method of transition from series to parallel without interrupting the torque or allowing any heavy rushes of current.

In shunt transition method, one motor is short circuited and the total torque is reduced by about 50% during transition period, causing a noticeable jerk in the motion of vehicle. The bridge transition is more complicated, but the resistances which are connected in parallel with or 'bridged' across the motors are of such a value that current through the motors is not altered in magnitude and the total torque is therefore held constant and hence it is normally used for railways. So in this method it is seen that, both motors remain in circuit throughout the transition. Thus, the jerks will not be experienced if this method is employed.

6.31.2.3. Series Parallel Control by Bridge Transition Method

    a. At starting, motors are in series with R_s, i.e. link P in position $= AA'$

    b. Motors in full series with link P in position $= BB'$ (No R_s in the circuit)

The motor and R_s are connected in the form of Wheatstone Bridge. Initially motors are in series with full R_s as shown in Fig. 6.36a. A and A′ are moved in direction of arrow heads. In position BB′ motors are in full series, as shown in Fig. 6.36b, with no R_s present in the circuit.

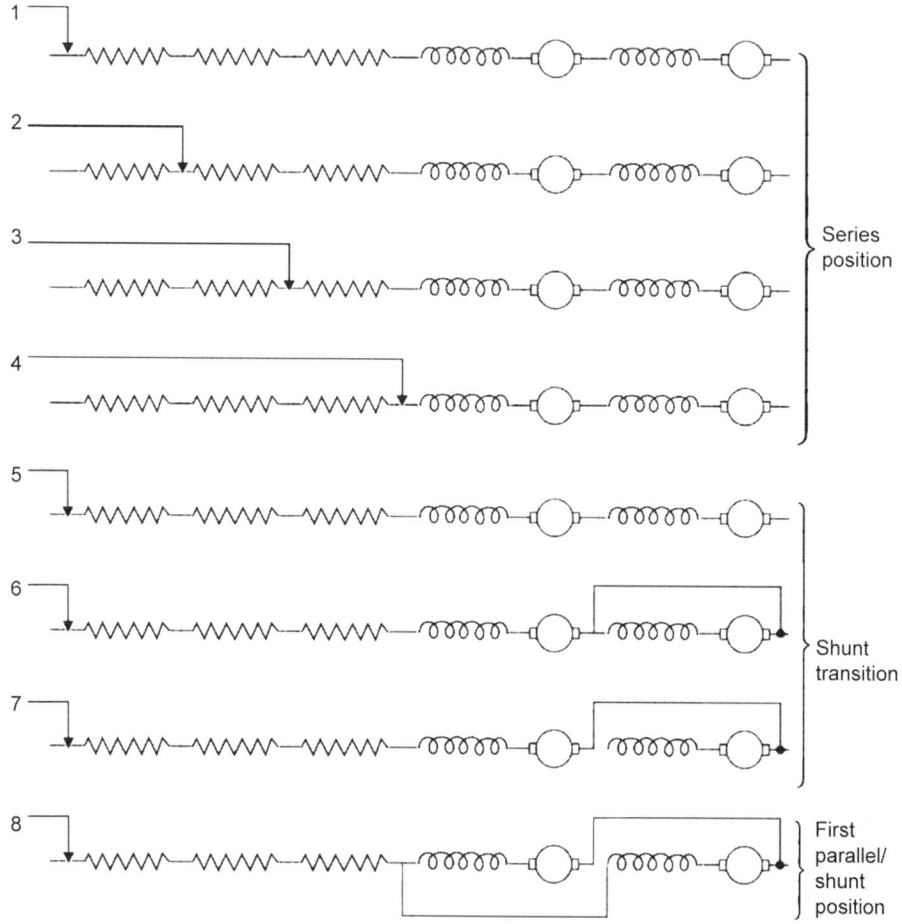

Fig. 6.35: Series parallel control by shunt transition method.

In transition step the R_s is reinserted. In 1st parallel step, link P is removed and motors are connected in parallel with full R_s as shown in Fig. 6.36c. Advantage of this method is that the normal acceleration torque is available from both the motors, through-out starting period. Therefore acceleration is smoother, without any jerks, which is very much desirable for traction motors.

6.32 ADVANTAGE OF SERIES PARALLEL STARTING

   i. It has higher efficiency than plain rheostat method of starting as proved above.
   ii. We get more than one economical speeds which are possible in plain rheostatic only by wasting energy in the rheostat.
   iii. Due to low energy loss in the starting resistance they are not of cumbersome size.

6.33 METALDYNE CONTROL

Metaldyne converter is a machine, which takes power at constant voltage and variable current and delivers the same at constant current and variable voltage. The main

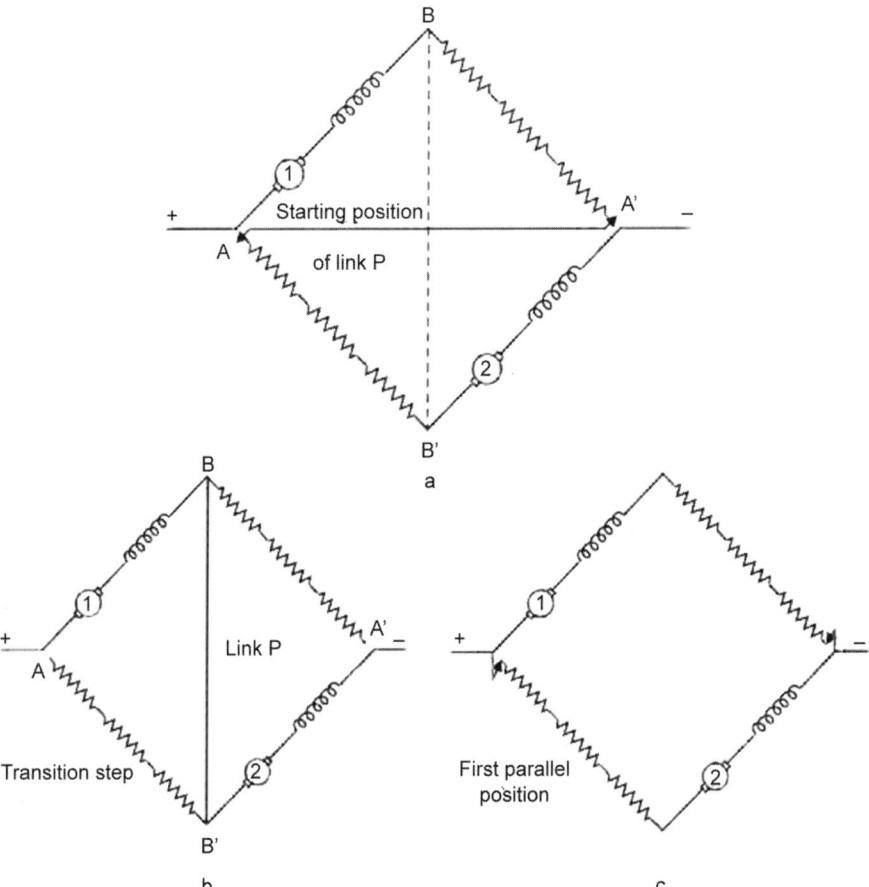

Figs 6.36a to c: Series parallel starting-Bridge transition: (a) Series position, (b) Transition step, (c) First parallel position.

advantage of using metaldyne converter in traction work is that it provides smoothly varying voltage to motors while maintaining predetermined current and, therefore, torque, during acceleration period, without incurring any loss of energy in the rheostats. In any method of motor control where notching is employed, motor current fluctuates between certain limits. Sudden growth of motor current at every notch produces sudden rise of tractive effort which results not only in jerky drive but at the same time coefficient of adhesion is also reduced. In metaldyne, control since current throughout the starting period remains constant uniform tractive effort is developed. This gives very smooth drive and high coefficient of adhesion.

It consists of DC armature, wound for two poles and provided with four brushes and a yoke with dummy poles as shown in Fig. 6.37. Constant voltage supply is given across brushes AB. Current Y, taken from supply, will produce armature reaction flux ϕ_s along the axis of brushes BA. Dummy poles will offer path of least reluctance to two parallel magnetic circuits. This armature reaction flux is called primary flux and is the cause for production of an emf E_2 across brushes CD. This current will produce secondary flux ϕ_s along the axis of brush CD. This flux in turn will produce an emf E_2 across brush BA so as to be equal and opposite of the applied voltage.

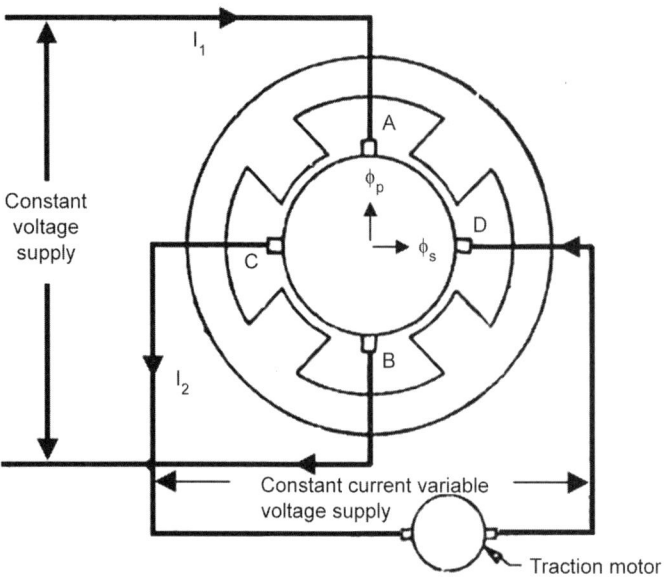

Fig. 6.37: Metaldyne convertert.

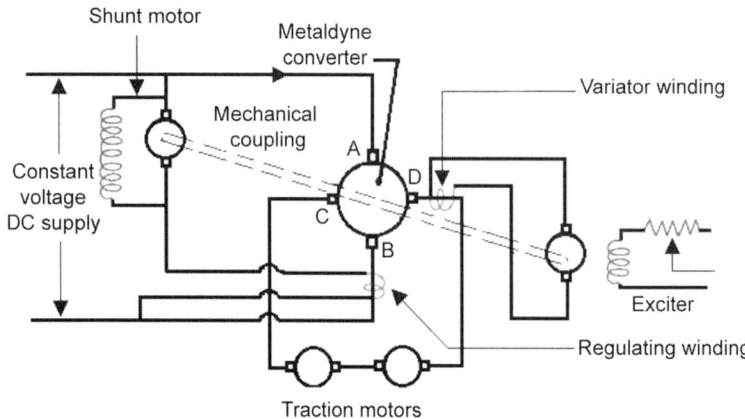

Fig. 6.38: Creating flux by providing variator winding.

$$\therefore \qquad\qquad E_1 = K \qquad\qquad (6.41)$$

Similarly, $\qquad\qquad E_2 = KI_1 \qquad\qquad (6.42)$

where K is a constant of machine and depends upon the construction of machine. Equation (6.41) suggests that if supply voltage is constant, load current will automatically be constant. Therefore, if load resistance varies and load current is constant, this will vary the load voltage E_2. Equation (6.42) suggests that E_2 can be varied, by corresponding change in the input current. In this way metaldyne converter, working as motor along brushes AB and as generator along brushes CD, acts like a transformer, transforming constant voltage variable DC current supply to constant current variable DC voltage supply.

For a given applied voltage, we will get only one constant current supply. In traction work, on the other hand, we require different tractive efforts during accelerating and constant speed running periods. Therefore, in order to develop different tractive efforts we require different magnitudes of constant currents during accelerating and constant speed running periods. Therefore, output current control is a necessity. This is achieved by providing variator winding which produces flux along brush axis CD as shown in Fig. 6.38. In order therefore, to produce constant magnitude of E_1 along BA axis, magnitude of flux ϕ_s along brush axis CD should tie constant. Flux ϕ_s is partly produced by variator winding and partly by load current I_2. As more and more flux is produced by variator winding, load current I_2 is correspondingly relieved of the responsibility of producing that much portion of ϕ_s. In other words magnitude of I_2 is reduced accordingly. If all the required magnitude of (k, is produced by variator winding, load current will become zero. At this stage even though motoring action is on, there is no generation action, i.e. there is input without output. This will give rise to increase of speed of the converter. On the other hand if flux produced by aviator winding is in opposite direction to that produced by I_2 magnitude of I_2 will increase. Now, if the back emf E_2 of traction motor being supplied by metaldyne converter is constant, this would mean more output than input I_1 as changes only when E_2 changes. Metaldyne converter will, therefore, require mechanical power to drive it. To generate given value of E_2 required amount of ϕ_s is to be produced. If some portion of ϕ_s is produced by regulating winding, then magnitude of I_2 decreases and power taken by metaldyne decreases. If, on the other hand regulating winding is producing flux in opposite direction to that produced by I_1 it would require more input current I_1 to produce given output voltage E_2. This would increase the input to metaldyne. We thus observe that any unbalance between output and input of metaldyne, brought about by variator winding, is restored automatically by regulator winding with the result that shunt motor is required only to meet the losses of metaldyne and drive exciter. Automatic action of regulator winding is explained as follows:

Any increase in the output of converter over and above the input to it, will tend to bring down its speed. With this, speed of shunt motor will also go down. This will reduce the back emf of shunt motor. This in turn will bring more current through regulator winding. If increase of regulator winding current brings about demagnetization of ϕ_s it will require more current I_1 to produce given amount of E_2. In this way balance is brought between input and output of metaldyne which now is run again at constant speed by shunt motor. With metaldyne converter, regenerative braking can also be achieved very easily by reversing the field of traction motor. This will change the direction of E_2 which in turn will change the direction of input current I_1. Magnitude of the regenerative braking can be regulated by controlling the magnitude of reversed excitation of traction motors supplied by metaldyne.

6.34 MULTIPLE UNIT CONTROL

Coaches which are fitted with traction motors are called motor coaches. For is usual to use motor coaches. Multiple unit trains suburban services it are better suited for high speed running than locomotive hauled trains. The space for traction motors in locomotive can only be extended by choosing wheels of larger diameter or by choosing wider gauge which cannot be practicable for economic reasons. There is then a limit to the capacity of locomotive hauled electric train: This is also so because number of driving axles becomes the number of locomotive axles. At the present state of technology of traction motor design, locomotive hauled train may be justifiable up to a certain speed. Beyond this speed multiple unit train becomes the solution as therein all the axles can be motored. Main advantage of using multiple trains is the flexibility of operation. We can have lengths of trains according to traffic density to be handled at different times of the day. At the time of heavy traffic a number of motor coaches along with trailer coaches are used. On the other hand at the times of light traffic, the train can be split up. Other advantages of multiple unit trains are: due to the fact that such trains, having no limitation of number of driving axles developing high tractive effort and, therefore, capable of achieving high acceleration rate, so much required by the suburban services. Another advantage is that the length of train being directly, proportional to the HP of traction motors ratio of weight of train to HP is constant. This enables lengthy trains, in rush hour periods, to operate without any difficulty to the same time table as shorter trains in off peak periods. Therefore, unlike in locomotive, whole of traction capacity is not tied to locomotive but can be distributed over a number of coaches required to handle the traffic. In such an eventuality it becomes necessary that all the traction motors of different motor coaches should be controlled from one point.

One such scheme is shown is Fig. 6.39 employing master controllers which control the sequence of closing of contactors of all the motors through control cables. There being more than one master controller, it becomes necessary to have interlock so that only one master controller can be operated at a time. Control cables running that the

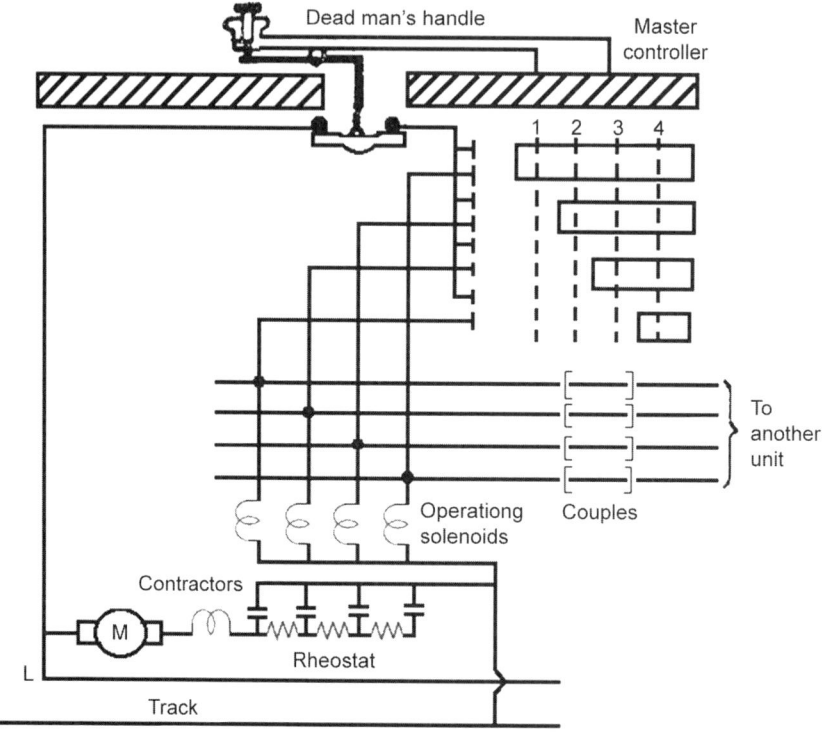

Fig. 6.39: Multiple unit control.

length of the coach are coupled at the end of the coach by means of special couplers so that only corresponding cables are connected together. One point which requires special mention here is the connections of reverser control wires.

Because the forward to the leading controller on the train is reverse to the trailing end controller, it becomes necessary to have the connections to the forward and reverse figures crossed. Only with a view to illustrate the principle of multiple unit control and to avoid complicated wiring, master controller shown in Fig. 6.39 embodies only the principle of plain rheostat starting. In actual embody not only rheostat starting practice master controller may but may incorporate series parallel control also as per control scheme. Master controller usually incorporates "dead man's handle", the function of which is to stop the train in case drier faints or becomes incapacitated. This consists of a contact attached to the knob of the controller handle shown in Fig. 6.39. In case driver releases the hold of the controller handle, it will open the circuit of the operating solenoids. This will lead to the opening of the motor circuit and bringing train to a stop by automatic application of brakes. It may be observed that the currents handled by the master controller are the solenoid currents and not the motor currents which are very heavy in case of locomotives. As such master controllers are not cumbersome or inconvenient to handle for controlling the motors.

6.35 WHEEL ARRANGEMENT

Tractive effort of a locomotive can be increased by not only increasing the HP of driving motors alone but we have to increase the adhesive weight also. Maximum axle load being limited by track design, therefore, total adhesive weight can be increased by increasing the number of driving axles. Number of driving axles can be increased in two ways.

1. By employing individual motor for each driving axle—this is called individual drive.
2. By employing one large motor and carrying the drive to the driving wheels through connecting rod. This is called collective drive. Recently monometer bogies, fully described in § 6.49.1 have been developed by French National Railways. These bogies employ only one motor which drives two or three axle units.

Main disadvantage of collective drive is the inertia forces which reduce vibrations at high speeds. Collective drive would, therefore, look one step backwards to steam locomotive practice. All the same, collective drive is preferred for low speed freight locomotives where heavy loads have to be hauled. This is due to the fact that tendency to slip with collective drive is lessened. Moreover in collective drive, since motor can be mounted up in the main body of the locomotive, there is freer band in the selection of the dimensions of the motor than in case of individual drive. Centre of gravity being high due to installation of motor in locomotive body, it produces less wear and tear of the track.

In case of DC locomotives, since weight to HP ratio is small, all the weight is put on the driving axles. As against this weight of AC motors is comparatively high. This has, therefore, to be taken, in addition to driving axles by non-driving axles or trailing axles. Hence in case of AC locomotive, provision of trailing axles is a necessity but in case of DC locomotive it is for improving the riding qualities.

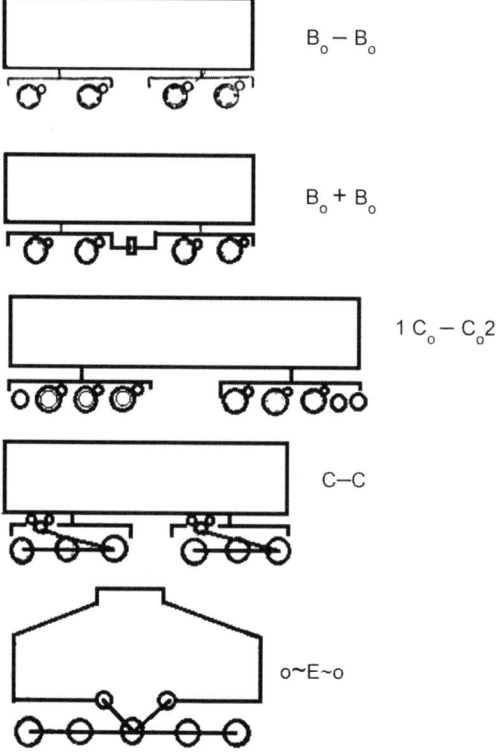

Fig. 6.40: Wheel arrangement.

Various arrangements of driving axles, trailing axles, individual or collective drive are expressed in a form of "short hand" as given in Fig. 6.40. Number of driving axles are denoted by alphabetical letters—A for 1, B for 2, C for 3 and so on. Number of trailing axles are denoted by figures. When axles are individually driven, small "o" is written after the letter indicating the number of axles. If there is no such letter, it indicates collective drive. Symbols for separate bogies are separated by dash line. If bogies are coupled by an articulating joint which transmits tractive effort between bogies, instead of dash, plus (+) sign is written.

6.36 BOGIE ARRANGEMENTS

Suspension arrangement of bogie with respect to axles is called primary suspension and that of body with respect to the bogie as secondary suspension. While the exact arrangement differs from bogie to bogie, we will state the desirable features of the bogie as follows:

1. Dynamic variation of vertical forces have to be kept to minimum to improve coefficient of adhesion. Reducing the unsprang weight and the use of helical springs in place of laminated springs weight this point of view fully suspended motors are preferable.
2. Effect of weight transfer should be minimised. This is achieved use of low traction bars in place of centre pin arrangement.
3. Loads on both the axles of the bogie should be equalized. This is achieved by the use of equalizer beam connecting axle boxes.

4. Resilient lateral connections between the body and the bogie should be provided to limit the side movements of the bogie. For this purpose use is made of friction dampers.

Having stated the desirable features, we will now give brief account of a few of the bogies used on Indian Railways which follow continental and Japanese designs.

In continental design, secondary suspension is not provided. The body is suspended at the end of a bolster through pendular suspension rods. This loading beam or bolster rests on the bogie frame entirely on side bearings. There is centre pin joining bolster with bogie frame. Its function is only to have lateral control and to avoid transmission of tractive effort through this pin as there is provided ample clearance in the longitudinal axis. Lateral oscillations between body and bogie are damped by two side bearer friction pads. Bogie frame on each side is supported on the equalizer beam through two pairs of double helical springs. Equalizer beam rigidly connects two axle boxes. Drag links provide longitudinal guidance of bogie frame with respect to equalizer. For damping out vertical oscillations, friction dampers are provided in between bogie frame and equalizer beam. Tractive effort is transmitted from bogie frame to the body by means of low traction bars which work only in tension. These traction bars are provided with adequate clearance in the elongated slot to allow for the bogie swivel on curves and to accommodate primary and secondary spring height variations due to load and vertical oscillations. The transmission of tractive effort through low traction bars instead of through centre pin reduces the weight transfer effect between the wheels of the same bogie. In Japanese design, both primary and secondary suspensions are provided. In BG MT locomotives (WAM_2 and WAM_3) the centre pivot and side bearers transfer the weight of the body to upper bolster and tractive effort is transferred through centre pivot. Upper bolster is further supported on the lower bolster through four sets of helical springs. Lower bolster is connected to the upper bolster by connecting rods and to the bogie frame by four articulated. The bogie frame is then supported on each axle box by means of two sets of helical springs. In FT locomotive (WAG_2), secondary suspension consists of helical springs carried on the brackets projecting from the side beams of the bogie frame. Body weight is supported on rubber seats fixed at the top of these helical springs. Tractive effort from the bogie to the body is transferred through "Oil centre pin" which is fixed to the cross beam on the bogie frame and engages with bolster beam on the body. Bogie frame is supported on each axle box by two sets of coaxial primary suspension helical springs. In both MT and FT locomotives the difference in the height of the coupler and centre pin is kept minimum to reduce weight transfer between bogies.

6.36.1 Monomotor Bogie

Recent development is the use of single motor in place of two nose suspended motors. This arrangement particularly makes it amenable to provide for changing the gear ratio so that locomotive can be made to haul equally well both slow heavy trains of goods traffic or high speed light trains of passenger traffic. The motor drives the motor pinion through resilient coupling. This motor pinion through anyone of the two reduction opinions drives the coupling wheel which is in engagement of the main gear wheels. The drive from main gear wheel to the wheel centre is resilient one. Gear change rocking device includes motor pinion and two reduction pinions, in permanent engagement with motor pinion. The changeover beam is pivoted about the axis of motor pinion to engage one or the other reduction pinions with the coupling wheel and so gives the two different gear ratios. The monomoter bogie arrangement has following advantages—over two motor bogie.

1. Due to coupling effect between two axles of 15% up to the speed of 10 mph and 20% for higher speeds. Improvement in the adhesion itself has resulted in the reduction in the weight of the locomotive to the extent of 18%. Apart from and this, use of one motor in place of two reduces wiring, shunting reversing equipment and traction motor blowers, etc. This yields considerable saving in the motor weight and therefore, the weight of the bogie.

2. Location of single motor at the centre of the bogie reduces the polar moment of inertia of the bogie. This is turn reduces the magnitude of the lateral forces and improves the stability.

3. Monomotor bogie construction is ideally suited for the mixed traffic. This increases the flexibility of operation of the locomotive. Because of this reason, for handling given pattern of traffic, we will require few types of locomotives. This leads to effective standardization of various parts and its consequent advantages in the operation and maintenance of the locomotives.

6.37 TRANSMISSION OF DRIVE

It is desirable to have as much of locomotive weight spring borne as possible. This reduces not only vibrations in the locomotive but at the same time damage to the track due to hammer blows is also reduced substantially. With this end in view, if motors are fixed on the locomotive or bogie frame, it will be seen that there will be relative motion between motor armature and driving axle. This will, therefore, need flexible drive between the two. There are various types of drives but we will consider only few to illustrate the principle underlying each. These are broadly classified as direct drive, partially suspended drive and fully suspended drive.

a. **Direct drive.** In direct drive shown in Fig. 6.41a armature of the motor is directly, mounted on the driving axle. Field poles, two in number along with yoke, are attached with the locomotive frame. In order that armature should not foul with pole faces due to vertical relative motion between the two, latter is made with flat faces. Disadvantages of this method of drive are: size of the armature of the

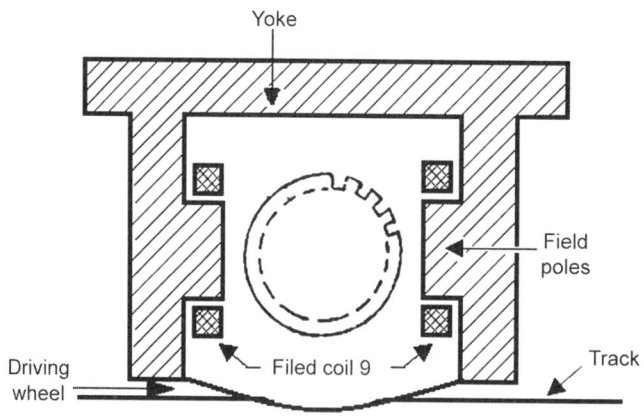

Fig. 6.41a: Direct drive.

motor is limited by the size of the driving wheel more of unsprang weight of the motor and low centre of gravity give rise to poor riding qualities besides large wear and tear of the track. In the absence of any reduction in speed, motor speed has to be low which makes the motor more bulky and costly. This method of drive is obsolete today and we have dealt with it for its historical interest only. There are, however, other examples of direct drive in which attempt has been made to do away with some of the disadvantages enumerated above.

i. **Direct quill drive.** In this method, armature is mounted on a hollow shaft called quill which is flexibly connected to the driving wheels by means of springs or link motion as in Fig. 6.41b. Main difference between direct drive and the quill drive is that in the latter case unsprang weight is reduced to minimum, which improves the riding qualities. In this method of drive also there is no reduction of speed and hence its motor speed is also limited.

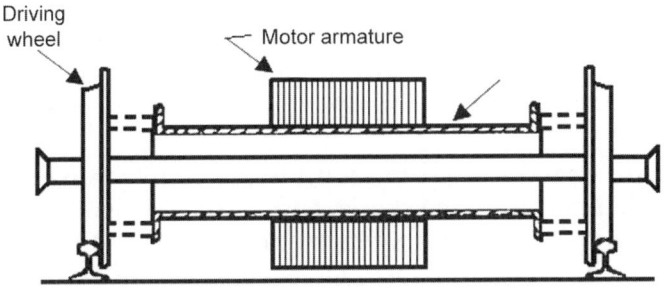

Fig. 6.41b: Direct quill drive.

ii. **Geared quill drive:** The only difference between direct quill drive and geared quill drive is that in the first case armature is mounted directly on the quill while in the latter case gear wheel is mounted on the quill. Traction motor is mounted on the bogie frame and its pinion engages with the gear wheel on the quill. With this method, we have free choice on the dimensions of the motor and, therefore, it is possible to employ high HP motors for the purpose of drive.

b. **Partially suspended drive.** In this drive the weight of the driving motor is partially borne on the axle and partially suspended from bogie. We will now take up various arrangements under this class of drive as follows:

i. **Nose suspension geared drive:** By employing speed reduction gears, it is possible to select a motor with high speed. This results in the reduction in physical size of the motor for given hip. The gear ratio depends upon the maximum safe speed of the motor and maximum designed speed of the vehicle. In this type of drive one end of the motor is supported on the axle by means of two plain bearings carried in the motor frame and the other end carries a nose or projection resiliently mounted on the bogie cross member. This arrangement maintains fixed centre to centre distance between armature shaft and axle and motor swings about the nose support as axle rises or falls in suspension on the bogie. Nearly 60% weight of the motor is spring borne. Remaining 40% unsprang weight on the axle gives rise to the hammer blow to the track. Motor itself is also subjected to direct shock from the track. Pinion on the motor engages with the gear wheel mounted on the axle. Resilient gear wheel is sometimes used to give some flexibility. In resilient gear wheel, loose rim is connected to the fixed hub through rubber or spring mounting. In this method of drive, limitation is imposed on the overall size of the motor due to its one end being supported. This method of motor suspension is simple and robust and is used for low speed traffic such as in motor coaches and tramways. For high speed it is necessary that whole of the weight of the motor should be spring borne for which more elaborate systems of motor mounting are employed.

  ii. *Oerlikon drive:* This combines the main features of nose suspended drive and gear quill drive. Instead of motor directly resting on the axle as in case of nose suspended motors, it rests on the quill. Main gear wheel is keyed to the quill which is connected to the axle by resilient crank pins arranged in the web of the wheel.

 iii. *Brown boveri drive.* In this drive, gearwheel is supported on a short quill shaft through roller bearings and revolves freely on it. Axle is fitted with a spider with arms which extend into the openings in the gear wheel. The torque is transmitted from the gear wheel spokes to the spider arms through springs.

c. **Fully suspended drive.** By making the motor fully suspended from bogie, we are isolating it from the shocks and abrupt accelerations to which axle is subjected. In this way not only mechanical duty of the motor improves and becomes smoother but it reduces the maintenance of permanent way and the motor. This further improves the adhesion characteristics of the bogie. This, however, creates difficulty of transmission of torque to the driving axle. Fully suspended gear system is that where resilient coupling is provided in between the road wheel and the main gearwheel. To this class of drive belong Jacqueline drive and WN drive.

  i. *Jacquemin drive.* In Jacquemin cardon transmission shown in Fig. 6.42, quill on one end is connected to the main gearwheel by resilient cardoon joint and on the other end to the axle wheel by similar joint. Main gear which is mounted on the roller bearing is supported on the extension of the transmission frame. Torque is transmitted from the main gearwheel through elastic arms perpendicular to the plane of cardoon ring. Thence from cardoon ring, torque is delivered to the quill by means of again elastic radial arm. Same is the arrangement at the other end of the quill. The whole arrangement is similar to the universal coupling shown in Fig. 6.43 where two nonaligned axes x_1 and x_2 are connected by the quill shaft 'y' and cardoon rings. This type of transmission is used in WAM_1 WAG_1 WAG_3 and WAG_4 locos. In WAG_2 quill is flexibily coupled to main gearwheel on one side and wheel centre on the other end, each through ten pieces of cylindrical rubber.

  ii. **W-N drive.** This drive makes us between motor use of 'VFW' coupling or and pinion. This permits ±17 mm vertical and ±12 mm horizontal misalignment between the two. 'WM' coupling with ring consists of two pinions with external gear teeth. These engage gears having internal teeth in each section of a two piece sleeve housings bolted together as single unit while the external teeth are crowned or rounded, the internal teeth are straight sided. Motor which is mounted on the bogie frame is directly coupled to

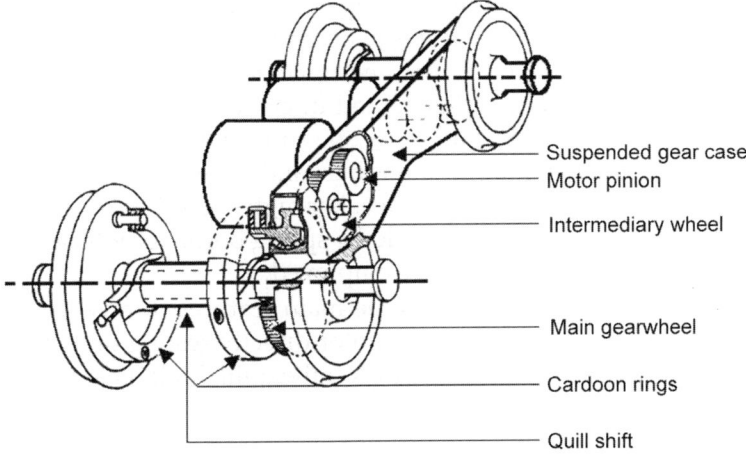

Fig. 6.42: Jacquemin cardoon transmission.

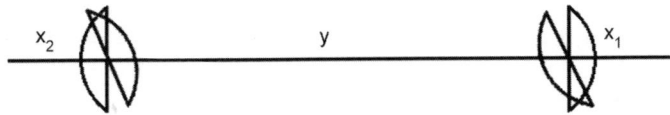

Fig. 6.43: Universal coupling.

the driving pinion through WN coupling. Driving pinion mounted on the roller bearing drives the main gearwheel. The gear case is supported from the bogie transom by a long suspension bolt. This type of drive is considered better than cardoon shaft drive from the point of view of easier maintenance in service.

6.38 BRAKING

To deal with heavy traffic requirements as a result of phenomenal increase of both the industrial activity and population, we have to resort to heavy and faster trains. With modern traction equipment, achievement of rapid acceleration does not present any problem and also failure during acceleration will not involve any danger. On the other hand resorting to heavy and faster trains involves the problem of dissipation of large amount of kinetic energy in shortest possible times as in the case of emergency braking. Any failure during emergency braking will be a catastrophe. As such the importance of braking system cannot be under estimated.

6.38.1 Requirements of a Braking System

Before we deal with various systems of braking we will first enumerate various desirable requirements which a braking system should satisfy. These are:
1. The braking system should be robust, simple and easy for driver to control and operate. It should require less maintenance and should be reliable.

2. The system should apply brakes simultaneously over all the vehicles.
3. Brake actuation time should be as small as possible.
4. To avoid damage to the goods and discomfort to the passengers, normal service application of brakes should be very gradual and smooth.
5. In case of emergency braking, safety consideration is the prime most consideration. As such retardation rate would be maximum consistent with the safety, so as to make unfailing halt in the minimum possible distance.
6. In order to obtain uniform deceleration, braking force applied to each axle should be proportional to axle load.
7. The braking system should be inexhaustible, i.e. repeated quick application of brake should be possible without needing any relaxation recuperation at normalising time in between consecutive operations.
8. Kinetic energy of the train should as far as possible be stored during braking which could subsequently be utilized for accelerating the train.
9. There should be automatic slack adjustment for constant piston stroke as a result of wear on the rim and the break block in the case of mechanical braking.

6.39 TYPES OF BRAKING

In broad sense, braking is of three types, viz. mechanical braking, electrical braking and hydraulic braking. In mechanical type of braking, the force is produced by the rubbing action of brake blocks on the tread of the wheels. Depending upon the method how braking force is produced and controlled, mechanical braking is further classified as pneumatic braking, vacuum braking, electropneumatic braking and disc braking. Electrical braking is similarly classified into plugging, rheostat braking, regenerative braking, electromagnetic braking, and eddy current braking. We will now discuss in details various types of braking in the following paragraphs.

6.39.1 Mechanical Braking

All types of electrical braking, as we will see later on, are effective only when the train is moving above certain speed. These are although helpful in reducing the speed of train, but cannot keep it stationary on down gradients. As such electric braking has to be supplemented with mechanical braking. Mechanical brakes are usually applied by brake block shoes pressed with force against the tyres of the wheels. Braking force is produced by the movement of piston which is transmitted to the brake blocks through a system of levers.

6.39.2 Air Brake System

In this system of braking for locomotive, brake shoe pressure is produced by admitting air under pressure behind piston as shown in Fig. 6.44 which moves piston rod against spring force in a direction in which brakes are applied. Releasing the pressure behind the piston enables spring to move the piston rod so as to release the brakes.

6.39.3 Vacuum Brake System

In this system of braking, force is applied by piston when vacuum underside piston is destroyed.

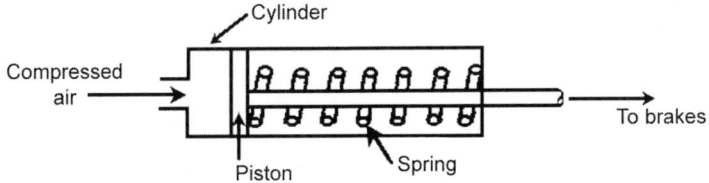

Fig. 6.44: Compressed air brake system.

The vacuum brake system, (Fig. 6.45) consists of a vertical cylinder having a piston and a piston rod which operates the braking arrangement through a system of levers. Vacuum is created on the top and underside of the system so that in normal condition, the piston rests at the bottom of the cylinder.

When brakes are to be applied, the vacuum is broken from the underside by admitting air at atmospheric pressure. The piston moves up and applies the brakes. The brakes may be released by either recreating the vacuum or by making the pressure equal on both sides of the piston.

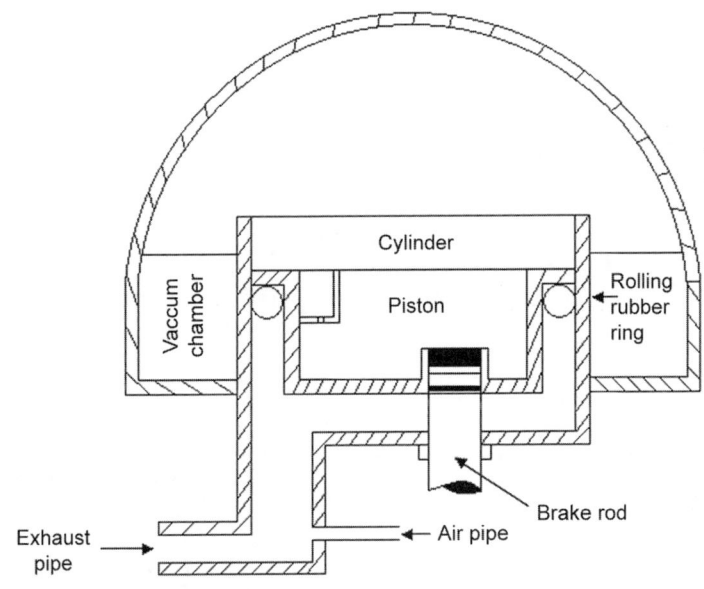

Fig. 6.45: Vacuum brake system.

☞ *Drawbacks*

The vacuum brake system suffers from the following drawbacks:

i. The air propagation time on long goods train is very long, this makes the response to the drivers control very much sluggish (and is also very much undesirable from the point of view of safety). To reduce this time of propagation, use is made of direct admission valves or accelerator valves.

ii. After the application of brakes, it is only after recreating the vacuum that brakes are released. This takes some time. Release of brakes on long trains can be speeded up by quick release valves.

6.40 HYDRAULIC BRAKE

It comprises an assembly resembling a hydraulic coupling the rotor being keyed to the axle and the stator being keyed to the bogie frame. The circulation of fluid (oil or water) in the assembly by the wanes of rotor generates a braking torque. The brake is applied by filling the coupling by means of a pump.

6.41 EDDY CURRENT BRAKES

There are two types of eddy current brakes-linear type and rotary type. In linear type, primary member is the shoe which carries excitation current and the rail itself forms the secondary element of the eddy current system. While braking, traction motors can be made to work as generators and supply power to the shoe of eddy current brakes. In rotary eddy current type brakes, the secondary member, i.e. the rotor is attached to the motor shaft and primary member which is supplied with DC excitation is fixed in space. This is a case similar to DC dynamic braking as applied to induction motor. The torque produced by the induced currents in the rotor will tend to reduce the relative speed between the rotor and the stationary field, i.e. it will be a braking torque. The only method of varying this torque is by varying the DC excitation of the stator. The main advantage of eddy current brake is same as that of the disc brake and is used satisfactorily in conjunction with the friction brakes for making the final stop.

6.42 MAGNETIC BRAKES

As shown in Fig. 6.46, magnetic brake consists of bipolar electromagnet with elongated pole faces a short distance apart and along with rails. Its body is made of cast steel and pole faces of soft steel pole faces are parallel to the rail. Passage of current through exciting coil produces magnetism which passes perpendicular to the rail face as shown

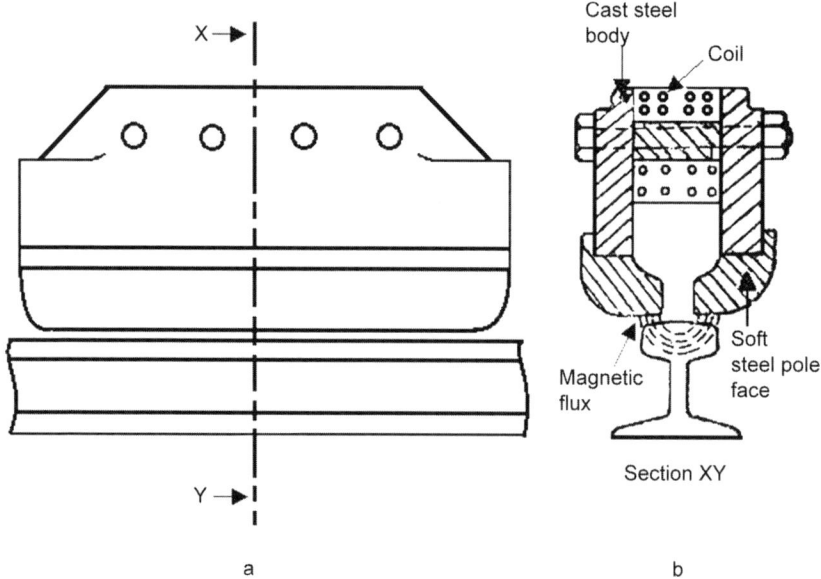

Figs 6.46a and b: Magnetic track break.

by dotted lines. This produces force of attraction between magnetic pole faces and rail given by equation 6.43.

$$F = \frac{B^2 a}{2\mu_0} \text{ newtons}$$

$$= \frac{B^2 a}{8\pi\, 10^{-7}} \tag{6.43}$$

This magnetic force in effect increases the weight on braking wheels with the result that braking force of the magnitude "μF" is produced. Magnetic brake is fitted in between wheels of the bogie and runs longitudinal along the track.

6.43 TYPES OF ELECTRIC BRAKING

There are three types of electric braking as applicable to common types of electric motors in addition to eddy-current brake which has been discussed in § 6.41.
1. Plugging or reverse current braking.
2. Rheostat or dynamic braking.
3. Regenerative braking.

6.44 MECHANICAL REGENERATIVE BRAKING

Energy is spent in imparting accelerating to the train. During coasting period, it is in imparting acceleration to the train. During coasting period, it is the kinetic energy of this moving train which is utilized in overcoming the track resistance. In suburban services, coasting period is about 20 to 50% of total run. After ceasing period, all of the remaining energy has to be dissipated by braking. Can be reduced by increasing the coasting period. This will, therefore, decrease the schedule speed. If, however, we have to stick to original schedule speed, prolonged coasting period can be achieved by adopting high acceleration which is possible with larger motors. One of the way to tackle this situation is to resort to mechanical regenerative braking. Essential element in achieving mechanical regeneration is that the stations are situated at certain height above level track. As shown in Fig. 6.47 slope of track while leaving station is 1 in 30 and while approaching station it is 1 in 60. While train is on level track, power is cut off and kinetic energy of train instead of being wasted in braking is utilized and stored as potential energy as train has to climb a slope of 1 in 60 while approaching the station. This potential energy is utilized in moving the train down gradient and imparting acceleration to it. This type of braking has been employed in certain section of London tube railways. It has been estimated that specific energy consumption has decreased to 75% of the value when mechanical regenerative braking was not employed.

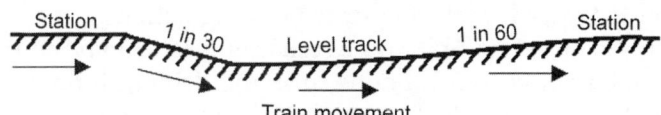

Fig. 6.47: Slope of track.

6.45 ELECTRICAL REGENERATIVE BRAKING

During electrical regenerative braking, traction motor are made to work as generators, utilizing kinetic energy of the moving train to run the generations. Electrical energy generated if fed back into the supply live. Main advantages of regenerative braking are reduced energy consumption, less wear and tear of brake blocks, wheels and track, consequently less maintenance required by those items, relatively small amount of brake block dust formed which increased the life of bearings, ease and safety with which heavy loads can be hauled over steep gradients. The disadvantages are: increased capital cost as motors of heavier size are employed, additional cost on control equipment with consequent complication in methods of operation and control. We will now find out the amount of energy recuperated between any two points during which regenerative braking has been applied. Equation 6.31 gives the specific energy output. During regeneration it is this energy which will be converted to electrical energy and fed back to the line. Since doting braking speed of train has to be reduced from V_1 to V_2 first term of equation 6.31 has to be modified accordingly.

Energy recuperated in W hrs.

6.45.1 How Electrical Regenerative Braking is Applied

Motor, without disconnecting it from the supply, is made to generate (instead of made to motor) and to feedback energy to supply. Magnetic drag, produced on account of generation action, offers the braking torque. This method of braking is most efficient. In many cases, the transition from motoring action to generating action is smooth and without any switching operation. As soon as overhauling load drivers the motor, it works as generator.

In case of DC machines, current can flow from machine to supply line only when emf generated by the machine is more than supply voltage. This condition of regeneration can be achieved by following two methods in traction practice.

1. With constant voltage supply, regeneration can be made possible by increasing the filed excitation.
2. Regenerations is also possible if load overhauls the motor and drives it at higher speed than no load speed.

In case of DC series motors, increase in speed results in decrease in excitation. The product of speed and excitation, therefore, remains constant. As such it is not possible to get emf more than the terminal voltage. Moreover it is not possible to make field current more than armature current, regeneration with series motors, as such, is not possible. We will further see that series motors for braking purpose are unstable in operation. Regenerative braking in the case of series motor can only be applied after connecting the field as separately excited and adopting measures to make it more stable.

In case of induction motor Fig. 6.48, we find that for speeds above synchronous, motor torque becomes negative. Machine is now working as induction generator. Depending upon the speed, motoring or generating action of the machine will be automatic. By this method of braking, overhauling load may be prevented from rising is speed much above synchronous. It is the driving torque of load exceeds the maximum braking torque which motor can develop, the motor will crossover to unstable operation. Rise in speed will then decreases braking torque. Motor in that case is heading far run away conditions.

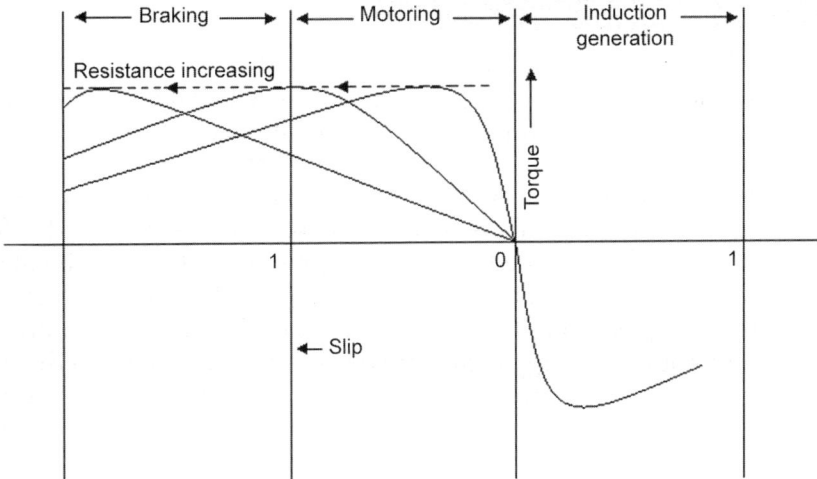

Fig. 6.48: Torque-speed curves of induction motor.

We can bring the speed below synchronous only where arrangement of pole changing available. If number of stator poles is increased, its new synchronous speed will be less than the actual running speed, machine will now work as induction generator and bring the motor below its first synchronous speed till operating speed will be little less than synchronous speed corresponding to increased number of poles. It will then continue to work as induction motor at this reduced. This method of electric braking is applicable to squirrel cage motors because rotor winding of slip-ring motor cannot be reconnected for different number of poles.

6.45.2 Conditions Necessary to Achieve Electric Regenerative Braking (ERB)

i. We know that in generators, generated emf is always more than terminal voltage. It is this difference between the generated emf and terminal voltage which is responsible for flow of current from generator armature to out side load. In case of motors, back emf is always less than the applied voltage and that is why it is taking current from supply. This, therefore, clearly suggests that in case motor is made to function as generator, its generated emf should be increased in such a regulated way that it is always more than the supply voltage by a given amount dictated by the requirement of braking, irrespective of the fact that its speed goes down as braking proceeds. This is possible only with the increase in excitation level.

ii. It is essential in the interest of safety that braking system should have mechanical stability. In other words, braking torque should be in a position to apply more and more in case powerful overhauling forces should try to increase the speed of the train.

iii. Braking system should have electrical stability as well, i.e. braking torque applied should be independent of line voltage fluctuations.

iv. In case recuperated energy happens to be surplus over and above the demands of other trains moving over that section, generating station equipment should be in a position to waste it, otherwise ERB would be ineffective.

6.45.3 Suitability of DC Shunt, Series and Induction Motors for ERB

a. **Shunt motor:** As the speed of motor increases due to say overhauling loads, its generated emf becomes more and more. Voltage differential between the generated voltage and line voltage will increase the braking current of the motor. Shunt motor as such has mechanical stability. Further shunt motors are electrical stable also, i.e. braking torque is independent of line voltage fluctuations. To clarify this, let us assume that a shunt motor is working as a generator and that line voltage has dropped. There being great potential difference between the motor and line, shunt machine will tend to give more current to the line. This is one side of the picture. Turning to other side of the picture, any reduction in the line voltage would reduce shunt field excitation. This in turn will reduce the generated voltage of the motor which will reduce the magnitude of regenerated current. Reduction of line voltage in this way, produces two opposite effects, on the magnitude of the regenerated current. Same is the case when line voltage instead of falling increases. Only limitation, shunt motor has got is that regeneration cannot be carried out to low motor speeds as this would need heavy field flux density which, due to saturation effect, would require heavy field current. This may bring about heating of the field winding of the motor.

b. **Series motor:** Speed torque characteristic of series motors is very much suited for the purpose of traction. It has further been shown that line voltage fluctuations have no effect on the torque developed, i.e. series motor are electrically stable in operation as motors. This, however, is not true when series motor is used for the purpose of regenerative braking. Series motor even tough high mechanical stability, suffers very badly from electrical instability. Any small fluctuation in the line voltage would conditions that tend to increase the magnitude of original disturbance. For instance, increase in line voltage reduces the voltage difference between motor generated emf and line voltage. This will reduce the magnitude of regenerated current. This reduced regenerated current in turn flowing through the series field will reduce the generated emf of motor. In this way disturbance of reducing the magnitude of regenerated current will go on extending till at the end, emf of both line and motor will be equal and there will be no regeneration action. Arguing on the same lines, initial disturbance of line voltage being reduced, will produce conditions of building up of regenerated current excessively. The root cause of this electrical instability is the flow of disturbance current through the series field. Therefore, first condition before series motors are adopted for regenerative braking, is to excite these fields separately and use stabilizing circuits.

It is, therefore, usual as shown in Fig. 6.49 to put armature of all the motors either in series or in parallel and also fields of all motor in series along with protective resistance across supply. Stabilizing action can be achieved either by the use of differentially compounding the exciter or by connecting circuit in parallel with stabilising resistance inserted in series with the main circuit. Circuit employing differentially excited exciter is shown in Fig. 6.50a. Any tendency of the drop in the line voltage will induce current surge from motors to the line. This recuperated current flowing through differentially excited field winding on exciter will demagnetize its fields. This will in turn reduce the excitation of the main traction motors whose generated voltage will also fall. In this way voltage

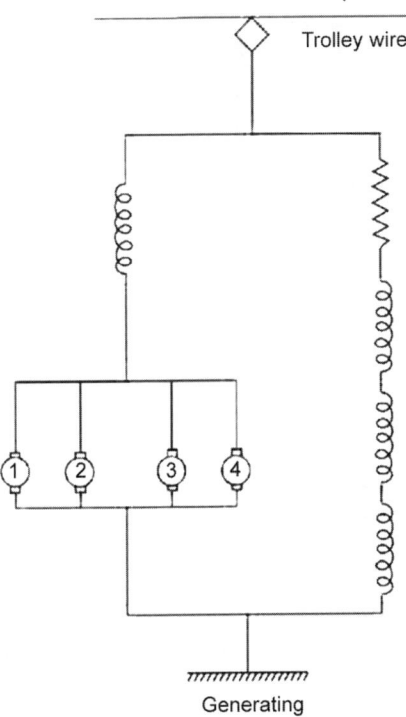

+

Trolley wire

Generating

Fig. 6.49: Electrical regenerative braking.

differential between the line voltage fluctuation. Exciter is driven by separate series motor which is coupled to a mechanical load also. Braking torque can be controlled by either varying the magnitude of series resistance or by regulating the excitation of the exciter by field rheostat.

Circuit employing stabilizing resistance is shown in Fig. 6.50b Current surge induced as a result of dip in the line voltage will cause more voltage drop across stabilizing resistance. Exciter armature supplying fields of traction motors in series will stabilizing resistance will now supply less exciting current consequent to increase in voltage drop in stabilizing resistance. As a result of this, emf generated in traction motor armatures will reduce thereby maintaining the required voltage differential between line voltage and motor voltage irrespective of line fluctuation.

Fig. 6.50c shows how one of the traction motors may be used as an exciter supplying fields of rest of traction motors and employing stabilizing resistance for ensuring electrical stability. Torque regulator resistance works diverter for series exciter field.

c. **AC induction motor:** ERB. with three phase induction motor is automatic and no addition apparatus is needed. Regenerative braking action depends upon the face that when motor is driven by overhauling load, such as encountered while train is descending a steep gradient above synchronous speed, it works as non synchronous generator, i.e. induction generator. Other condition necessary for obtaining generation action, in view or induction machine being none self-exciting, is that it needs supply system which supplies excitation and

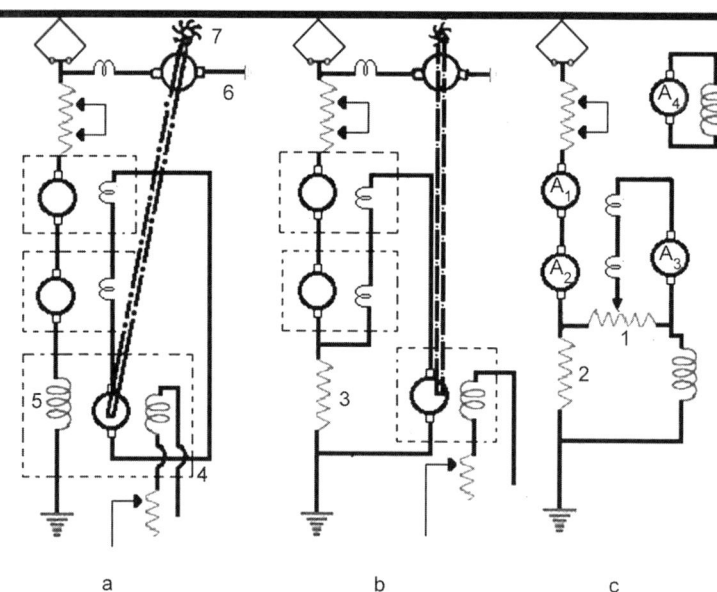

1. Torque regulator
2. Stabilizing rheostat
3. Stabilizing resistance
4. Exciter
5. Differential winding
   of exciter
6. Series motor
7. Mechanical load

Figs 6.50a to c: Circuit employing (a) Differentially excited exciter, (b) Stabilizing resistance, (c) Use in traction motor as an exciter.

fixes frequency. It is, therefore necessary that supply system should be at least connected with one synchronous alternator in addition to load dissipation arrangement, speed torque curves for both motor and generator operation are shown in Fig. 6.48. It is observed there from that if no resistance in the rotor circuit is inserted, speed of the train will be practically unaffected by this steepness of the gradient or weight of the train. If increased speeds are required with the light loads, these can be obtained by inserting resistance in the rotor circuit.

WORKED EXAMPLES

☙ **EXAMPLE 6.1**: The speed–time curve of a train consists of:
    i. Uniform acceleration of 6 km/h/s for 25 s;
   ii. Free running for 10 minutes;
 iii. Uniform deceleration of 6 km/h/s to stop the train;
  iv. A stop of 5 minutes.
Find the distance between the stations, the average and schedule speeds.
[AMIE Sec B Electric Drives and Their Control Winter 1996]

Solution: Acceleration, $= 6$ kmphps.
Accelerating period, $t_1 = 25$ seconds.
Maximum speed, $V_m = \alpha \times t_1 = 6 \times 25 = 150$ kmph.
Time for free running, $t_2 = 10 \times 60 = 600$ seconds.
Retardation, $\beta = 6$ kmphps

Time for retardation, $t_3 = \dfrac{V_m}{\beta} = \dfrac{150}{6} = 25$ seconds

Distance travelled during acceleration period, $S_1 = \dfrac{1}{2}\dfrac{V_m t_1}{3600} = \dfrac{1}{2}\times\dfrac{150\times25}{3600} = \dfrac{25}{48}$ km

Distance travelled during free running, $S_2 = \dfrac{V_m t_2}{3600} = \dfrac{150\times600}{3600} = 25$ km

Distance travelled during braking period, $S_3 = \dfrac{V_m t_3}{7200} = \dfrac{150\times25}{7200} = \dfrac{25}{48}$ km

Total distance between stations, $S = S_1 + S_2 + S_3 = \dfrac{25}{48} + 25 + \dfrac{25}{48} = 26.042$ km

Average speed, $V_a = \dfrac{S\times3600}{T} = \dfrac{26.042\times3600}{25+600+25} = 144.23$ kmph

Schedule speed, $V_S = \dfrac{S\times3600}{T + \text{stop line}} = \dfrac{26.042\times3600}{650 + 5\times60} = 98.68$ kmph

❧ **EXAMPLE 6.2**: A schedule speed of 45 km per hour is required between two stops 1.5 km apart. Find the maximum speed over the run if the stop is of 20 second duration. The values of acceleration and retardation are 2.4 kmphps and 3.2 kmphps respectively. Assume a simplified trapezoidal speed–time curve. [BTE UP Electric Traction. 2003]

Solution: Acceleration, $\alpha = 2.4$ kmphps
Retardation, $\beta = 3.2$ kmphps
Distance of run, $S = 1.5$ km
Schedule speed, $V_3 = 45$ kmph

Schedule time, $T_s = \dfrac{s\times3600}{V_s} = \dfrac{1.5\times3600}{45}$

Actual time for run, $T = T_s - \text{stop duration} = 120 - 20 = 100$ seconds
Maximum speed,

$$V_m = \dfrac{T}{2K} - \sqrt{\dfrac{T^2}{4K^2} - \dfrac{3600\,S}{K}} \quad \text{where } K = \dfrac{1}{2\alpha} + \dfrac{1}{2\beta} = \dfrac{1}{2\times2.4} + \dfrac{1}{2\times3.2} = 0.3646$$

$$= \dfrac{100}{2\times0.3646} - \sqrt{\dfrac{100^2}{4\times(0.3646)^2} - \dfrac{3600\times1.5}{0.3646}} = 74\ kmph$$

❧ **EXAMPLE 6.3**: A train is required to run between two stations 1.6 km apart at an average speed of 40 kmph. The run is to be made to a simplified quadrilateral speed–time curve. If the maximum speed is to be limited to 64 kmph, acceleration to 2.0 kmphps and coasting and braking retardation to 0.16 kmphps and 3.2 kmphps respectively, determine the duration of acceleration, coasting and braking periods.

[Gorakhpur Unit), Utilisation of Electrical Power with Traction 1980; Pb. Univ,. Electric and Utilization January 1991]

Solution. Distance of run, $S = 1.6$ km

Average speed, $V_a = 40$ km per hour

Maximum speed, $V_1 = 64$ km per hour

Acceleration, $\alpha = 2.0$ kmphps

Coasting retardation, $\beta_c = 0.16$ kmphps

Braking retardation, $\beta = 3.2$ kmphps

Duration of acceleration, $\dfrac{V_1}{\alpha} = \dfrac{64}{2} = 32$ seconds

Actual time of run, $T = \dfrac{3600\,S}{V_a} = \dfrac{3600 \times 1.6}{40} = 144$ seconds

Let the speed before applying brakes be V_2

then duration of coasting, $t_2 = \dfrac{V_1 - V_2}{\beta_c} = \dfrac{64 - V_2}{0.16}$ seconds

Duration of braking, $t_3 = \dfrac{V_2}{\beta} = \dfrac{V_2}{3.2}$ seconds

Since actual time of run, $T = t_1 + t_2 + t_3$

$$144 = 32 + \frac{64 - V_2}{0.16} + \frac{V_2}{3.2}$$

$$V_2\left(\frac{1}{0.16} - \frac{1}{3.2}\right) = 32 + 400 - 144$$

$$V_2 = \frac{288}{6.25 - 0.315} = 48.5 \text{ km per hour}$$

Duration of coasting, $t_2 = \dfrac{V_m - V_2}{\beta_c} = \dfrac{64 - 48.5}{0.16} = 96.85$ seconds

Duration of braking, $t_3 = \dfrac{V_2}{\beta} = \dfrac{48.5}{3.2}$

⊙ **EXAMPLE 6.4**: The speed–time curve of an electric train on a uniform rising gradient of 1 in 100 comprises:

    i. Uniform acceleration from rest at 2 kmphps for 30 seconds

    ii. Coasting with power off for 70 seconds

    iii. Braking at 3 kmphps to stand still.

The weight of the train is 250 tonnes, the train resistance on level track being 5 kg/tonne, and allowance for rotary inertia 10%.

Calculate the maximum power developed by traction motors and total distance travelled by the train. Assume transmission efficiency as 97%.

[Agra Univ. Traction & Utilisation of Elec. Power 1979]

Solution: Weight of train, W = 250 tonnes

Effective weight of train, $W_e = \left(1 + \dfrac{10}{100}\right)W = 1.1\,W$

Train resistance, r = 5 kg/tonne = 5 × 9.81 = 49.05 N/tonne
Gradient, G = 1 percent
Maximum velocity, $V_m = \alpha t_1 = 2 \times 30 = 60$ km/hr
Tractive effort required, $F_t = 277.8\,W_e\alpha + 98.1\,WG + Wr$
= 277.8 × 1.1 × 250 × 2 + 98.1 × 250 × 1 + 250 × 49.05
= 189577 newtons
Maximum power output from the driving axles

$$= \frac{F_t V_m}{3600} = \frac{189577}{3600} \times 60 = 3160 \text{ kW}$$

Maximum power developed by traction motors

$$= \frac{3160}{0.97} = 3258 \text{ kW}$$

Let the coasting retardation be β_c then
Ft = $277.8W_e\,(-\beta_c)$ + 98.1 WG + Wr
Or   0 = 277.8 × 1.1 W × β_c + 98.1 W × 1 + W × 49.05

Or $\beta_c = \dfrac{98.1 + 49.05}{277.8 \times 1.1} = 0.4815$ kmphps

$V_2 = V_1 - \beta_c t_2 = 60 - 70 \times 0.4815 = 26.3$ km/hour

Braking period, $t_3 = \dfrac{V_2}{\beta} = \dfrac{26.3}{3} = 8.77$ seconds

Total distance travelled by the train,

$$S = \frac{V_1 t_1}{7200} + \frac{(V_1 + V_2)t_2}{7200} + \frac{V_2 t_3}{7200}$$

$$= \frac{60 \times 30}{7200} + \frac{60 + 26.3}{7200} \times 70 + \frac{26.3 \times 8.77}{7200} = 1.12 \text{ km}$$

☻ **EXAMPLE 6.5:** Two 600 V motors each having a resistance of 0.1 are started on the series parallel system, the mean current per motor throughout the starting period being 300 A. The starting period is 15 seconds and the train speed at the end of this period is 29 km per hour. Calculate (i) the rheostatic losses (in kWh) during (a) the series and (b) the parallel combinations of motors (ii) the train speed at which transition from series to parallel must be made.

Solution: Line voltage, V = 600 volts
Current per motor, I = 300 amperes

Starting period, T_s = 15 seconds

Motor resistance, R = 0.1 Ω

Maximum speed, V_m = 29 kmph

Back emf of each motor in full series position,

$$E_{bse} = \frac{V}{2} - IR = \frac{600}{2} - 300 \times 0.1 = 270 \text{ volts}$$

Back emf of each motor in full parallel position,

$$E_{bp} = V - IR = 600 - 300 \times 0.1 = 570 \text{ volts}$$

Assuming acceleration smooth, back emf will be built up at constant rate.

Since motors take 15 seconds to build up 570 volts, therefore, time taken to build up 270 volts emf

i.e.
$$T_{series} = 5 \times \frac{270}{570} = 7.1 \text{ seconds}$$

$$T_{series} = 15 - 7.1 = 7.9 \text{ seconds}$$

(i) (a) Voltage drop in the starting rheostat in series combination at the starting instant

$$= V - 2IR = 600 - 2 \times 300 \times 0.1 = 540 \text{ volts.}$$

which reduces to zero in full series position.

Energy dissipated in starting resistance during series combination

$$= \frac{(V - 2IR) + 0}{2} \times I \times \frac{T_{series}}{3600}$$

$$= \frac{540 + 0}{2} \times 300 \times \frac{7.1}{3600} = 160 \text{ watt hours} = 0.16 \text{ kwh}$$

(b) Voltage drop across the starting resistance in first parallel position is equal to $\frac{v}{2}$, i.e. 300 watts which gradually reduces to zero.

Energy dissipated in starting resistance during parallel combination

$$= \frac{\frac{V}{2} + 0}{2} \times 21 \times \frac{T_{parallel}}{3600}$$

$$= \frac{\frac{600}{2} + 0}{2} \times 2 \times 300 \times \frac{7.9}{3600} = 198 \text{ watt hours} = 0.198 \text{ kwh}$$

ii. Acceleration, $\alpha = \frac{\text{Maximum speed}}{\text{Starting period}} = \frac{V_m}{T_s} = \frac{29}{15} \text{ kmphps}$

Speed at the end of series period = $\alpha T_{se} = \frac{29}{15} \times 7.1 = 13.7 \text{ km/hour}$

☺ **EXAMPLE 6.6**: A motor coach weighing 150 tonnes is equipped with four 600 V motors for series-parallel control. If during series-parallel starting the current per motor is maintained at 300 A.

Calculate:

    a. The durations of the starting period.

    b. The speed of the train at transition.

    c. The rheostatic losses during (i) series (ii) parallel steps and start.

At 300 A 600 V, the tractive effort is 15,000 N per motor and the train speed is 30 kmph. Assume the train is started up a gradient of 1 percent and train resistance is 10 N per tonne. Allow 10 percent for the effect of rotational inertia. Each motor has a resistance of 0.1 Ω.

Solution: Dead weight of train, W = 150 tonnes

Accelerating weight of train, W_e = 1.1 W = 1.1 × 150 = 165 tonnes

Average current per motor, I = 300 A

Maximum speed, V_{max} = 30 kmph

Supply voltage, V = 600 volts

Train resistance, r = 10 N per tonne

Motor resistance, R = 0.1 Ω

Tractive effort, F_t = 4 15,000 = 60,000 N

$F_t = 277.8 + W_e × α + Wr + 98.1\ WG$

    ∴ 60,000 = 277.8 × 165 × α + 150 × 10 + 98.1 × 150 × 1

or $α = \dfrac{43785}{277.8 \times 165}$ = 0.955 kmphps

(a) Duration of starting period, $T_s = \dfrac{V_m}{α} = \dfrac{30}{0.955}$ = 31.4 seconds

    Back emf of each motor in series (2 circuits, each consisting of two motors in series)

$$E_{bse} = \dfrac{V}{2} - IR = \dfrac{600}{2} - 300 \times 0.1 = 270\ V$$

    Back emf of each motor in full parallel combination,

$$E_{bp} = V - IR = 600 - 300 \times 0.1 = 570\ V$$

    Duration for series running,

$$T_{series} = T_S \times \dfrac{E_{bse}}{E_{bp}} = 31.4 \times \dfrac{270}{570} = 14.9\ seconds$$

    Duration for parallel running, $T_{parallel} = T_S - T_{series}$ = 3.14 − 14.9 = 16.5 seconds.

(b) Speed of the train at transition = $α × T_{series}$ = 0.955 × 14.9 = 14.23 $kmph$

(c) (i) Rheostat losses during series operation

$$= \dfrac{(V - 2IR) + 0}{2} \times 2I \times \dfrac{T_{series}}{3600} = \dfrac{(600 - 2 \times 300 \times 0.1) + 0}{2} \times 2 \times 300 \times \dfrac{14.9}{3600}$$

$$= 270 \times 600 \times \dfrac{14.9}{3600} = 670.5\ watt\text{-}hours$$

(ii) Rheostat losses during parallel operation

$$= \frac{\frac{V}{2}+0}{2} \times 4I \times \frac{T_{parallel}}{3600} = \frac{300+0}{2} \times 4 \times 300 \times \frac{16.5}{3600} = 825 \; watt\text{-}hours$$

Rheostat losses during start = 670.5 + 825 = 1,495.5 watt-hours

☻ **EXAMPLE 6.7**: A train has schedule speed of 40 Kmph over a level track, distance between stations being 2 km. Station stopping time is 25 seconds. Assuming braking retardation of 4 kmphps and maximum speed 30% greater than average speed, calculate acceleration required to run the service.

Solution: Schedule time of run

$$= \frac{1}{40} \times 3600 = 90 \; seconds$$

Actual time of run = 90 – 25 = 65 sec.

$$V_a = \frac{1}{65} \times 3600 = 55.38 \; kmph$$

$$V_m = 1.3 \times 55.38 = 72 \; kmph$$

Substituting various values in equation 6.3, we get

$$\frac{1}{\alpha} + \frac{1}{4} = \frac{7200 \times 1}{72 \times 72}(1.3-1) = \frac{5}{6}$$

∴
$$\frac{1}{\alpha} = \frac{5}{6} - \frac{1}{4} = 0.583$$

∴
$$\frac{1}{\alpha} = \frac{5}{6} - \frac{1}{4} = 0.583$$

$$\alpha = \frac{1}{0.583} - 1.71 \; kmphps$$

☻ **EXAMPLE 6.8**: A 300 tonne motor coach having 5 motors , each developing 5000 NM torque during acceleration, starts from rest. If upgradient is 25 in 1000, gear ratio 3, gear transmission efficiency 90%, wheel radius 50 cm, train resistance 60 N/tonne, addition of rotational inertia 10%, calculate time taken to attain speed of 60 kmph.

If line voltage is 3000 VDC and efficiency of motors 80%, find the current taken during notching period.

Solution. Tractive effort is obtained by substituting various values in equation 6.34

$$F = \eta \frac{T}{r_w} \gamma$$

Putting various values in equation 6.20, we get

$$= 0.9 \times \frac{5000 \times 5}{0.5} \times 3 = 135000 \text{ newtons}$$

$$F_t = 277.8 \, W_e \alpha + 9.81 \, WG + Wr$$

$$\therefore \qquad 13500 = 277.8 \, (1.1 \times 300)\alpha + 9.81 \times 300 \times 25 + 300 \times 60$$

$$\therefore \qquad \alpha = \frac{135000 - 91575}{91674} = 0.47 \text{ kmphps}$$

$$t_1 = \frac{60}{0.47} = 126.66 \text{ seconds}$$

Maximum power of motors is obtained by inserting values in equation 6.26

$$P_m = \frac{0.278 \, F_t V_m}{\eta}$$

$$= \frac{0.278 \times 135000 \times 60}{0.8} \text{ watts}$$

$$= 2814.75 \text{ KW}$$

$$\text{Total current drawn} = \frac{2814.75 \times 1000}{3000} = 938.25 \text{ Amp.}$$

$$\text{Current drawn per motor} = \frac{938.25}{5} = 187.65 \text{ Amp.}$$

☙ **EXAMPLE 6.9**: 500 tonne goods train is to be hauled by a locomotive up a gradient of 25% with acceleration of 1 kmphps. Coefficient of adhesion is 20%, track resistance 50 N/tonne and effective rotating masses 10% of dead weight. Find the weight of locomotive and number of axles, if axle load is not to increase beyond 24 tonne.

Solution: Applying equation 6.22, we get

$$F_t = 277.8 \, W_e a + 9.81 \, WG + Wr$$
$$= (277.8 \times 1.1 \times 1 + 9.81 \times 25 + 40) \, W$$
$$= 600.83 \, W \text{ newtons}$$

If weight of locomotive $= W_L$

$$F_t = 600.83 \, (500 + W_L)$$

$$= 3000415 + 600,83 \, W_L \qquad \text{(i)}$$

Maximum value of tractive effort can be obtained by applying:

$$F_t = 9.81 \times 1000 \times 0.2 \times W_L \qquad \text{(ii)}$$

Equating eq (i) and (ii) we get

$$1962 \, W_L = 300415 + 600.83 \, W_L$$

$$W_L = \frac{300415}{1962 - 600.83} = 220.70 \text{ tonnes}$$

$$\text{Number of axles} = \frac{220.70}{24} = 9.19 \approx 7$$

QUESTIONS

1. What various traction systems you know of?
2. Discuss merits and demerits of steam engine drive.
3. What are the advantages and disadvantages of internal combustion engine drive?
4. What are the advantages of electric drive and state the limitations to its use?
5. What are the advantages and disadvantages of diesel electric traction?
6. What are the main characteristics of diesel engine with special reference to its application for traction purposes?
7. What is the scope of application of battery drive?
8. Why tramways are losing ground to other systems of traction?
9. Where will you recommended hybrid drive?
10. What is special advantage of flywheel drive? Will you recommend this for long runs?
11. What are different systems of track electrification?
12. What are the merits and demerits of DC system of track electrification?
13. What are the disadvantages of third rail system of track electrification?
14. What are the advantages of single phase low frequency system of track electrification? What are the factors due to which its wide spread application remains limited?
15. Why 3 phase traction system employing induction motors is now obsolete; give reasons?
16. What is Kando system? Explain the potentialities of this system for its being adopted in future?
17. What are the advantages of composite system of traction employing 25 KV arc. Supply and DC traction motors?
18. What are the disadvantages of 25 KV arc. Traction system?
19. What steps are taken to reduce the interference in the telecommunication circuits?
20. Compare AC and DC systems of traction.
21. How many types of train services railways have to cater for and what are their distinguishing textures?
22. What types of train service correspond to trapezoidal and quadrilateral speed–time curves?
23. What are the factors affecting the schedule speed of a train?
24. What are different types of functions performed by the attractive effort developed by a traction unit?
25. What is the difference between dead weight and accelerating weight of a locomotive?
26. What do you understand by train resistance and on what factors this depends?
27. What precautions will you take to reduce starting friction of a train?
28. Distinguish between the overloading due to excessive hip developed and overloading due to excessive armature current to the motor.
29. What do you understand by the specific energy consumption and what factors affect the same?
30. What is the coefficient of adhesion? How the value of coefficient of adhesion affects the slipping and skidding of the driving wells of traction unit?
31. What are the factors affecting slip of traction unit?
32. What are various supply arrangements of feeding at traction substation?
33. What are the disadvantages of 25 kV AC traction system?
34. Write a short note on negative booster.
35. Compare the DC and AC systems of railway electrification from the point of main line and suburban line railway service.
36. Why we provide neutral section in OHE? What considerations determine the length of neutral section?
37. Explain the working of a remote control center.

38. What are the functions of a DC substation?
39. What is the major equipment of a DC substation?
40. What are different systems of current collection and give their merits and demerits?
41. Explain the construction of OHE employing single catenary and supported by the bracket assembly of swiveling type and explain its superiority ore other methods of construction.
42. What are the advantages of automatic weight tensioning and temperature compensation arrangements in the OHE?
43. Where insulated overlap is used?
44. Explain why neutral section is provided in the OHE in arc? Traction system and not in DC traction system.
45. What precautions are taken in connection with the location of the neutral section?
46. What are relative merits and demerits of employing insulated overlaps and section insulators?
47. What are various types of construction of polygonal OHE and give their scope of application?
48. What precautions are taken to prevent corrosion of underground installations?
49. What are various current collection systems?
50. Describe in details the conductor rail system of feeding traction unit. What precautions are taken in this system at level crossing?
51. What are different current collection gears and the scope of their application?
52. What is the main drawback of pole collector?
53. What are the disadvantages of bow collector?
54. What are different types of pantographs and where they are used?
55. What are the advantages of single ended naively type pantograph over those of conventional diamond type pantograph?
56. What are the characteristics that a traction motor should possess?
57. What types of motor find application in traction work?
58. Discuss the suitability of series motor for traction duties.
59. What are various methods of speed control of series motors and their scope of speed range?
60. Discuss the effect of feeding undulation DC to a traction motor on its performance.
61. Discuss the unsuitability of shunt motor for traction duties.
62. Explain the construction difference between single phase series motor and deck series motor.
63. Discuss the merits and demerits of the induction motor for traction duties. What is your view regarding its adoption for future traction?
64. Explain the principle of linear induction motor and how its tractive effort control is exerted.
65. What are the advantages and disadvantages of linear induction motor as compared to the rotary induction motor?
66. What is the function of a motor starter?
67. What is the essential difference in restricting the starting current of motor in arc and DC traction practice?
68. What is the main advantage of series parallel control of motors over rheostat method of starting and speed control?
69. What are main transition connections from series to parallel operation of traction motors and what are their relative merits?
70. How direction of rotation of a traction motor is reversed?
71. Explain the working principle of Metaldyne control of traction motor. What are merits and demerits of this control?
72. What are the advantages and disadvantages of thermistor control of traction motors?
73. What is multiple unit control and for what applications will you suggest this?
74. What is the function of a circuit breaker employed on the locomotive and give reason why only are blast type breaker is used and not the oil break type?

75. What is the function of a tap change? Explain how it is better method of speed control and starting than that obtained by series resistance method combined with series parallel method?
76. Compare tap changing on the HT and LT side of the transformer.
77. Explain what special precautions are taken in tap changer to avoid burning of main contracts?
78. What are various rectifier connections and give their scope of application?
79. What are various traction motor connections and state their advantages and disadvantages?
80. Explain the function of a reactor when used in series with traction motors?
81. What are the advantages and disadvantages of collective driver in a locomotive?
82. What are the advantages of mono-motor bogie?
83. Explain various systems of transmission of drive bringing out their merits and demerits.
84. What are the requirements which an ideal braking system should possess?
85. Explain vacuum brake system and give the drawbacks this system of braking suffers from.
86. How electropneumatic braking system works and what are its advantages?
87. Explain how hydraulic and eddy current brakes work?
88. Explain how magnetic brake works?
89. What is mechanical regenerative braking and what are its advantages?
90. Discuss the suitability of DC shunt and series machines for regenerative braking.

ELECTROLYTIC PROCESSES

7.1 INTRODUCTION

An **electrolytic process** is basically using of **electrolysis** industrially to purify metals or compounds at a relatively higher purity and lower expense. Electrolytic processes is widely used for the extraction of pure metals from their ores (such as aluminium, zinc, copper, magnesium, sodium, etc.) manufacturing of several chemicals (such as caustic soda, potassium permanganate, hydrogen, oxygen, chlorine, etc), electro-deposition of metal including electroplating, electrotyping electroforming, patching up of worn-out parts in metallurgical, and related industries. Though the various processes mentioned are different in apparent detail, they all are based on the principle of electrolysis.

7.2 ELECTROLYSIS-BASIC PRINCIPLE

Figure 7.1 shows the principle circuit of electrolysis bath or salt bath. Figure 7.2 detail circuit of practical salt bath.

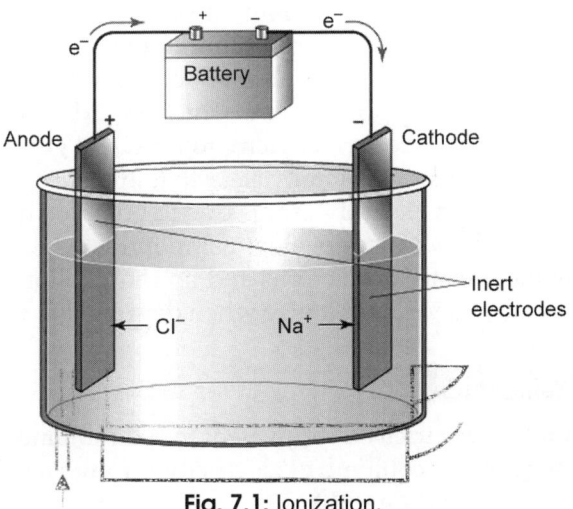

Fig. 7.1: Ionization.

In case of compound formed by electrovalent bond when it is dissolved in water, which has high dielectric constant, results in weakening of the electrostatic force of attraction between the ionized atoms. This makes the charged ions to lead an independent existence. For example, Nancy molecule in water solution will dissociate into constituent atoms of sodium and chlorine, both of these atoms are no longer electrically neutral. Chlorine atom will receive one more electron from sodium and sodium atom in turn will have one electron less as it had given to chlorine. Chlorine atom will, therefore, has net negative charge of one electron and sodium atom has net positive charge equal to that of one electron. Atoms in these charged phases are called ions. If now solution is subjected to electric field, positive ions will be acted upon by force in the direction of electric field and negative ions will be acted upon by the force in opposite direction of the field as shown in Fig. 7.1 Negative ions which are collected at Battery positive electrode often called anode, and give out excess electron to anode and get converted to atoms.

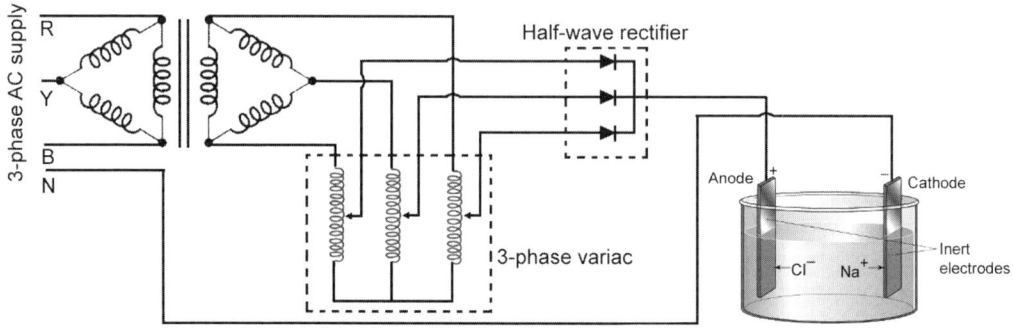

Fig. 7.2: Practical circuit of salt bath.

Similarly positive ions which are collected at negative electrode often called cathode, receive excess of electrons from cathode and get converted to atoms. Number of electrons extracted—to those given to anode from cathode are equal. Total number of electrons in external circuit, therefore, remain same. Thus, the objective of source of supply seems only to serve as an electron pump, pumping electrons from positive side of supply to negative side of supply. Negative ions which are to be collected at anode are called anions. Similarly positive ions which arc to be collected at cathode are called cations, solution which allows electricity to flow and gets dissociated is called electrolyte. From above analysis, it is totally clear that in metals electric current is due to flow of free electrons but in electrolytes it is due to both positive and negative ions flowing in opposite directions. Since covalent compounds which are usually organic are not produced by the transfer of electrons and no charged ions are present in solid state in these compounds, they do not ionize when dissolved in water.

7.3 LAWS OF ELECTROLYSIS

Charge of one coulomb of electricity is equivalent to the combined charge or 6.3×10^{18} electrons. For a passage of one coulomb of electricity through an electrolyte, 6.3×10^{18} number of univalent atoms, $\left(\dfrac{6.3 \times 10^{18}}{3}\right)$ number of divalent atoms, and $\left(\dfrac{6.3 \times 10^{18}}{2}\right)$

number of trivalent atoms are required and so on. Number of atoms taking part in electrolysis for a given amount of electric charge is inversely proportional to the valency. Weight of the element deposited will of course be proportional to the atomic weight of the element.

If Q = Quantity of electricity in coulombs

I = Current in amperes

t = Time in seconds

a = Atomic weight of element

v = Valancy of the element

W = Weight of element deposited

Then $W \propto \dfrac{a}{v} . It$

$\therefore \qquad W = B\dfrac{a}{v} . It$ \hfill (7.1)

where B is a constant. In case of hydrogen, $v = 1$, $a = 1$ and if $It = 1$ coulomb, then $V = W$. Therefore, value of B is the weight of hydrogen deposited per coulomb of electricity which is 0.000010446 gm/coulomb. We can write equation 7.1 in the following form

$$W = \frac{1}{F}\frac{??}{v} . It$$

where F is called Faraday constant, its value being 96487.

Faraday's first law of electrolysis

Equation 7.1 can be written as

$$W = Z.It \qquad (7.2)$$

where

$$Z = B\frac{a}{v}$$

$$Z = 0.000010446\frac{a}{v}$$

$$Z = \frac{1}{96487}\frac{a}{v}$$

Faraday's first law of electrolysis states that the weight of the element deposited is proportional to the quantity of electricity passed through it. In equation 7.2 constant of proportionality Z is called electrochemical equivalent. It is defined as the weight of the element in grams liberated per coulomb of electricity. Value of B is numerically equal to the electrochemical equivalent of hydrogen. Actual quantity of substance deposited is little less than that given by equation 7.2 on account of some quantity of impurities is being liberated. This is taken into account by the use of a factor know as current efficiency. It is defined as the ratio of actual quantity of substance liberated to that which would be liberated theoretically by employing equation 7.2. Its value lies between 90 and 98%.

☞ *Faraday's Second Law of Electrolysis*

For the passage of given quantity of electricity, weights of different elements deposited are proportional to their chemical equivalent, where chemical equivalent is defined as the weight of that substance in grams which will combine or displace one gram of hydrogen or 8 grams of oxygen and is equal to $\left(\dfrac{a}{v}\right)$ Table 7.1 gives values of electrochemical equivalent and chemical equivalent of various elements.

Table.7.1:	Values of atomic weight, chemical equivalent, ECE, current and temperature of solution of various elements				
Elements	Atomic weight	Valency	Chemical equivalent	Electrochemical equivalent kg/ coulomb	Current in amp/sq.m	Tem. of sol.
Aluminium	27.00	3	9	9.4×10^{-8}	-	-
Cadmium	112.41	2	56.2	58.2×10^{-8}	100–150	Cold or warm
Calcium	40.08	2	20.04	20.75×10^{-8}	-	-
Chlorine	35.457	1	35.07	36.74×10^{-8}	-	-
Chromium	52.01	6	8.75	9.1×10^{-8}	1500–2000	35°C
Cupric	63.18	2	31.59	32.9×10^{-8}	250–300	Cold or warm
Cuprous	63.18	1	63.18	65.62×10^{-8}	30–40	50°C
Gold	197	3	65.4	67.68×10^{-8}	100–300	60–80°C
Hydrogen	1.008	1	1	10.45×10^{-8}	-	-
Ferrous	55.9	2	27.95	29.02×10^{-8}	100–200	90°C
Ferric	55.5	3	18.63	$19. \times 10^{-8}$	-	-
Lead	207.21	2	103.2	107.16×10^{-8}	100–200	Cold
Magnesium	24.32	2	11.97	12.43×10^{-8}	-	-
Nickel	58.71	2	29.3	30.43×10^{-8}	100–200	50–60°C
Nitrogen	14.00	3	4.67	4.85×10^{-8}	-	-
Oxygen	16.00	2	7.98	8.296×10^{-8}	-	-
Potassium	39.10	1	39.04	40.54×10^{-8}	-	-
Silver	107.88	1	107.67	111.81×10^{-8}	30–60	Cold
Sodium	22.99	1	23	23.87×10^{-8}	-	-
Tin	118.70	2	59.3	61.4×10^{-8}	100–200	70°C
Zinc	65.38	2	32.45	33.88×10^{-8}	200–300	Cold

7.4 TERMS CONNECTED WITH ELECTROLYTIC PROCESSES

Following terms are used in electrolytic processes:
  i. Current efficiency   ii. Voltage   iii. Energy efficiency
  i. **Current efficiency:** Due to impurities, which cause secondary reactions, the quantity of substance or substances liberated is slightly less than that theoretical value from Faraday's laws. This is taken into account by employing a factor, called the Current efficiency.

The current efficiency is defined as the ratio of the actual quantity of substance liberated or deposited to the theoretical quantity, as calculated from Faraday's laws.

$$\text{i.e., Current efficiency} = \frac{\text{Actual quantity of substance liberated or deposited}}{\text{Theoretical quantity of substance liberated or deposited}}$$

Its value usually lies between 90 and 98%.

In certain cases this efficiency is very low. For example, in chromium plating it is roughly 12 to 15 percent. It is because only 15 percent of the total current passed through the electrolyte consisting of some chromium acid solution, is used in depositing chromium and the rest is wasted in producing oxygen and hydrogen gases, which for the purpose in hand, are useless.

ii. **Voltage:** The voltage required to pass the current through any electrolytic cell is equal to the sum of voltage drops at the resistance of the electrolyte and the voltage drops at electrodes (anode and cathode). From the resistivity of the electrolyte and the X-sectional area and the length of the electric current path, the resistance of electrolyte can be determined. For economic purpose, the electrolytic resistance should be minimum and to achieve it, in many cases special conducting agents are added to the electrolyte. The addition of sulphuric acid to copper sulphate solution in copper plating is an example of this. There is some potential difference between the cathode and the electrolyte and between the electrolyte and anode. This potential difference is a measure of the tendency of the metal to go into the solution and is known as electrode potential. The electrode potential depends upon some conditions (i.e. temperature and concentration) and also upon the nature of the metal and the electrolyte. Under ideal conditions the value of electrode potential for most of the substances lie between 0.5 and 1.0 volt, the total potential difference required to pass the necessary current through electrorlytic cell is about 1 or 2 volts.

iii. **Energy efficiency:** On account of various secondary effects and reactions the substance deposited by a given quantity of electricity is less than that determined theoretically from Faraday's laws. Voltage required is also higher than that determined theoretically. Hence actual energy consumption will be higher than that determined theoretically for depositing a given quantity of the substance.

The ratio of theoretical energy required to the actual energy required for depositing a given quantity of metal is known as energy efficiency.

$$\text{i.e., Energy efficiency} = \frac{\text{Theoretical energy required}}{\text{Actual energy required}}$$

7.5   APPLICATIONS OF ELECTROLYSIS

Following are the major applications of electrolysis.
  i. **Electrodeposition.** This includes:
     a. Electroplating
     b. Electrodeposition of rubber

   c. Electrometallisation

   d. Electrofacing

   e. Electroforming

   f. Electrotyping

  ii. Manufacture of chemicals

 iii. Anodizing

 iv. Electropolishing

  v. Electrocleaning or pickling

 vi. Electroparting or electrostripping

 vii. Electro metallurgy. This includes:

   a. Electroextraction

   b. Electrorefining.

7.6   ELECTRODEPOSITION

We have seen in article 7.2 that compounds in the solution dissociate into positive and negative ions when subjected to electric field and travel towards respective electrodes. One of the following events may then take place:

  i. If an ion, after giving away electric charge to electrode, has stable existence and does not have chemical reaction with electrode material, it will be deposited on the electrode itself. This is the principle of electrodeposition and electro-extraction.

 ii. If ion after giving away electric charge to electrode, has chemical reaction with the material of electrode, the product of reaction in turn is soluble in the electrolyte. In this way the electrode is gradually eaten away. This is the principle employed in electrorefining.

iii. Ion, if after giving away charge to electrode, does not neither react with the material of electrode, nor has any independent and stable existence, will react with the water of solution, thereby liberating oxygen and hydrogen. This is what takes place on the electrolysis of acidulated water.

Factors on which quality of electrodeposition depends

Following are the factors on which the quality of electrodeposition depends:

1. **Nature of electrolyte**. The electrolyte from which complex ions can be obtained (e.g., cyanides) provides a smooth deposit.

2. **Current density**. The deposit of metal will be uniform and fine-grained if the current density is used at a rate higher than that at which the nuclei are formed. The deposit will be strong and porous if the rate of nuclei formation is very high due to very high current density.

3. **Temperature**. A low temperature of the solution favours formation of small crystals of metal; and a high temperature, large crystals.

4. **Conductivity**. The solution of good conductivity provides economy in power consumption and also reduces the tendency to form trees and rough deposits.

5. **Electrolytic concentration**. By increasing the concentration of the electrolyte, higher current density can be achieved, which is necessary to obtain uniform and fine-grain deposit.

6. **Addition agents**. The addition of add or other substances to the electrolyte reduces its resistance. Addition agents like glue, gums, dextrose, dextrins, etc. influence the nature of deposit. The crystal nuclei absorb the addition agent added in the electrolyte; this prevents it to have large growth and thus deposition will be fine-grained.

7. **"Throwing power"**. It is defined as the ability of the electrolyte to produce even irregular surfaces. Due to irregular shape of the cathode the distance between the various portions of the cathode and anode will be different. Due to unequal distance, the resistance of the current path through the electrolyte for various portions of the cathode will be different but the potential difference between the anode and any point on the article to be plate (cathode) will be, of course, be the same and the result will be that the current density will be more on the portion nearer to anode and it will cause uneven deposit of the metal.

Throwing power can be improved by two methods:

1. By increasing the distance between the anode and cathode.
2. By reducing the voltage drop at cathode surface. In some cases decrease of current density causes a decrease in voltage drop at the cathode leaving more voltage available for overcoming the resistance of the electrolyte, thus tending to counteract any changing current concentration.

Copper cyanide bath is therefore better suited for electroplating intricate articles. Zinc cyanide bath has better throwing power than zinc sulphate solution for zinc plating.

8. **Polarisation:** Due to increase in the electroplating current density, rate of metal deposition also gets increased up to a certain limit, but after some time, electrolyte surrounding the base metals becomes so much depleted of metal ions that rate of deposition does not increase with increase in current density. If current density crosses the prescribed limit, it will result in electrolysis of water and hydrogen deposition on the cathode. This hydrogen evolved, blankets the base metal which diminishes the rate of metal deposition. This phenomenon is called polarization. Blanketing effect can be minimized by agitating the electrolyte.

With "reverse current electroplating", in which at regular intervals plating current is reversed for a second so that sufficient electron concentration is established around the base metal and the polarisation effect becomes negligible even with very high overall speed of plating. The other advantages of reverse current plating are:

i. During reverse current period the unsound nod inferior metal is depleted and the fiat level surfaces are produced.

ii. Metal surface is brightened causing elimination of buffing or polishing operation.

7.6.1 Electroplating

Electroplating is defined as electrodeposition of metal upon metallic surfaces.

This is done for

i. Protection of metals against corrosion.

ii. Giving a shiny appearance to articles.

   iii. Giving reflecting properties to reflectors.

   iv. Replacing worn out material.

Various operations involved in electroplating are degreasing and surface cleaning, deposition of metal by passing current and polishing.

7.6.1.1   Cleaning Operation

If electroplating is done on an ungreased and shabby surface, the deposit formed is not well adherent to the base metal and is likely to peel off. Therefore, to reduce percentage of spoilage, proper care is given to cleaning operation. Surface contaminants may be dirt and grease sticking to the surface, oxide and scale formed as a result of hot working or heat treatment of metal, or sand of the mould sticking to the surface.

Depending upon the type of contaminant, cleansing operation can be of sand blasting, hydraulic jet, chemical and electrolytic pickling. Chemical means employed for removing dirt and grease makes use of trichloroethylene. Alternatively, it may employ alkaline cleansers which include caustic soda, sodium carbonate, sodium silicate and tri sodium phosphate. To remove oxide and scale from the surface, acid dip is used. Acid used for iron, steel, brass and copper is dilute sulphuric acid; for aluminium, pickling solution consists of 10% nitric acid, 10% hydrofluoric acid and 80% water. In electrolytic pickling, iron and steel articles are made anode and solution is made up of $\frac{1}{2}$ kgm of each of sulphuric acid and sodium sulphate is 5 kg. of water contained in lead lined vat. Current density of 300 to 400 A/m^2 is employed. Cathode is made up of lead sheets.

7.6.1.2   Electrolytic Bath

The electrolyte used in the electrolytic bath depends upon the nature of the metal to be deposited. Details of preparation of solutions and current densities employed for electrolytic bath of various metals are given as follows:

- For copper plating two types of electrolytic baths are used. In acid type bath, solution is made of 150–200 gm of copper sulphate and 25–35 gm of sulphuric acid per 1000 cc of solution. Current density used is 150–400 A/m^2 and temperature of 25–50°C. Deposit obtained is thick and rough requiring polishing.

- In cyanide bath, solution is made of 25 gm of copper cyanide, 28 gm of sodium cyanide, 6 gm of sodium carbonate, and 6 gm of sodium bisulphate per 1000 cc of solution. Current density used is 50–150 A/m^2 and the temperature required is 25–40°C. It provides thin and smooth deposits. Copper anodes are used in both of the baths.

- For silver plating solution consisting of 24 gm of silver cyanide, 24 gm of potassium carbonate and 36 gm of potassium cyanide per 1000 cc is used. The required current density and temperature are 50–150 A/m^2 and temperature of 20–35°C respectively.

- For gold plating solution used consists of 18 gm of potassium gold cyanide, 12 gm of potassium cyanide, 6 gm of potassium sulphate and 12 gm of caustic potash per 1000 cc. Anode employed is stainless steel. Current density of 50–150 A/m^2 and temperature of 50–70°C are used.

- For nickel plating the solution consists of 180–240 gm of nickel sulphate, 36 gm of nickel chloride and 25 gm of boric acid per 1000 cc. The current density used is 100–200 A/m^2 and temperature of 25–40°C. Anode employed is pure nickel.
- For chromium plating solution consisting of 180–300 gm of chromic acid, and 2–3 gm of sulphuric acid per 1000 cc is used. The required current density and temperature are 1500–2500 A/m^2 and temperature of 35–50°C. Current density used is higher for hard chromium plating than for decorative plating. Anode employed is antimonial lead. Vats used for chromium plating of steel with lead lining. Chromic acid is added in the solution when required. Arrangements for removal of the fumes are also to be provided.

7.6.1.3   Polishing and Baffling

Silver, nickel or chromium plating requires polishing by mops driven at peripheral speed of 2000 to 3000 metres/minute. It is required to run these mops at high speeds. Lower speeds than this will cause the mop to drag and toe article becomes hot and burnishing, rather than polishing. At higher speeds metal flows, from tops to the valleys of the scratches as though it were a thick liquid, under high pressure and temperature, usually developed. Top thin film of surface material which is called Bailey layer, has been found to be a mass of cemented metal powder and debris of mop, hard durable and lustrous.

7.6.2 Electrodeposition of Rubber

Colloidal solution contains charged particles in case of rubber latex, as obtained from tree and as for colloidal solution, there exist very fine negatively charged particles of rubber. When the solution is subjected to electrolysis, rubber particles are deposited at anode. Current density of 100 A/m^2 is employed.

7.6.3 Electrometallisation

It is a process of depositing metal on conducting base for decoration and for protective purposes. Nonconductive base is made conductive by a coating of graphite which is made the cathode.

7.6.4 Electrofacing

It is a process of coating of metallic surface with a harder metal by electrodeposition to make it more durable.

7.6.5 Electroforming

It is the production or reproduction of objects by electrode position. First step is to prepare a mould of wax using the object to be reproduced as pattern. The surface of the wax which bears exact impressions of the object, is made conductive by giving graphite coating. Thick coating of copper is then deposited on this surface. This copper shell is then stripped off from the mould and filled with low melting point tin lead alloy to give it strength and rigidity.

Production of gramophone records is another instance of electroforming. Original record which carries sound track is made of wax. Negative metallic records which are used as dies for moulding, are made in the same way as described above. These negative record dies are the used as moulds in the mass production of ebonite material recorder's. These dies are sometimes chromium plated (electrofacing) to give it long life.

Production of seamless tubes is another application of electroforming. Revolving mandrel is made the cathode. For producing copper tubes, copper sulphate acid solution is used, cast copper being made as anode. For iron, ferric chloride ($FeCl_2.4$ H_2O) solution with cast iron anodes is used. Mandrel is coated thinly with material which renders easy withdrawal of tube of material deposited.

Copper and gold foils are deposited on cathode belt, moving slowly through the solution. Copper and gold deposition is removed as belt emerges from the vat. In case of gold foil, cathode belt is made of silver.

7.6.6 Electrotyping

This is a special application of electroforming where reproduction of printer's type and engravings is made. Whole process is same as electroforming.

7.7 MANUFACTURE OF CHEMICALS

Various industrial applications of electrolysis in the production of chemicals are listed in Table 7.2. Out of these, production of caustic soda by electrolysis of brine has commercial value. On electrolysis of brine, chlorine ions move towards anode where they are collected as chlorine gas. Sodium ion on the other hand travels towards cathode where it reacts with water producing caustic soda, liberating hydrogen. The brine solution becomes richer in caustic soda. The by-products—chlorine and hydrogen, in no way, are less important. These by-products are further used to produce hydrochloric acid. If chlorine and hydrogen evolved are not separated, they produce perchlorates, chlorates and hypochlorates.

Table 7.2	Manufacture of Chemicals	
Product	*Solution*	*Consumption of electricity*
Caustic soda	Brine	3000–4000 kwhrs/tonne
Chlorin, compounds of sodium	Brine	3000–7000 kwhrs/tonne
Potassium permanganate	Potassium manganate	70–80 kwhrs/tonne
Ammonium persulphate	Ammonium sulphate	2000–2500 kwhrs/tonne
Hydrogen and oxygen	Water	45–60 kwhrs/one cubic metre of hydrogen and $\frac{1}{2}$ cubic metre of oxygen.

7.8 ANODIZING

Many metals, when exposed to atmospheric conditions get corroded, the product of corrosion being oxides of the metal. Whether or not process of corrosion will be continuous, depends upon the nature and texture of the first film of oxide formed and its solubility with surroundings. For instance, if the film is rapidly dissolved by the surrounding medium, fresh metal will be exposed to corrosive atmosphere. This will continue till the metal has been eaten away totally. If film is cracked or is discontinuous, severe local corrosion may occur. If film formed is tough, thin, uniform and impermeable, it will protect the base metal from further corrosion. Such protective films are called passive films.

Study in the process of corrosion revealed that it is produced by the surface electrolytic currents. Due to impurities in the metal, large number of thermocouples are produced in the metal. It is this thermocouple emf which seems responsible for the production of surface electrolytic currents. Taking this clue, question arises as to why not produce a thin uniform passive film artificially by the passage of electric current. This is actually what is attempted in the process of anodic oxidation which is often called anodizing.

The process of anodizing consists of passing electric current of 10 to 30 A/m^2 through an electrolyte of 3% solution of chromic acid for a period of 30 to 60 minutes. The article to be anodized is thoroughly cleaned and made the anode, carbon rod is made the cathode. Aluminium and magnesium are capable of producing such passive films and, therefore, anodizing is applicable to them. In case of aluminium great feature of these coatings is their sub microscopic porosity. This enables the passive films to take dyes which are absorbed and affect the entire depth of the film.

7.9 ELECTROPOLISHING

Electropolishing has one advantage over buffing. Unlike buffing electropolished surfaces are free from cold worked surface cemented layer embedding abrasive particles of mop in the metal. In principle, process of polishing consists of making the work as anode in a suitable solution. This produces insoluble compounds. These compounds are broken down by more anodic action on the hills than on valleys of the surface. In this way smoothening of surface results. Aluminium work piece which is to be polished after being properly buffed is subjected to two anodic treatments in series. In the first treatment job is made anode in the bath of flu boric acid. This removes thin coating of metal from the surface uniformly. This surface is then given anodizing treatment in sulphuric acid bath as described in article 7.8. This process produces an oxide film clear and transparent with reflectivity as much as 90%.

Polishing in case of silver plated article is done after plating is over by reversing the current through bath at 4 to 5 times the current strength used for plating and at intermittent intervals every few seconds.

7.10 ELECTROCLEANING OR PICKLING

Electrolytic solution of sodium phosphate is contained in iron tank which is made anode. Work is suspended as cathode. Passage of current produces caustic soda on cathode which has cleaning action. Also large volume of hydrogen evolved at cathode quickly removes grease. This is called catholic cleansing and is applicable to zinc and aluminium. In anodic cleansing, work is made anode. Dirt particles are positively charged in alkaline bath and by electrophoresis process they migrate to negative pole.

7.11 ELECTROPARTING OR STRIPPING

Separation of two or more metals is done electrically. For instance if copper is to be taken off from steel, the work piece has to be made anode in a solution of 75 gm of sodium cyanide, 25 gm of caustic soda in 1000 cc of water. Sheet of iron is made cathode and pressure used is 6 volts.

7.12 ELECTROEXTRACTION

There are two types of processes adopted for electroextraction of metals.

i. In one of the process, electrolyte is the solution of the salt of metal obtained by treating the ore with acid.

ii. In second process, electrolyte is ore in the molten state.

The methods adopted for extracting zinc and aluminium is explained below:

1. **Extraction of zinc:** This is an example in which an aqueous solution of the salt is used. The ore, consisting mainly of zinc oxide, is treated with concentrated sulphuric acid, roasted, and passed through various chemical processes in order to remove impurities (such as cadmium, copper, etc.) by precipitation. The zinc sulphate solution, so obtained is then electrolysed. The electrolysis of zinc sulphate is performed in large lead-lined wooden boxes having a number of aluminium cathodes and lead anodes. Zinc gets deposited on the cathodes and is removed periodically (once or twice a day). The current density on the cathodes is about 1,000 A/m². The voltage per cell is about 3.5 V and usually 100 to 150 cells are used in series requiring a pressure of roughly 500 V. Energy consumption is 3,000 to 5,000 kWh/tonne.

2. **Extraction of aluminium**: This is an example of fused electrolyte process. Aluminium is produced from bauxite containing aluminium oxide or alumina (70% in case of high grade bauxite), silica (silicon oxide) and iron oxide. The bauxite ore is first reduced to aluminium oxide by chemical treatment and it is dissolved in fused cryolite. Cryolite is a solution of aluminium fluoride and fluoride of either of sodium, potassium or calcium. The mixture thus obtained is electrolysed. The fusion and electrolysis are accomplished in a large shallow rectangular steel bath lined with carbon; carbon anodes projecting downwards into the bath and the bottom of the bath forms the cathode. The charge is melted by the arc struck between the carbon anodes and cathode and is then maintained in a molten state by the heating action of the electric current flowing through the charge. The liquid metal deposits at the cathode and settles at the bath bottom and is periodically siphoned out into large ladles from which it is poured into pig or ingot moulds. Fresh alumina is fed into the bath at short intervals to replace that which has been decomposed by the current and the process is, therefore, a continuous one. The aluminium obtained by this process is 99.5% pure. A furnace having an area of about 15 square meters will need a voltage of about 6 volts and a current of about 40000 amps. Energy consumption is 20,000–25,000 units/tone. Almost the whole of the aluminium requires in the present days industry is produced in this way. As the electrolytic process requires large amount of electric power and process is continuous, so such plants are installed near hydroelectric power stations. The high temperature (1000°C)

necessary to keep the ores in a fused state is maintained by the ohmic losses due to the current flowing through the electrodes and electrolyte.

7.13 ELECTROREFINING

By electrorefining, it is possible to get metal of almost 100% purity. This is one of the most important prerequisite expected out of copper and aluminium in order to have high electrical conductivity. Electrorefining process in essence is same as electroplating, anode being made of impure metal and electrolyte being made of the salt of the metal to be refined. Pure metal is deposited at cathode.

- Copper refining requires copper sulphate solution with 150 to 300 kWhrs/1000 kgs. of copper refined.
- Silver is refined requiring solution of nitric acid and silver nitrate with electric consumption of 400 to 420 kWhr/1000 kgm.
- Solution of iron ammonium sulphate is required for refining iron with electric consumption of 1000 to 1500 kWhrs/1000 kgm.

7.14 POWER SUPPLY FOR ELECTROLYTIC PROCESSES

Power supply required for electrolytic processes is direct current and at very low voltage. The power required for electrodeposition is usually very small (between 100 and 200 amperes at 10 or 12 volts) and can be obtained either by employing a motor-generator set consisting of a standard induction motor driving a heavy-current low-voltage DC generator (preferably separately excited) or by employing the copper oxide rectifier. The latter is preferred because of low maintenance cost, occupying less space and higher operating efficiency. Mercury arc rectifier cannot be used because it has low efficiency at low output DC voltage on account of constant voltage drops at cathode and anodes. The plate rectifier unit is usually placed along with its transformer in the oil so that it may be protected from the corrosive fumes of the electrolyte. Recently the solid state rectifyng devices employing germanium and silicon diodes have been developed for use. These solid state devices occupy very small space even as compared to metal rectifiers. Output DC voltage can be controlled by controlling the excitation of the generator in case of motor-generator set and by means of continuously variable autotransformer in case of rectifier supply. This method of control is suitable where only one bath is being supplied. In case, more than one bath are supplied, a variable resistance is connected in series with each bath so that the supply to each bath can be controlled independently. With the development of SCRs or thyristors, which can control output voltages of power supplies, their use in power supplies for electro-chemical processes has increased. They are also compact and light in weight, even cooling attachments are included. Voltage control is by using output transformers.

Power supply required for extraction and refining of metals and large scale manufacture of chemicals is in very large amounts. Since most of the processes are continuous, therefore, have a load factor of 100 percent. Because of power requirements in huge amount and at 100 percent load factor, such plants are located near the hydro-electric power stations or atomic power stations even if extra transportation of raw material is necessitated. The advantage of a high load factor is greater with such stations than with steam stations and also transmission costs are eliminated. Nangal Fertilizer Factory producing calcium ammonium nitrate and heavy water and utilizing power of

180 MW from left bank Bhakra Power House and Sri Ram Fertilizer Factory located at Kota (Rajasthan) are instances in the support of the above statement.

The voltage of each cell is about 10 volts, but if many cells are connected in series, current of the order of several thousand amperes will be required at the voltage of the order of 500 to 800 volts. Thus by employing heavy current motor-generators, rotary convertors or even mercury arc rectifiers, the required supply may be obtained from the modern grid network.

WORKED EXAMPLES

EXAMPLE 7.1: If 22.092 g of nickel is deposited by 110 A current flowing for 11 minutes, how much copper would be deposited by 55 A current in 7 minutes? Atomic weights of nickel and copper are 58.6 and 63.18 respectively and valency of both is 2.

Solution. Given : Nickel : m_{Ni} = 22.092 × 10^{-3} kg, I = 110 A, t = 11 minutes;
Atomic wt. = 58.6 Valency = 2
Copper: m_{cu} =? I = 55 A, t = 7 minutes, Atomic wt. = 63.18, Valency = 2.

$$\text{ECE of nickel, } Z_{Ni} = \frac{m}{I \times t} = \frac{22.092 \times 10^{-3}}{110 \times (11 \times 60)} = 30.43 \times 10^{-8} \frac{kg}{C}$$

$$\text{ECE of copper, } Z_{cu} = Z_{Ni} = \frac{\text{Chemical equivalent of copper}}{\text{Chemical equivalent of nickel}}$$

Mass of copper deposited, $m_{cu} = Z_{cu} \times I \times t$
= 32.81 × 10^{-8} × 55 × (7 × 60) = 7.58 × 10^{-3} kg = 7.58 g.

EXAMPLE 7.2: Calculate the quantity of aluminium produced from aluminium oxide in 24 hours if the average current is 2800 A and current efficiency is 95 percent. Aluminium is trivalent and atomic weight is 27. The chemical equivalent weight and ECE of silver are 107.98 and 111 × 10^{-8} kg / C respectively.

Solution. Average current   = 2800 A
Current efficiency   = 95%
Valency of aluminium   = 3
Atomic weight of aluminium   = 27
Chemical equivalent weight of silver   = 107.98
∴ ECE of silver   = 111 × 10^{-8} kg/C

$$\text{Chemical equivalent weight of aluminium} = \frac{\text{Atomic weight}}{\text{Valency}} = \frac{27}{3} = 9$$

∴ ECE of aluminium,

$$Z = \frac{\text{ECE of silver} \times \text{chemical equivalent weight of aluminium}}{\text{Chemical equivalent weight of silver}}$$

$$= \frac{111 \times 10^{-8} \times 9}{107.98} = 0.252 \times 10^{-8} \text{ kg} / C$$

Mass of aluminium produced, m = Z It × current efficiency

= 9.252 × 10^{-8} × 2800 × (24 × 60 × 60) × 0.95

= 21.26 kg.

QUESTIONS

1. Discuss the part played by the electron in the outer orbit of an atom in promoting chemical bond and electrical conductivity.
2. Explain the nature of electric current flow through metals and electrolytes.
3. Explain electrovalent and covalent bonds. Why compounds formed by covalent bonds are not ionisable?
4. Discuss various laws of electrolysis.
5. What are different applications of electrolysis?
6. Discuss the advantages of reverse current process of electroplating.
7. What is polarisation and how its bad effects on electrodeposition process can be reduced?
8. Why positive side of DC supply meant for electroplating is earthed?
9. What are special constructional features of electroplating dynamos?
10. Why it is preferred to supply large chemical works from hydroelectric or atomic power stations?

REFRIGERATION AND AIR CONDITIONING

Refrigeration and air conditioning have become a need in every sector of our civilization and is almost irreplaceable in industrial set ups whether it is in houses, school, colleges, departmental stores, cinema houses, restaurants, cars, hospitals, power houses, etc. In the early phase of twentieth century it was considered to be a luxury but with the change in the environments and standards of leaving it has become a necessity. Some fortunate souls take birth in air conditioned delivery rooms and equally unfortunate souls in incubators which are meant for premature born babies. Refrigeration is even applied after death. In between these applications at two extreme ends of human life, there are various applications which require brief account.

8.2 BASIC TERMINOLOGY

The important terms used in refrigeration are defined below:

1. **Refrigeration:** Refrigeration means cooling or simply removal of heat from a substance under controlled condition.

2. **Refrigeration system:** The equipment employed to maintain the system at a low temperature is called refrigeration system.

3. **Refrigerator:** Theoretically, a refrigerator is a reversed heat engine or a heat pump which takes out heat from a cold body and delivers it to a hot body.

4. **Refrigerant:** The refrigerant is basically a heat conducting medium which during their cycle (i.e., compression, condensation, expansion, evaporation) absorbs heat from a low temperature system and delivers it to a higher temperature system (hot body).

5. **Tonne of refrigeration (TR):** It is a practical unit of refrigeration and is defined as the amount of refrigeration effect produced by the uniform melting of one tonne (1000 kg) of ice from and in 24 hours. It is equivalent to heat extraction rate of 50 kcal/minute.

8.3   APPLICATIONS OF REFRIGERATION

Broadly speaking, refrigeration applications can be classified into comfort refrigeration, food preservation refrigeration and industrial refrigeration. Important refrigeration applications are given below:

1. Ice making
2. Transportation of foods above and below freezing
3. Industrial air conditioning
4. Comfort air conditioning
5. Chemical and related industries
6. Medical and surgical aids
7. Processing food products and beverages
8. Oil refining and synthetic rubber manufacturing
9. Manufacturing and treatment of materials
10. Freezing food products
11. Plumbing
12. Building constructions.

8.4   REFRIGERATION SYSTEMS

There are three refrigeration systems:

1. Vapour compression refrigeration system.
2. Vapour absorption refrigeration system.
3. Thermoelectric refrigeration system.

8.4.1 Vapour Compression Refrigeration System

When a volatile matter evaporates, it absorbs the latent heat from its surrounding, e.g. when we take petrol or ether in hand we can feel the cold. A pure compression refrigeration system runs on the principle of absorbing latent heat of evaporation required by the refrigerant form the surrounding which is required for cooling purposes. The schematic diagram of Vapour compression refrigeration system is shown in Fig. 8.1.

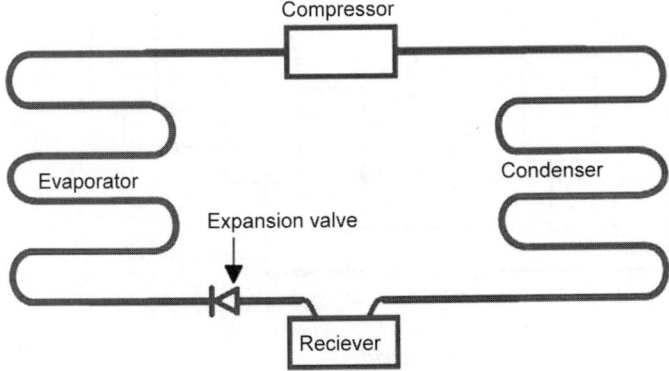

Fig. 8.1: Vapour compression refrigeration system.

Compressor takes out refrigerant vapour from evaporator as soon as it is formed and we can maintain a low pressure in the evaporator. Saturation temperature of refrigerant corresponding to low pressure will be low as compared to that surrounding the evaporator. Latent heat of evaporation of the refrigerant will, therefore, be extracted from the surroundings. Compressor delivers the refrigerant vapour to the condenser under high pressure. On compression, vapour gets heated up so that its temperature is higher than the surroundings of condenser coil. Latent heat of evaporation is extracted from the heated vapour by the cooling medium in the condenser. This converts refrigerant vapour into liquid which comes to the receiver. From receiver, refrigerant liquid goes to expansion valve which regulates the rate of flow of refrigerant from high pressure to low pressure. It is, therefore, observed that system works as a heat pump, extracting heat at low temperature from evaporator and delivering the same at high temperature to condenser. According to the laws of thermodynamics, this can be done only on the expenditure of energy which is supplied to the system in the form of electric energy which is supplied to the system in the form of electric energy driving the compressor. It may be noted that compressor and expansion valve constitute the two dividing points between high and low pressure sides in the system.

8.4.2 Absorption Refrigeration System

The vapour absorption refrigeration system is one of the oldest methods of producing refrigerating effect. Main advantage of this system of refrigeration is the absence of moving parts and universal power source which may be a gas burner or an electric heater. It works on the principle that refrigerant which is ammonia or lithium bromide, is given off from strong solution at high pressure when it is heated and gets absorbed in cold and weak solution. Different elements employed in this system are shown

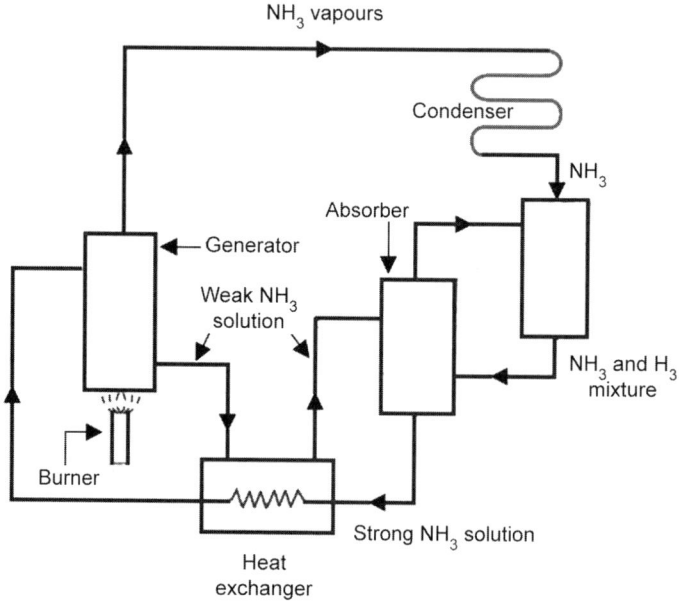

Fig. 8.2: Vapour absorption refrigeration system.

in Fig. 8.2. It essentially consists of an absorber, a pump, a generator and a pressure reducing valve, in addition to condenser, receiver, expansion valve and evaporator, as in the vapour absorption refrigeration system. The refrigeration cycle work as follows.

On heating the generator, containing strong solution of ammonia and water, it evolves ammonia gas at high pressure. This gas as usual goes to condenser where it is liquefied. Liquid ammonia is passed to the expansion valve through the receiver and then goes to evaporator by gravity which contains hydrogen say at 12Kg/sq. cm. Whole system is charged to 15Kg/sq. cm pressure. Thus according to Dalton's law of partial pressure, liquid ammonia evaporates at 3Kg/sq. cm pressure and (–). Latent heat of evaporation is abstracted from the surroundings of evaporator. Mixture of NH_3 and H_2 then goes to absorber. Here NH_3 is absorbed in water and H_2 rises and goes back to evaporator. The function of H_2 is to reduce the partial pressure of NH_3 while maintaining the whole system at high pressure. Strong solution of NH_3 flows to generator via counter flow type heat exchanger. Hot and weak solution from generator flows back to absorber giving its heat in the exchanger to strong solution of ammonia. This completes the cycle.

8.4.3 Thermoelectric Refrigeration System

This system of refrigeration works on the principle of pettier effect. If direct current is passed through a junction of two dissimilar metals like antimony and bismuth, one junction heated up and other junction is cooled. As shown in Fig. 8.3, it consists of a number of thermoelectric module assemblies in series joined by copper strips. Each thermoelectric module is built-up of a large number of thermoelectric couples. Junctions at the top of module assembly, in Fig. 8.3, are cooled and those at bottom heated up. Top junctions, therefore, abstract heat from the surroundings thereby producing refrigerating effect. Bottom junctions of module require cooling by water. Module assembly, therefore, downs the job of abstracting heat from the medium at top and rejecting the same to the medium at bottom. By reversing the current flow, heat production at the junctions is reversed, i.e. now top of the modules will be heated and bottom cooled. If it is an air conditioner, air will be heated up instead of being cooled. Thus summer and winter operating of air conditioner is easily obtained just by reversing the current flow. Main advantages of this system of refrigeration are absolutely no moving parts and ease of automatic control by varying the magnitude of current. Only drawback is its prohibitive initial cost.

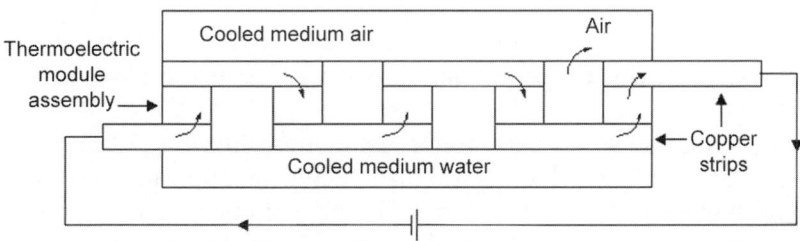

Fig. 8.3: Thermoelectric refrigeration system.

8.5 COEFFICIENT OF PERFORMANCE

The performance of refrigeration system is expressed by a term known as the co-efficient of performance which is defined as the ratio of heat absorbed by the refrigerant while passing through the evaporator to the work input required to compress the refrigerant in the compressor; in short it is the ratio between heat extracted and work done (in heat units).

If R_n = net refrigerating effect, and W = work expanded in by the machine during the same interval of time, then,

$$= \frac{111 \times 10^{-8} \times 9}{107.98} = 0.252 \times 10^{-8} \text{ kg/C}$$

where actual COP = ratio of R_n and W actually measured during a test and theoretical COP = ratio of theoretical values of R_n and W obtained applying laws of thermodynamics to the refrigeration systems.

8.6 UNIT OF REFRIGERATION

The unit of refrigeration is 'Ton' it is the rate of cooling obtained by melting 1 tonne or (2000 lbs) or ice at 0°C in 24 hours. It is equivalent to heat extraction rate of 50 Kcal/min. While large commercial refrigeration plants are designated in tonnes, domestic refrigerators are designated by the capacity (volume) of the cabinet. One tonne capacity of the refrigeration plant requires approximately one HP motor.

8.7 REFRIGERANTS

For safe, economical and thermodynamically efficient production of cold, refrigerants should be possessed as may as practicable of the following desirable properties:

1. It should be non-corrosive
2. It should be non-inflammable
3. It should be non-toxic
4. It should be free from objectionable odour
5. Working pressures should not be high
6. Low boiling point
7. High latent heat of vapourisation
8. Low specific heat of liquid
9. Low cost
10. High critical temperature.

There is no one refrigerant which can be used for all types of application, i.e., there is no ideal refrigerant. It is one refrigerant has certain advantages; it will has some drawbacks also. Hence a refrigerant has chosen which has more advantages and less disadvantages.

Various refrigerants are freon, ammonia, methyl chloride, SO_2 and CO_2 freon (F) is trade name given to gas which is mixture of chlorine, fluorine and carbon. F_{12}, F_{22} and F_{114} are extensively used for domestic and commercial refrigeration. NH_3 is used for commercial and industrial refrigeration for ice manufacture, cold storages, etc.

8.8 DOMESTIC REFRIGERATOR

Domestic refrigerator employing vapour compression system of refrigeration described in § 8.4.1 is shown in Fig 8.4. Here condenser coil, provided with metallic fins to augment the cooling surface, employs natural air cooling. Instead of expansion value domestic refrigerator employs capillary tubing, about 15 inches long and of 0.03 to 0.07 inches in diameter. The resistance offered by the tubing results in pressure drop. As a result of this, refrigerant after capillary expands and extracts its heat from the evaporator coils wound round freezer compartment. Main advantage of capillary tubing over expansion value is the absence of moving parts and hence it is very much reliable in operation. Because of limited range, capillary tubing is used in small size of domestic refrigerators, air conditioners and water coolers. Capillary tubing and condenser coil (in black) both are located outside and at the back of refrigerator. While natural air cooling is adopted for condenser cooling in domestic refrigerators, in larger units either fan is provided for circulating air or city water is made use of depending upon the design. In order to provide sufficient ventilation to the condenser coils and sealed unit, refrigerator should invariably be placed about 20 to 25 cm away from the wall. Also accumulated dust be blown off periodically. Another specialty about domestic refrigerator is that compressor and motor are assembled in a single unit and is called "hermetically sealed unit". This is installed at the bottom of refrigerator cabinet. Main advantages of hermetically sealed unit are that. The problem of gas leakage is

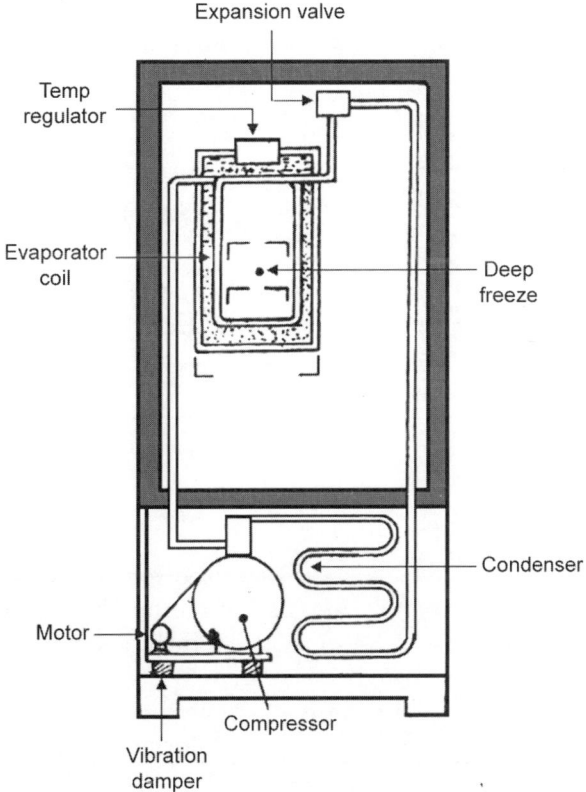

Fig. 8.4: Domestic refrigerator.

minimised as no moving part extends though the sealed housing, it has relatively silent operation and it does not need any maintenance. In open type compressor, the motor drives the compressor using pulley and belt arrangement. This is used in large commercial refrigerators.

Presence of even small traces of moisture in refrigerant can choke-up the capillary and adversely after the cooling. This is avoided by installing dehydrator containing silica gel in the refrigerant piping. In order to prevent leakage of atmospheric heat inside the refrigerator, it is provided with double walled cabinet-packed with materials having high thermal insulation such as fiber glass, cork or expanded rubber. All around the inside of door flap, soft rubber seal is used which makes the cabinet air tight. Also the door is provided with automatic closing mechanism. Door hinges are provided in such a way that door flap when left in open position automatically comes to closing positions due to gravity and as it approaches closing position, magnetic strip fitted behind the sealing rubber ring is attracted to the cabinet closing the door with snap action. If hot thins are kept in refrigerator, it will quickly evaporate the refrigerant in evaporate coils producing lot of vapour pressure, this will increase the duty of compressor and as such motor has to work for longer time to bring down the temperature in the freezer compartment. The motor may get damaged as it is shorts time rated. Also moisture released from hot foods will freeze around the freezer compartment which has to be defrosted.

Electrical circuit of a refrigerator is shown in Fig. 8.5. When we open the refrigerator, door push switch closed and refrigerator lamp gives light. Temperature inside the refrigerator is to be maintained between to and inside freezer to. This is made possible by the use of thermostat switch shown in Fig. 1.22 and described in § 1.7.4. The thermal phial is clamped to the evaporator. Vapour pressure is developed

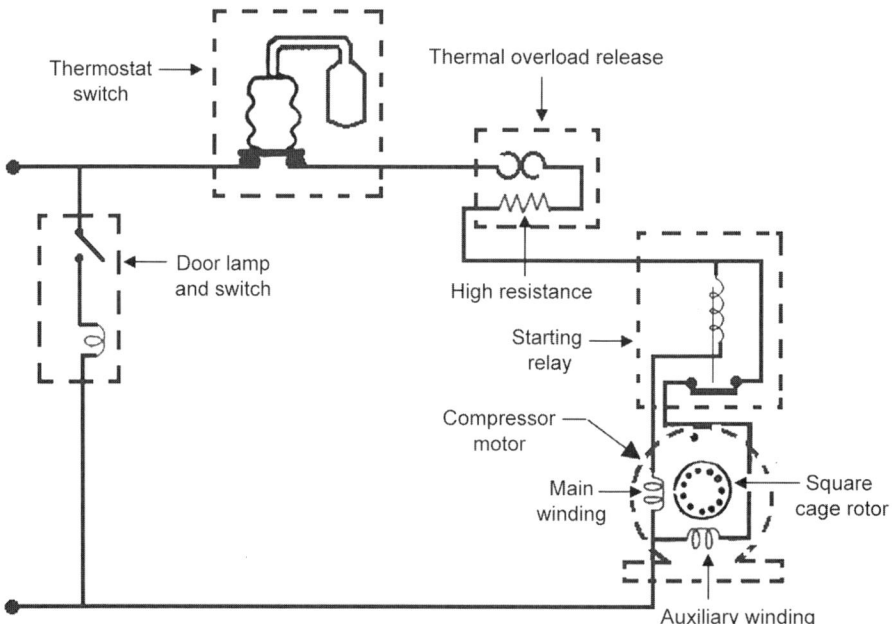

Fig. 8.5: Electrical circuit of a refrigerator.

depending upon the temperature of the freezer. The bellows expert pressure on the lever against the action of adjustable spring. Temperature inside the refrigerator can be varied by means of temperature control screw. Greater the distance between contracts, greater will be the temperature and vice versa. If thermostat is taken out and kept in the room, the motor will run continuously because the temperature outside is more and the temperature inside will go no decreasing. Refrigerator imposes heavy duty on the motor which has to develop high starting torque. Single phase motors of capacitor start type are used in open type refrigerators. In sealed unit refrigerator split phase motor is used. When under the action of thermostat switch, motor is connected to line, heavy current flows through the main winding which operates the current relay and connects the auxiliary winding of motor also to the supply. As motor start and initial in such of current subsides the ready drops out and auxiliary winding gets disconnected. Thermal overload relay is provided to protect the motor from damage. If starting relay fails to close, motor may fail to start and if it fails to open then it may result in blown fuse or tripping out of overload release or failure of capacitor if any or auxiliary winding. Torque developed by all induction motors including refrigerator motor is proportional to square of applied voltage. In case of less applied voltage. In case of less applied voltage, motor will drew heavy current to develop the required torque and will become hot. Thermal overload relay will repeatedly disconnect and connect the motor to supply, eventually burning it out. It is therefore necessary to supply refrigerator through automatic voltage regulator.

8.8.1 Troubleshooting of Refrigerator

Various common defects, cause and remedies of refrigerator are listed below:

	Defect	Cause	Remedy
1.	Motor does not start on giving supply	i. Power not reaching the motor due to (a) blown fuse (b) thermostat contacts open (c) overload release open	i. (a) Check for blown fuses in the main switch (b) thermostat wiring may be defective or improperly adjusted (c) relay may be defective
		ii. Motor defective	ii. Main winding may be open or shorted or grounded
2.	Motor runs hot	i. Overload	i. Reduce load
		ii. Low voltage	ii. Install automatic voltage regulator
		iii. Insulation failure of motor winding developing short circuit between turns or with ground	iii. Replace or repair the sealed unit
		iv. Bearings worm out	iv. – do –
3.	Motor does not start and gives humming noise	i. Either relay contacts being not closed or auxiliary winding being open	i. Replace relay or replace or repair sealed unit
		ii. Low voltage	ii. Install automatic voltage regulator

Contd.

Defect	Cause	Remedy
	iii. Motor overloaded accompanied with tripping of overload relay	iii. Choke in the refrigerant circuit which requires purging
4. Motor slow	i. Low voltage	i. Correct the voltage
	ii. Overload of motor	ii. Check for choke of the refrigerant piping
5. Motor keeps on running even through it is very cold inside the refrigerator	i. Defective thermostat due to short in its wiring or sticking contacts	i. Check and repair
	ii. Wrong setting of thermostat	ii. Set correctly

6. Motors keeps on running but
 	
  i. Cooling is insufficient.
	i. Less refrigerant gas in the system due to
	  a. leakage of gas or
	  b. partial choking of capillary (indicsated by hot capillary up to point of choke and cold beyond it)
		i.a. Replenish the gas lost due to leakage
		  b. Purge the system and recharge it

ii. Cooling as nil	ii. (a) No refrigerant in the system (indicated by the unit, tubing and condenser being at ambient temperature)	ii. a. Replace gas
	b. Complete choke of capillary.	b. Purge the system and recharge
7. Motor starts at short intervals otherwise cooling is good	i. Improper setting of thermostat	i. Read just the switch
	ii. Bad door seal	ii. Replace the door seal
8. Motor runs normal with normal cooling in freeze but cooling in the rest of compartment is not satisfactory.	i. Defective door seal or warped door or opening of refrigerator door too frequently	i. Replace or repair door seal or door and reduce the frequency of opening or the refrigerator door
9. Too much frosting around the freezer	i. High atmospheric humidity or steaming hot liquids stored	i. Avoid keeping hot liquids and defrost frequently
10. Motor works normal with good cooling but defrosting starts all of a sudden. Cooling again starts after some time	i. Presence of moisture in the refrigerant cycle	i. Install drier on hip side or if already installed replace silica gel
11. Noisy operation	i. Bearings faulty	i. Repair sealed unit
	ii. Loose mounting bolts	ii. Tighten the bolts

8.9 WATER COOLER

Water is one of the most needed things for a person. In summer season cold water gives life to a thirsty person. At water is most refreshing. Thus cooling of water in summer season becomes necessary. Water coolers are used to produce cold water at about to. The temperature of water is controlled with the help of thermostatic switch.

Water coolers may be classified as follows:
1. Instantaneous type water coolers:
    i. Bottle type cooler
    ii. Pressure type cooler
    iii. Self controlled or remote type cooler
2. Storage type water cooler

8.9.1 Instantaneous Type Water Coolers

In instantaneous type of water cooler, the cooling coil is wrapped round the pipe line such that by the time water reaches the tank it is cooled to desired temperature. The description of various types of instantaneous type water cooler is given below:

8.9.1.1 Bottle Type Water Cooler

In this type of water cooler, water to be cooled is stored in a bottle or reservoir. For filling glass tumblers or containers faucet or similar means are provided. The dripping water from the faucet is collected in the waste water basin or water drip as shown in Fig. 8.6. Its usual size is 25 liters and is suitable for places where plumbing installations is expensive and drains are available.

8.9.1.2 Pressure Type Water Cooler

Schematic diagram of pressure type cooler is shown in Fig. 8.7. Here water is supplied under pressure. For filling glass tumblers or container faucets or similar means are

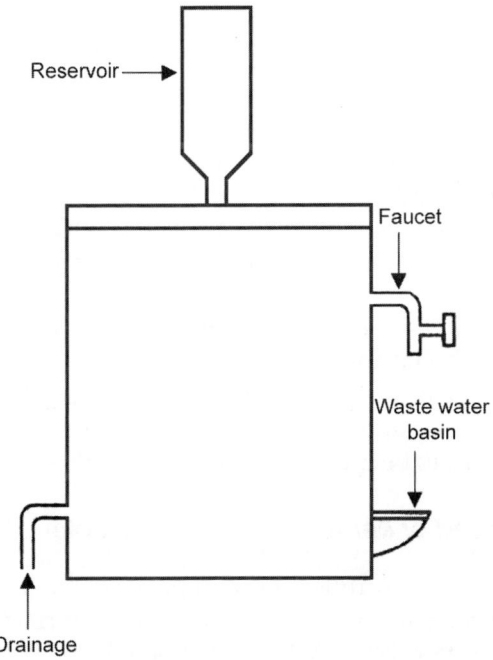

Fig. 8.6: Bottle type water cooler.

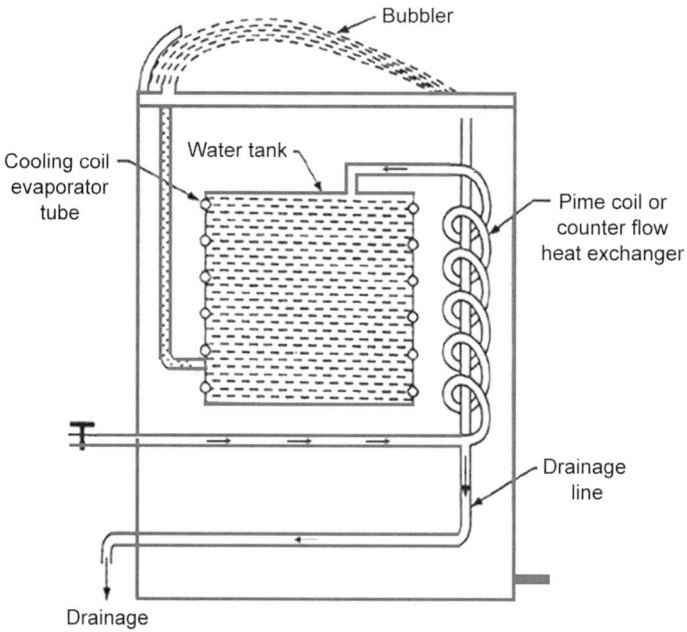

Fig. 8.7: Pressure type water cooler.

provided. A valve is employed to control an appropriate flow of water or projected stream of water from a bubbler. An arrangement should be made to collect water and allow complete collection of water spreading from the bubbler. The temperature of waste water is low; it is used for cooling the supply water by passing through a pipe coil wrapped round the drainage line. By doing so, the cooling load for cooler is reduced. Since the water is supplied under pressure the cold water can be obtained from the top mounted at any height of the cooler. In case of bottle type, faucet has to be at a height up to which siphoned water can be obtained from the tank of the cooler. The refrigeration system is usually mounted at the bottom structure of the cooler body and a cooling coil is wrapped round the water tank, to ensure good surface contact between the evaporative tube and tank, either the tank surface is corrugated to accommodate pipe or pipes are secured using soft solder to give metal contact. Some times a helical or U type coil is immersed in the water tank.

Although this arrangement gives high heat transfer from water to the coil yet formation of undesirable salt due to chemical reaction between water contaminant and the copper surface proves to be great disadvantage in this system.

8.9.1.3 Self-Controlled or Remote Type Water Cooler

Schematic diagram of self-controlled or remote type cooler is shown in Fig. 8.8. This type of water cooler employs mechanical refrigeration system and is a factory assembled unit. A remote cooler cools the water which is supplied to the desired drinking place (away from the system). It is quite a useful unit since it does not require extra space near the place of work.

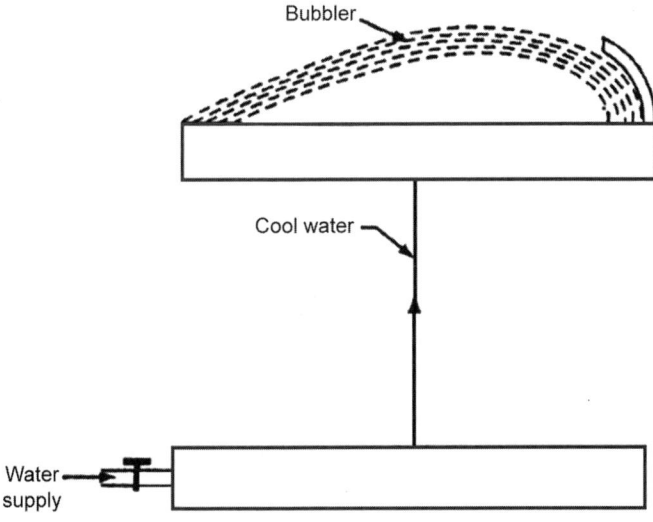

Fig. 8.8: Self-controlled or remote type water cooler.

8.9.2 Storage Type Water Cooler

Such type of coolers are used where continuous supply of water is not available. Fig. 8.9 shows a schematic storage type water cooler which is self-explanatory. In storage type water coolers, evaporator coil is wound around the water tank which

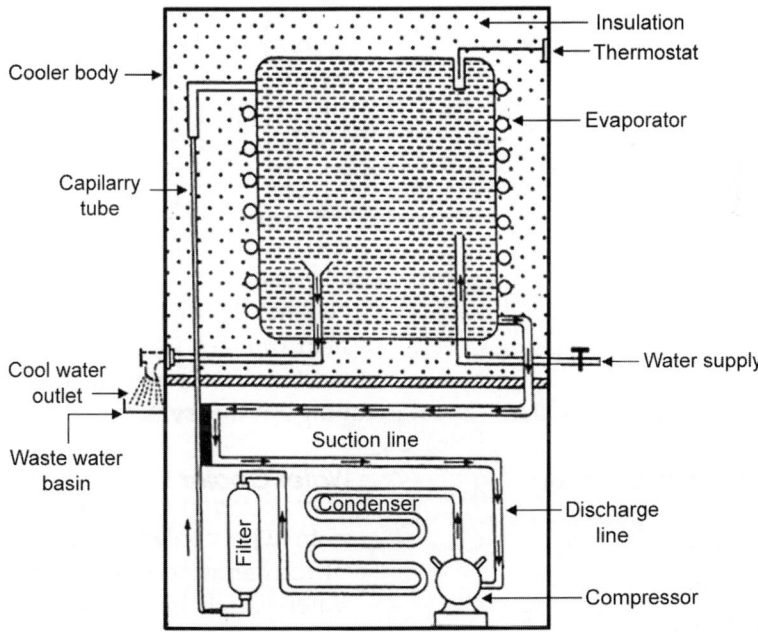

Fig. 8.9: Storage type water cooler.

is kept full of water by float valve. A low pressure liquid refrigerant which is flows through the evaporator coil takes away the heat of water and thus makes it cold. When the water attains desired temperature the thermostat operates and disconnects the power supply to the motor. The motor used here is capacitor-start capacitor-run single phase induction motor.

8.10 DESERT COOLER

As the name implies, it is suitable for an places where the humidity is quite low and temperature quite high. Such conditions are in conformity with desert areas. Hence the coolers are called the desert coolers. The principle on which a desert cooler operates is "evaporative cooling". Evaporative cooling is a process is which sensible heat is removed and moisture added to the air. When the air passes through a spray of water it gives up heat to water, some of water evaporates and picks up heat from the air equivalent to its latent heat. The vapours thus formed are carried along in air stream. In this way air is cooled and humidified. Fig. 8.10 shows a schematic diagram of desert cooler.

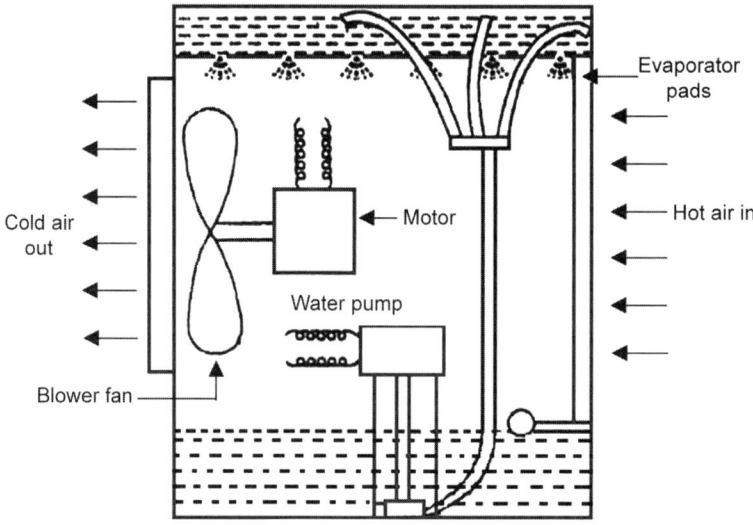

Fig. 8.10: Desert cooler.

☞ *Construction and Working*

A desert cooler consists of the following main parts:

1. Blower/fan
2. Water circulating pump
3. Water wetted pads
4. Water tank
5. Float valve
6. Water traps.

   The water is filled in the sump of the cooler from water supply mains, the level of which is controlled by a float valve. A water pump lifts the water and supplies it at the top of the cooler to the water distribution system which consists of small branches of copper pipe or so equipped with orifices which deliver equal amount of water to the troughs which in turn supply water to the wetted pads. The water which drops back from the pads is re-circulated. The pump may be made of brass, stainless steel or even plastic.

The blower pulls the air through the wetted pads and delivers it to the space to be cooled through and opening in the fourth side of the desert cooler cabinet. The air sucked through the pads is cooled by the principle of evaporative cooling. The blower provides adequate velocity to the air before it is delivered to the space to be cooled.

To have long-life of the desert cooler and better performance, pads should be changed every year and holes for water distribution system should be cleaned. The tank should be cleaned just after the season and coated with the corrosion resisting paint like red oxide and some other paints.

The efficiency of desert cooler is given as

$$\eta_{desert} = \frac{t_{db} - t_{dbg}}{t_{db} - t_{wb}}$$

where t_{db} is dry bulb temperature of ambient air, t_{dbg} is the temperature of air leaving the grille, and t_{wb} is the wet bulb temperature of ambient air.

According to ISS, the velocity across the pads should be around 40 m/minute. The desert cooler efficiency should be about 75%. Though the high efficiency of desert cooler corresponds to cooled air, the large humidity causes discomfort. The desert cooler efficiency above 80% is not desirable. For good performance the efficiency of a desert cooler should not fall below 70%.

8.11 AIR CONDITIONING

Air conditioning is a combined process that performs many functions simultaneously. It conditions the air, transports it, and introduces it to the conditioned space. It provides heating and cooling from its central plant or rooftop units. It also controls and maintains the temperature, humidity, air movement, air cleanliness, sound level, and pressure differential in a space within predetermined limits for the comfort and health of the occupants of the conditioned space or for the purpose of product processing. More often it involves following operations.

  i. To supply duly cleaned fresh air to keep down body odours, concentration of CO_2 and smoke within low limits.
  ii. To cool the air in summer and to heat it in winter in order to maintain proper temperature inside the premises air conditioned.
  iii. To control the humidity of air to the required level.
  iv. To obtain air movement without producing annoying draught.

8.12 TYPES OF AIR CONDITIONING

The air conditioning systems may be classified as follows:
  i. Unitary type.
  ii. Central type.
  iii. Unitary-central type.

8.12.1 Unitary Type

Window type air conditioner is an example of this type. It is factory encased, self-contained unit with compressor, condenser, evaporator, filter and refrigeration piping. It is made in capacities ranging from ½ to 2 tonnes employing motor up to 3 HP. These units are usually fitted flush with window sill on inside having attractive frontage. On outside it is supported on brackets and projects beyond the wall. These units are suitable for rooms having at least one exposed wall as hot air from condenser has to be discharged into open. Unitary type air conditioning for large premises costs about 50 to 100% more than central type but has advantage of flexible operation. Standby capacity required in unitary type air conditioning need not be high. In case of any trouble with any unit it can be immediately replaced, whereas in central type air conditioning, this is not possible. Very few structural modifications become necessary, i.e. only wall opening of say is needed in unitary type air conditioning. No duct work is required as in case of central type air conditioning. This type of air conditioning is best suited where only few rooms are to be air conditioned.

8.12.2 Central Type

In this type, air conditioning plant is situated at some central place usually in basement from where conditioned air is led through ducts to the rooms to be cooled. There are return ducts for carrying air from these rooms back to the central plant where it is dehumidified, cooled, and recharged with fresh ventilating air. Advantages of this type of air conditioning are high efficiency and robustness of the plant. This type of air conditioning is especially suited for large buildings and rooms which do not have exposed walls. Main disadvantages of this type of air conditioning are: absence of any adjustment of individual room temperature during being costly and taking lot of space and mixing of odours, cigarette smoke and bacteria present in the return air from infected rooms and re-distributing them to healthy room.

8.12.3 Unitary Central Type

In air conditioning, only 15% fresh air is required, rest of 85% air is recalculated. This type of air conditioning employs both the above types, central type supplying 15% fresh air. All the disadvantages of central type are done away with. There will be no return air duct, no mixing of odours and bacteria of one room and redistributing the same among other room unit. Fresh air duct will be very much small in size. Fresh air interdicted will displace corresponding quantity of air which will cases through openings act.

    In this type of air conditioning, each unit cooler will be connected by means of lagged piping to central type air conditioner caring either refrigerant or chilled water. This piping should be so placed that there should be no direct exposure to sun.

8.13   WINDOW AIR CONDITIONER

It is a common type of air conditioner used to condition the air of a particular space occupied by human beings, i.e., office room, residential room, etc. It has automatic operation to cool and humidify the air.

A window type air conditioner is shown in Fig. 8.11. It consists of a case divided into two parts by a partition with a small opening at the top as the outdoor part and indoor part.

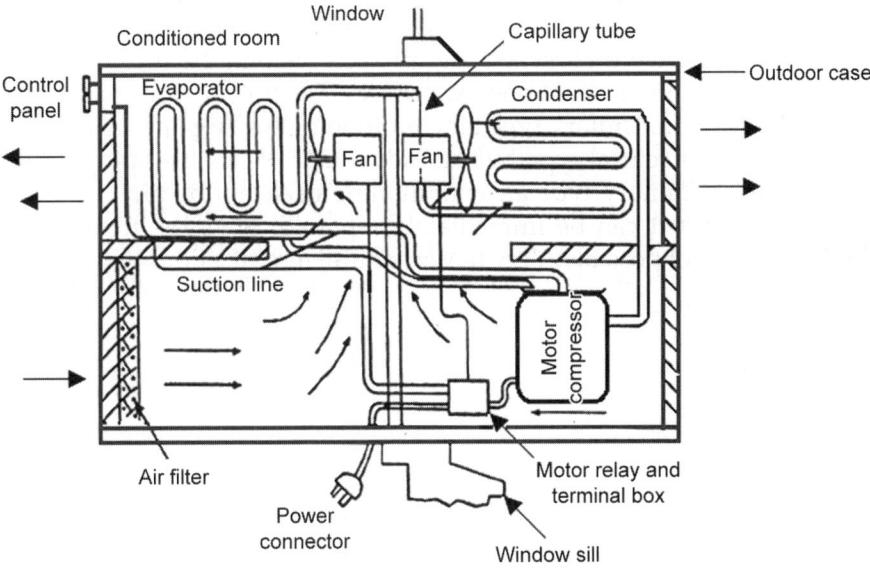

Fig. 8.11: Window type air conditioner.

The outdoor portion consists of a hermetically sealed motor compressor unit, condenser, motor driven fan and a tray. This portion partitioned by a portion L into two parts is provided with a left hand side opening.

The indoor portion consists of evaporator, motor driven fan, remote bulb, refrigerant, control, a control panel, an air filter, power connector and a tray. This portion is further subdivided into two parts by opening on the right hand side. A pipe line connects the two trays in the inner and outer parts. A capillary line control through a refrigerant filter connects the condenser to the evaporator. Evaporator is connected to the compressor by a suction pipe line. These units are usually fitted flush with window still on inside having attractive frontage. On outside it is supported on brackets and projects beyond the wall. The front and back of the inner and outer portions of the cooler is fitted with shutters. These shutters are adjustable at different inclinations according to the requirements.

The working substance is a vapour which readily evaporates and condenses. The most popular substances are ammonia, freon, carbon-dioxide, etc.

As the unit is put into operation, the low pressure vapour through the suction pipe line is drawn from the evaporator and passed to the compressor. The compressor delivers it at high pressure to the condenser. In the condenser, the vapour gets condensed and the heat is removed from the refrigerant vapour. The liquid refrigerant collected at its lower coils is passed through the filter into the capillary tube control before flowing back at low pressure into the evaporator coils.

The low pressure liquid refrigerant inside the evaporator coils quickly picks up from its surface. The air is drawn from inside the room through a filter from the lower portion of the unit by a motor driven fan. This air is passed over the evaporator coils.

The cooled air from the evaporator coils surface is supplied back into the room. Due to dehumidification of moisture from the circulating air, the moisture after flowing downward from the surface of the evaporator coils is collected at the bottom of the evaporator into a pan. This moisture in the pan flows due to gravity into another pan placed in the outdoor portion at the bottom of the condenser. The moisture helps to cool the compressor and condenser due to evaporation.

In the compressor and condenser compartment, the fan draws in the outside air from the lower portion of the unit, circulates it over the condenser and discharges from the upper portion of unit to the outside due to mounting of the compressor and condenser in a particular design. The heat of the vapour refrigerant is taken away by the air during its passage through the condenser and it is cooled into the liquid form. The unit automatically stops as soon as the desired temperature is reached inside the room. The automatic operation of the unit is accomplished by a valve at the control panel.

☞ *Advantages*

1. Saving in installation and field assembly labour.
2. Exact requirement of each separate room is met whereas a central system cannot meet the individual needs of separate rooms
3. Zoning and duct work is eliminated.
4. Only those rooms which need cooling will have their units running, whereas the central plant will have to operate all the time for the sake of only a few rooms.
5. Failure of unit affects a single room whereas all the rooms are affected when failure occurs in the central system.
6. Low initial cost.
7. Flexibility of operation.
8. Starting with only a few rooms, the air conditioning may be progressively extended to other rooms without spending a large sum from the very beginning.

Table 8.1	Comparison between desert cooler and window type air conditioner	
Particulars	*Desert cooler*	*Window type air conditioner*
Initial cost	Low	Quite high
Main components	Motor and fan	Complete refrigeration unit, compressor, condenser, evaporator, and controls, etc.
Humidity control	Not possible	Possible
Power required per unit cooling	More	Less
Location	Inside or in the window of the room	Fitted either in the window or wall
Utility	Can be used to cool the air and that too in the dry season	Can be used for both cooling and heating purposes

QUESTIONS

1. Is refrigeration a luxury or a necessity?
2. How does vapour absorption refrigeration cycle differ from vapour compression refrigeration cycle?
3. Explain vapour compression refrigeration cycle and show how cooling effects is produced?
4. Describe the electric circuit of a refrigerator.
5. What will happen if the thermostat is taken out and placed in the room?
6. Why is it necessary to maintain constant voltage for the refrigerator?
7. Why it is not advisable to put very hot things in the refrigerator?
8. Why is an automatic door closing mechanism provided in modern refrigerator?
9. How can the temperature inside the refrigerator be adjusted?
10. What is the main difference between a refrigerator and water cooler and between water cooler and air conditioner?
11. Write a short note on following:
    - i. Thermoelectric refrigeration.
    - ii. Air conditioning.
    - iii. Window air conditioner.

BIBLIOGRAPHY

1. A. Chakrabarti, M.L. Sony, P.V. Gupta, U.S. Bhatnagar, A Textbook on Power System Engineering, Dhanpat Rai & Co., India, 2012.
2. Beam. A.R. and Simons, R.H., Lighting Fittings, Performance and Design, Pergamon Press, UK, 1968.
3. Boast, W.B. Illumination Engineering, Mc Graw-Hill Book Co., New York, INC, 1953.
4. C.L. Wadhwa, Generation Distribution and Utilization of Electrical Energy, New Age International Publishers, India, 2010.
5. Chilikin, M., Electric Drive, Mir Publishers, Moscow, 1978.
6. Chute, G.M. & Chute R.D., Electronics in Industry, 4th Edition, McGraw-Hill, Koga Kusha Ltd., 1971.
7. Davies, A.C., The Science and Practice of Welding, Seventh Edition, Cambridge University Press-Cambridge, London, 1977.
8. Dover. A.T. Electric Traction, Sir ISAAC Pitman & Sons Ltd., London, Fourth Edition, 1963.
9. Fitzgerald, A.E., Kingsley, C., Kusko, A., Electric Machinery, International Student Edition McGraw Hill, New York, 1971.
10. H. Partab, Art and Science of Utilization of Electrical Energy, Dhanpat Rai & Co., India, 2012.
11. Hancock, N.N., Electric Power Utilization, Isaac Pitman, London 1967.
12. Henry Horwitz P.E., Welding Principles and Practice, Houghton Mifflin Co., Boston 1979.
13. J. Wesley Cable, Induction and Dielectric Heating, Reinhold Publishing Corporation, New York, 1954.
14. J.B. Gupta, A Course In Electrical Power, S.K. Kataria & Sons., India, 2010.
15. J.B. Gupta, Rajeev Manglik, Rothit Manglik, Utilization of Electrical Energy and Traction, S.K. Kataria & Sons, 2012.
16. N.V. Surayanarayana, Utilization of Electric Power & Electric Traction- Including Electric Drive & Traction, New age International Publishers., India, 2005.
17. P.S. Bimbhra, Electrical Machinery, Khanna Publishers, India, 2012.
18. Parry Moon, The Scientific Basis of Illuminating Engineering, Dover Publication Inc., New York 1996.
19. Phillip T. Krein, Elements of Power Electronics, Oxford University Press, New York, 1998.
20. Pillai, S.K., A First Course on Electrical Drives, Wiley Eastern Ltd., India, 1983.
21. R.K. Rajput, Utilisation of Electrical Power, Laxmi Publications (P) Ltd., India, 2012.
22. S. Sivanagaraju, M. Balasubba Reddy, D. Srilatha, Generation & Utilization of Electrical Energy, Pearson, 2010.
23. Taylor, E.O., Utilization of Electric Energy, Orient Longman, India, 2006.
24. Thereja BL, A Textbook of Electrical Technology (Volume 3) 23rd Edition, S. Chand & Company Ltd, New Delhi, 2004.

INDEX

AC Commutator motors 40
AC Timer circuit 238
Absorption factor 116
AC electric locomotive 277
AC rectified welding unit 219
AC welding machines 219
Air brake system 308
Air conditioning 355
Amplidyne 16
Anodizing 336
Atomic hydrogen welding 208
Automatic control 55
Automatic speed control 54
Average speed and schedule speed 258

Bare metal arc welding 205
Battery drive 245
Beam factor 117
Bearings 82
Bogie arrangements 302
Booster control 46
Bottle type cooler 351
Bow collector 282
Braking 307
Butt welding 224

Candela 113
Capacitor discharge circuit 239
Carbon arc welding 215
Carbon-arc lamp 122
Chopper control 53
Coated electrodes 205
Coefficient of adhesion 267
Coefficient of utilisation 117
Compact fluorescent lamp 142
Composite system 249
Contactors 23
Cooling time constant 68
Coreless furnace 190
Current collection system 280

DC motor 40
DC welding machines 218
Depredation factor 116
Desert cooler 354
Diac 9
Dielectric heating 193
Diesel-electric drive 243
Diode 1
Direct arc furnace 174
Direct core-type furnace 188

Direct current system 248
Direct resistance heating 165
Domestic refrigerator 347
Duty cycles 68

Eddy current brakes 310
Electric arc furnace 174
Electric arc welding 202
Electric braking 59
Electric drive 244
Electrical regenerative braking 312
Electrocleaning or pickling 337
Electrodeposition 332
Electroextraction 338
Electroforming 335
Electro gas welding 230
Electrometallisation 335
Electroparting or stripping 338
Electropolishing 337
Electrorefining 339
Electro slag welding 229
Electrofacing 335
Electrolysis 327
Electrolytic bath 334
Electron beam welding 227
Electron bombardment heating 184
Electronic control 48
Electronic welding control 236
Electroplating 333
Electrotyping 336
Enclosures 81
Energy storage welding 239

Factory lighting 148
Feeding and distribution system for DC tramways 276
Feeding and distribution system on AC traction 275
Field control 42
Field excitation 52
Flame-arc lamp 123
Flash butt welding 224
Float switches 21
Flood lighting 150
Fluorescent tubes 135
Flywheel drive 245
Flywheels 77
Frequency method 56
Furnaces 166

Gaseous discharge lamps 128
Gas-filled lamps 126

Gate-turn-off switch 10
Globes and reflectors 144
Group drive 31

Halogen lamp 128
Heat control unit 237
Heating time constant 66
High frequency heating 186
High-pressure mercury-vapour discharge lamp 131
Hybrid drive 245
Hydraulic brake 310

Ignitron 15
Ignitron contactor 236
Illumination 113
Incandescent lamps 124
Indirect arc furnace 183
Indirect core-type furnace 190
Indirect resistances heating 165
Individual drive 32
Induction heating 186
Inert gas metal arc welding 212
Instantaneous water coolers 351
Integrating sphere 156
Internal combustion engine drive 243

Kando system 250

Lamp efficiency 115
Laser welding 232
Led 142
Light energy 112
Light fittings or luminaries 146
Limit switches 21
Linear motor 288
Load equalisation 77
Luminance 114
Luminous efficiency 116
Luminous flux 112
Luminous intensity 113

Mag welding 214
Magnetic brakes 310
Magnetic storage welding circuit 240
Magnetic-arc lamp 124
Maintenance factor (MF) 116
Master controller 23
Mean hemispherical candle power (MHSCP) 115
Mean horizontal candle power (MHCP) 115

Mean spherical candle power (MSCP) 115
Mechanical braking 59, 308
Mechanical regenerative braking 311
Mercury iodide lamps 133
Metal inert-gas (MIG) 213
Metaldyne control 48, 296
Monomoter bogie 303
Mountings 82
Multiple unit control 299

Negative boosters 276
Neon lamp 133
Noise 84

Overhead equipment (OHE) 278
Overload capacity 73

Pantograph collector 282
Percussion welding 226
Photometry 152
Pilot devices 20
Plane angle and solid angle 111
Plasma arc welding 230
Plugging 60
Plugging switches 22
Polar curve 118
Pole changing 57
Polishing and baffling 335
Power transistor 4
Pressure switches 21
Pressure type cooler 351
Projection welding 224
Push button 20

Radiating heating 173
Railway electrification 247
Rating of machines 71, 73
Reduction factor 115
Reflection factor 117
Refrigerants 346

Refrigeration systems 343
Regenerative braking 63
Repulsion motor 41
Resistance heating 164
Resistance welding 221
Rheostatic or dynamic braking 61
Rotor copper losses 58
Rousseau diagram 118

Saturable reactor (magnetic amplifier) 18
Schrage motor 40
Seam welding 223
Self-controlled or remote type cooler 352
Semiconductor devices 1
Series parallel control 293
Series parallel starting 296
Silicon-controlled rectifier (SCR) 6
Single-phase low frequency AC system 248
Single-phase series motor 41, 288
Single-phase induction motors 38, 41
Slip coupling 58
Smoothing reactors 283
Sodium-vapour discharge lamp 129
Space height ratio 115
Specific consumption 115
Specific energy consumption 265
Speed control 41, 55
Speed governors 26
Speed–time curve 253, 254
Spot welding 221
Steam engine drive 242
Storage water cooler 353
Street lighting 149
Submerged arc welding 210
Substations 274
Synchronous and synchronous induction motor 41
Synchronous motors 39

Thermostats 22
Three-phase low frequency system 249
Three-phase series motor 41
Three-phase thyristor rectifiers 51
Three-phase induction motors 33, 40, 288
Thyratron 12
Thyristor 5
Thyristor control 49
Traction mechanics 253
Traction motor 282, 284, 285
Traction motor control 292
Tractive effort 258
Tractive effort-speed characteristic 262
Train movement 258
Tramways 246
Transmission of drive 304
Triac 8
Trolley bus 246
Trolley collector 281
Troubleshooting of refrigerator 349
Tungsten inert-gas (TIG) 212

Ultrasonic welding 233
Undulating DC 286
Unijunction transistor (UJT) 11
Upset butt welding 226
Utilization factor (UF) 115

Vacuum brake system 308
Ventilation and cooling 75
Vertical core-type furnace 189

Ward-Leonard method 46
Waste light factor 116
Water cooler 350
Wheel arrangement 301
Window air conditioner 356

Zener diode 3